Advanced Applications of Micro and Nano Clay II
Synthetic Polymer Composites

Edited by

Amir Al-Ahmed[1] and Inamuddin[2]

[1] Interdisciplinary Research Center for Renewable Energy and Power System (IRC-REPS), King Fahd University of Petroleum & Minerals (KFUPM), Dhahran-31261, Kingdom of Saudi Arabia

[2] Department of Applied Chemistry, Zakir Husain College of Engineering and Technology, Faculty of Engineering and Technology, Aligarh Muslim University, Aligarh-202002, India

Published by **Materials Research Forum LLC**
Millersville, PA 17551, USA

Published as part of the book series
Materials Research Foundations
Volume 129 (2022)
ISSN 2471-8890 (Print)
ISSN 2471-8904 (Online)

Print ISBN 978-1-64490-202-8
eBook ISBN 978-1-64490-203-5

Distributed worldwide by

Materials Research Forum LLC
105 Springdale Lane
Millersville, PA 17551
USA
https://www.mrforum.com

Manufactured in the United States of America
10 9 8 7 6 5 4 3 2 1

Table of Contents

Preface

Due to the layered structure, clay materials have many potential applications. Here, the layered silicates are made-up of silicon and oxygen bonds with some other elements. Incorporation of clays in a polymer to fabricated clay-polymer composite found to provide unique properties to the new materials, which opened many advanced application opportunities. It was observed to provide better absorption, and thermal/mechanical stabilities, carrier mobility, electric, magnetic, and dielectric properties and so on. In this book clay based composites with different synthetic polymers have been covered, which contains 10 state of the art articles covering different aspects of the clay and synthetic polymer based composites and their applications.

Chapter-1: The primary focus of this chapter is the composite formation by in-situ polymerization of hectorite/clay materials. The chapter also covers the application of this composite in different areas, like, food packaging, rheological control agent, wastewater treatment, biomedical applications and drug delivery.

Chapter-2: This chapter is about the use of cost-effective clay-based bio-polymeric composites in a variety of biomedical applications, such as; drug delivery. The main focus is to explore the compositional-structural properties, route of action of clay-based micro and nanocomposites in in-vitro drug delivery system, and administration route of different clay-based composites in drug delivery system

Chapter-3: This chapter primarily focuses on the synthesis, material characterization and testing of nanoclay-based conducting polymeric nanocomposites for electromagnetic interference shielding application. Employing nanoclay is found to improve mechanical strength, dielectric properties, thermal stability, barrier properties and shielding efficiency of nanocomposites through better dispersions and exfoliation of fillers.

Chapter-4: This chapter provides detailed information about the flame retardancy of micro/nano clay polymeric materials. Methods and advantages of the use of micro/nano clay over conventional flame retardant have also been discussed in this chapter. Characterization techniques of polymer micro/nano clay composites have been studied in detail.

Chapter-5: This chapter is designed to be source for nano-clay reinforced thermoplastic nano-composite research, including synthesis, characterization, structure/property relationship and applications considering optical and electrical conductivity of the composite films. Further, the UV-A shielding behaviors of the composite films are also presented.

Chapter-6: This chapter focuses on the application of nano-clay/polymer composites in removal of inorganic, organic pollutants from wastewater. The chapter elucidates characterization, occurrence, types and synthesis of nanoclay polymer composites. The feasibility and efficiencies of few nano clay polymer composites for the removal of specific water pollutants is discussed in detail.

Chapter-7: In this chapter attempts have been made to provide a brief overview of synthesis and applications of nanocomposites based on nontronite, an iron rich smectite clay; starch, a common polysaccharide. Being naturally abundant, affordable, non-toxic and biocompatible, clay-based minerals and biopolymers are advantageous to afford eco-friendly nanocomposites, especially useful for biological applications.

Chapter-8: This chapter discusses the numerous approaches for preparing clay-based PET nanocomposites as well as their physico-mechanical features, morphological and structural depiction, and crystallization and rheological behavior. In addition, the applications of PET-based nanocomposites are thoroughly covered in this chapter.

Chapter-9: This chapter deals with various types of fabrication of the polypropylene/clay nanocomposites such as extrusion process, solution blending method, melt-blending method, in-situ polymerization method, direct melt compounding, ultra sound-aided extrusion and master batch dilution. The mechanical, thermal, tribological, optical, creep, hygrothermal, rheological and morphological properties of these composites are also discussed.

Chapter-10: This chapter describes the various aspects of polymer nanocomposite synthesized through sonochemical in-situ polymerization intercalation and nanoparticles dispersion. It also explained how the different effects of ultrasound and cavitation improve the thermal, mechanical and electrical properties of polymer nanocomposites even at low filler loadings ($<=1.0$ wt.%).

At the end, we thankfully acknowledge all the authors and co-authors for their valued contribution and also express our sincere gratitude to the other publisher and authors for granting us the copyright permission to use their illustrations. Every effort was made to obtain the copyright permissions from the respective owners to include the citation for the reproduced materials, still we also express our sincere apology to any copyright holder, if, unknowingly, their right is being infringed. We also thankfully acknowledge the sincere efforts of Mr. Thomas Wohlbier and his team for evolving this book into its final shape.

Dr. Amir Al-Ahmed
Interdisciplinary Research Center for Renewable Energy and Power System (IRC-REPS)
King Fahd University of Petroleum & Minerals (KFUPM)
Dhahran-31261, Kingdom of Saudi Arabia

Dr. Inamuddin
Department of Applied Chemistry
Aligarh Muslim University
Aligarh-202002 (UP), India.

Adv. App. of Micro and Nano Clay II – Synthetic Polymer Composites   Materials Research Forum LLC
Materials Research Foundations 129 (2022) 1-23   https://doi.org/10.21741/9781644902035-1

Chapter 1

In-situ Composite Formation by Polymerization on the Hectorite or other Clay Materials

Madhur Babu Singh[1], Prashant Singh[2], Pallavi Jain[1]*

[1]Department of Chemistry, SRM Institute of Science and Technology, Delhi-NCR Campus, Modinagar, Ghaziabad, 201204, India

[2]Department of Chemistry, Atma Ram Sanatan Dharma College, New Delhi-110021, India

*palli24@gmail.com

Abstract

Clays are the naturally existing mineral having layered structures with at least one dimension in the nano-range that are economical and environment friendly. There exist two types of nanoclays, anionic and cationic, depending on the surface charge layered. Nanoclays have wide application in different areas for improving physical properties like heat resistance, mechanical strength and anticorrosion quality of the polymer matrix. Clay and its composite have promising applications including tissue engineering, petroleum, drug delivery, food packaging and enzyme immobilization. Due to their superior properties like flame retardancy, non-toxic, magnetic properties and large surface areas; hectorites and their composite are of great interest. The primary focus of this chapter is Composite Formation by in-situ polymerization of hectorite/clay materials and its application in different areas.

Keywords

Hectorite, In-Suit Polymerization, Nanoclays, Economical, Environment Friendly

Contents

1.   Introduction

Clays are of different classes, like smectite, chlorite, kaolinite, illite and halloysite depending upon their particle morphology and chemical composition. They are highly efficient and readily available hence developed for different applications. There has been increasing use of clay minerals as natural nanomaterials with rapid nanotechnological advancements. Some common types of clays are given in table 1. Octahedral sheets are made up of magnesium/aluminum having six-fold coordination with the tetrahedron oxygen and with the hydroxyl [1]. The various aspects that define and differentiate nanolayers are influenced by the location of these plates, about 30 various types of nanoclays exist, that are utilized in various applications depending on their properties. As shown in Table 1, in conventional nanoclay materials, three main types of sheets are observed 1:1, 2:1 and 2:1:1 on their mineralogical composition. Each tetrahedron is connected to an octahedron in a 1:1 structure; in a 2:1 structure, each octahedron is connected to two tetrahedral plates (one for each face); and, each octahedral plate adjoins another octahedron and connects with two tetrahedra in the 2:1:1 structure [2,3]. For example, halloysite nanoclay has an average size of 15 nm × 1000 nm which is a natural aluminosilicate nanotube [4]. Halloysite nano-coating (1:1 nanotube) comes in the use of medicine, packaging of food and applied science due to their hollow tube structure. The

Adv. App. of Micro and Nano Clay II – Synthetic Polymer Composites   Materials Research Forum LLC
Materials Research Foundations 129 (2022) 1-23   https://doi.org/10.21741/9781644902035-1

most common lamellar nanoclay of montmorillonite (smectite) (MMT) consists of layers of aluminosilicates about 1 nm thick, the surface of which is replaced by metal cations and is stacked in several layers about 10 μm in size. In the polymer matrix stacking agents can be dispersed as a filler/additive for forming polymer/composites of nanoclay, which has various applications as increased mechanical strength thickening and gelling conditioners, flame retardant materials, wastewater treatment and air permeability control [5–8]. MMT nanoclays (2:1 subclass silicates) with the high exchanging ability of cations such as biological or organic molecules have sites for the exchange of cations on the siloxane surface that could be bound to various substances [9]. MMT nanoclays stacks can be dispersed as a filler/additive in a polymer matrix that can make polymer/nanoclay composites that are extensively focused upon because of their high capacity of cationic exchange, edema properties and cross-sectional area. The large area of surface [10–12] halogenizite nanolayers are easily dispersed in various polymers without delamination. This is because that there are fewer hydroxyl groups on the nanoclay surface than in MMT. In addition, tubular nanolayers of halloysite are excellent nano controllers for various chemical molecules [13]. Therefore, to enhance their mechanical and thermal properties functionalized nanolayers of halloysite are used as effective fillers for polymers. To get continuous delivery of active molecules like flame retardants, antioxidants, anticorrosive and antibacterial agents, these nanolayers are also used as a means [14,15]. In recent years R&D of new polymer/nanoclay materials is an area of progress in material chemistry [3,6–8,16–19]. A rigid nanolayer can be used as a filler and can strengthen the macromolecular structure and prevent freely moving polymer chains around the filler [13]. In addition, this works as a carrier component when surface adhesion is achieved between the filler and the chain [20]. It is necessary to solve two important problems in the field of synthesis:(1) at nanoscale, chemical compatibility of the polymer matrix and nanofiber; and (2) uniform distribution of nanofibers in the macromolecular matrix. Between nanoclay filler and the polymer matrix, the surface interaction, also the nanoclay dispersion's quality has a considerable effect on the characteristics of polymer/ nanoclay composites. Interrelated characteristics decide the morphology of polymer matrixes and hence its basic majority characteristics like heat distortion temperature, elastic modulus, thermal stability, strength, ability to get self-heated, gas barrier and shape memory ability [21,22]. The functionalization of nanolayers on a surface is a practical method to improve the surface interaction of the nanoclay fillers with polymer substrates, which makes it possible to transfer the surface tension of polymer to nanolayers [23]. For example, tensile properties and thermal stability of the polymer/nanoclay composites can be enhanced by altering (covalent) the outer layer of halloysite nanotubes, which enhance the degree of dispersion in the matrix of the polymer. The synthetic approaches can lead to various combinations of polymer matrix and nanolayers like immiscible, interstitial and flocculent structures

[24]. The polymers get detached from the layers of clay. In an immiscible structure, the nanolayer dispersion is aggregated into the polymer matrix [25]. Sandwich structure is made in between the layer of clays by the polymer, due to which the shapes of the clay layers get changed. This change comprises changing the layering method of the layers, changing the distance between the layers and reducing the electrostatic force acting between the layers of clay, which leads to a greater enhancement in the thermal and mechanical behavior of the material [26,27]. The layers of nanoclay are completely separated by polymer chains in a layered structure, which provides high mechanical properties and the possibility of polymer processing [28]. There are three main ways by which polymer/nanoclay composites can be synthesized, including melting, mixing in solution, and in-situ polymerization [1]. Undoubtedly, the most frequently used synthesis method is in-situ polymerization, in which the amount of grafted organic matter is controlled, and the distance between and by changing the conditions for polymerization, the clay layers are controlled [29]. The association of in-situ polymerization with effective splicing methods, consisting of chemical activation [30], radical-mediated polymerization [31,32], tandem preparation [33], photo genesis [29], and small emulsions [34], made it possible to efficiently disperse nanoclay in the form of single platelets in macromolecular matrices, represents a serious problem inherent in the synthesis of composites polymer/nanoclay. Every method is applied successfully for the transformation of the surface's chemical of the clay with grafted materials with high or low molecular weight [35]. Polymer/nanoclay composites have extended to other materials. Various hybrid and enhanced materials include nanoclays/conductive polymers(polypyrrole (PPy), poly (2,3-dihydrothieno-1,4-dioxin) (PEDOT), polyaniline and (PANI), Thiofuran, biocomposites and hybrid organic clays, membranes have much better properties compared with traditional composites [36]. These new compounds are manufactured with less content modified nanoclay fillers than conventional filler systems and, therefore, have less weight [37]. The polymer/nanoclay composites are come into use in numerous fields such as construction (structural panels), aerospace (fire-resistant panels and high-performance component), automotive (bumpers, fuel tanks, etc.) inside and outside, chemical processes (catalytic), pharmaceuticals (as penetrants and carriers), wrapping of food and textiles because of their unique properties [38]. The scarcity of fossil resources and the urgent environmental protection have drawn our attention towards the generations and polymer/nanoclay composite's application. It is expected that innovative bio air composites polymer/nanoclay will find application in different fields of science and technology [39]. This chapter provides an overlook of the latest progress in preparation and various new applications of polymer/nanoclay compounds. Provides an updated overview of various preparation methodology and its utilization on modified-surface nanoclay fillers and possible strength and upcoming opportunities for uses of clays

composites. Individual wafers formed in a polymer matrix represent a major challenge inherent in polymer/nanoclay synthesis. Grafted clay surfaces with molecular weight [35].

Table 1. Types of clay commonly used

Ratio of the Clay	Examples
1:1	Halloysite, Kaolinite, Rectorite
2:1	Laponite, Bentonite, Saponite, Montmorillonite, Vermiculite, Sepiolite, Hetorite
2:1:1	Chlorite

2.   Synthesis

Different methodologies are possible for the three synthesis processes, namely (a) In-Situ Polymerization (ISP), (b) Melt-Blending and (c) Solution-Blending. However, achieving a homogeneous dispersion of clay-nanoparticles into the polymeric metrics is an important move in the synthesis of nanoclay/polymeric compounds. On comparing the solution-blending and melt-blending processes, the former process regularly acquiesces better dispersion of clay-nanomaterial into the polymer matrix than melt-blending because of the strong agitation power and low viscosity. Alternatively, the melt-blending method is regarded as an environmentally beneficial process and industrially feasible with significant financial potential. The ISP process is generally exercised preparation approach that offers consistent dispersion and is simple to change by altering the polymerization situations [29]. Numerous innovative synthesis techniques for generating nanoclay/polymer composites with unique characteristics have recently been proposed. The different emerging synthesis methods are described in this chapter, as well as correlations between the synthesis pathways, formation of structure, physical and mechanical properties, as well as the features of nanoclay/polymer materials are described.

2.1   In-situ polymerization method

Insufficient dispersion of the filler, causing agglomeration and entanglement, especially at a high filler content is obtained by the melting method [40]. Based on data on silicate dispersion [41] in-situ polymerization is more efficient in complexation and can circumvent the strict thermodynamic demands related to polymer alternation [42]. In addition, in-situ polymerization provides a flexible molecular structure to the matrix of the polymer. It provides an efficient way for the preparation of various polymer/nanoclay compounds having an extended reach of properties and allows the design of interfaces between nanolayers and polymers through flexible adjustment of nanolayers, parts, and background structure [43] on the preparation of new polymer/nanoclay compounds using the in-situ polymerization and demonstrated the advantages of this methodology over the various class of synthesis methods [44]. For example, poly(2-ethyl-4,5-dihydrooxazole)/nanoclay compounds were first synthesized with the help of the in-situ method [26]. This research developed an opening ring polymerization method that triggers the separation of the clay layers in the matrix of the polymer and is directed towards the production of a composite. The results showed that the layers underwent mixed/intercalated exfoliation and thermal stability was improved and analysed with poly-(2-ethyl-4,5-dihydrooxazole). Later, a conductive composition enclosing PPy, silver and attapulgite clay was developed by use of the in-situ morphology of granular fibres acquire in 10 min [45]. The in-situ polymerization provides a fine-dispersed clay filler in a matrix of polymer, which imparts exceptional antimicrobial properties to the biodegradable compound, materials and also an improved storage module. Herrero et al. [46] searched about the effect of the area and shape of nanolayers utilized in in-situ polymerization. Their research shows that needle clay contributes to higher molecular weight and it has better mechanical properties than sheet clay, improved in-situ polymerization due to non-isothermal temperature profile and extra catalyst. Kheroub et al. [47] applied in-situ polymerization for the preparation of poly (furfural alcohol)/MMT compounds. The resulting composite has upgraded mechanical and thermal properties due to an intercalated or peel-able structure as compared to original polymers [48]. By use of in-situ polymerization and solution of trichloromethane in an environment of an ultrasonic probe Prado et al prepared many PMMA/MMT composites. The resulting compounds have varied and intercalated morphology, that provides favorable properties for optoelectronic devices. Sharma and his colleagues [49] improved the mechanical and thermal properties of poly (MMAcoBA)/Cloisite 30B composites by polymerization in-situ emulsification with ultrasonic assistance. This innovative method provides greater resistance to thermal degradation and mechanical strength as compared to the synthesis of composites by common techniques. Because of micro-convection caused by cavitation and ultrasonic waves resulting composite possesses a greatly peeling and dispersed structure. Compared with the use of the smelting method, the improved mechanical properties and increased clay base distance in a PP/MMT polymer matrix of compounds were obtained with the use of the in-situ polymerization method [50]. Similarly, a rubber/clay compound (polybutadiene/MMT, cis-1,4) was developed with the use of the same process, and overcome its mechanical defects such as the problems of tear resistance, thermal stability and low tensile strength [51]. The research shows that for a butadiene polymerization, Ni-based catalytic systems have high potency and selectivity under the presence of MMT. Cherifi et al. [28] also used ultrasound in-situ polymerization to form poly(2,3-epoxyproply methacrylate)/MMT compounds. In the comparison of solution-blending process, ISP method employing ultrasound resulted in better composite having better reaction time, clay dispersion and compound yield. Several recent ISP processes for polymer/nanoclay composites have particular stress on advanced synthesis schemes like active/controlled radical polymerization, photo genesis, emulsion polymerization and coupling chemistry, chemical modification of the clay surface with high or low molecular weight grafts. The systematic scheme of the in-situ polymerization method is shown in Figure 1.

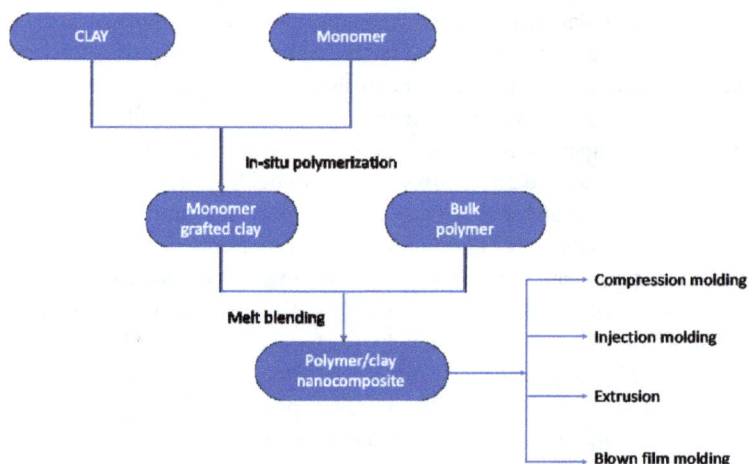

Fig. 1 In-situ polymerization method scheme

2.1.1   Mediated controlled radical photopolymerization (P-CRP)

For the fabrication of novel nanoclay composite, light is being used very smartly and strategically to mediate P-CRP. Generally, P-CRP is categorized into two, namely the photo redox method and intramolecular photochemical method [52]. Based on the catalyst, reaction mechanism and specific reagent there are various synthesis subcategories. In the P-CRP method, polymer nanoclay composites were prepared by utilizing cost-efficient and universal light sources as external regulators. This method has much potential supremacy in different utilization such as particle preparation and surface fabrication, photo-responsive gel design [53]. By using this method, nanoclay composites are made with well-organized and controlled molecular weights and topology, and it also made the polymerization rate fast at relatively low temperatures [33]. The exfoliated poly(glycidyl methacrylate) was prepared by Jiassi et al. [54] through the P-CRP method and thiolated-bentonite was employed as a chain transfer agent and he was able to create a composite with the best mechanical and thermal properties. Shanmugam et al. [55] went through advanced photo-induced electron transfer reversible addition-fragmentation chain transfer (PET-RAFT) method for the preparation of different stereo-polymers. In comparison to thermal polymerization, this method was found to be better in control of molecular weight, dispersity and tacticity. Better development was made by Chen et al. [56] in the fields of P-CRP process, he simply made a simple flow reactor that facilitates the action to create tri-thiocarbonates (TTCs) having an ambient increase in expansibility and rate of reaction when compared to batch reactions. P-CRP method approach for the synthesis of clay composite is widely explored now a day's using light as an external stimulus to control and unprecedented levels of function.

2.1.2 Click coupling chemistry

For chemical modification of nanoclays surface, the most versatile and effective way is the use of click chemistry. When we are talking about the synthesis of polymer nanoclays [39] by using the click coupling reaction, there must be the need for a complementary functional group at polymer and nanoclays [57]. The incorporation of click chemistry with another different method of preparation resulted in the grifted and potential of the two methods as well as the good and potent alternative path for designing advanced nanoclay composite having better physicochemical properties.

Nowadays, click chemistry is being greatly explored to functionalize the end group of polymers to design a better polymer by following controlled living radical polymerization (CLRP) [58]. Yadav et al. [59] have used the click coupling way for linking cellulose and nanoclays covalently. He used Cu(I) catalyzed azide-alkyne for achieving covalent linkage that is established between nanoclays and modified cellulose. This way of approach increases the interfacial interaction between nanoclays and cellulose, which leads to the betterment of the mechanical properties of the composite. Zou et al. [60] gave the conclusion that with the help of click chemistry having refined and orthogonal reactions in a weak environment have possibilities for the creation of stimuli for developing bioactive materials and versatile output like polymer nanoclay composite scaffolds can also be made to the industrial scale.

2.1.3 Mini-emulsion polymerization (MEP) method

MEP can produce compounds from polymerization within monomeric droplets (50-500 nm) [61]. In the polymerisation process, ultrasonication is mainly used but this method is energy consuming, nowadays many low-energy consuming methods have been come into knowledge introducing in situ surfactant creation with the utilization of CO_2 [62]. The MEP has many supremacies over other methods of polymerization like it is free from many harmful solvents, high molecular weight and high monomer conversion. When we compare the nucleation step of mini-emulsion and emulsion polymerization, mini-emulsion has the least complex step for encapsulating both nonexfoliable and exfoliable nanoclays, to get excellent dispersion, mini-emulsion has to be incorporated into the CLPR method [63]. The MEP method was applied for the creation of the hybrid latex composite film with the use of a composition of poly(methylmethacrylate-co-butyl acrylate) (PMMBA) with MMT nanoclays [64]. A large amount (>8.0 wt percent) of nanoclays in a mini-emulsion will provide a constructive path for the MMT to diffuse, resulting in high latex viscosity and a large volume of coagulum. The prepared film has great efficiency of assisting board and paper factories which are making excellent quality barrier paper that is being used in food packaging. Ultrasound-assisted MEP is being used by Buruga et al. [65] for the fabrication of PMMA/halloysite nanoclay composites. For improving the dispersion of nanoclays into the polymer matrix there should be the inclusion of ultrasonication in the MEP, as a result of this, we achieve increase mechanical and thermal properties.

3.   Application

Polymer/nanoclays composites have very excellent properties such as high fatigue endurance, high specific strength, high damping, low density and increased thermal behavior have attracted researchers to explore their application in the different fields (Figure 2) [66].

In this section, there is a focus on its application in food packaging, wastewater pre-treatment and biomedical industries.

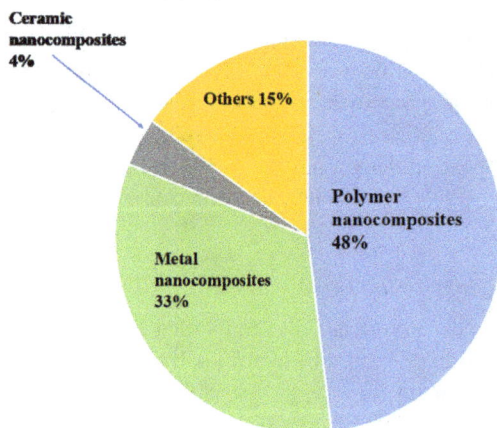

Fig. 2 Statistics of applications of different nanocomposites

3.1   Food packaging

Having barrier properties against the permeability of many gases like CO_2, oxygen, water vapor and many volatile components having additional properties such as thermal, mechanical and optical properties make polymer/nanoclays composite an excellent substance for food packaging materials [66]. In comparison to older traditional materials (plastic), nanoclays have more superior properties, so different nanoclay materials have devolved and investigated with better mechanical property and reduced permeability towards gases. To control the toxic gases such as CH_4, and CO_2 selective and intelligent polymer/nanoclays composite are being designed which extend the survival of preserved food [67]. NASA and the US military had made a joint effort to develop nanoclays that work as a barrier enhancer for ethylene vinyl alcohol (EVA) which work as a long-shelf-life wrapping material. The material which was developed by utilizing composites of EVA or nanoclays matrix has a long shelf existence of 3-4 years with no use of refrigeration. A food packaging material with antimicrobial properties was created to limit the growth of *Escherichia Coli* and *Staphylococcus aureus* and to reduce water vapour permeability to a large amount by employing poly(6-hexanolactone) (PCL)/nanoclay composites possess intercalated structures. Multilayer packaging film was developed by utilizing ethanol/vermiculite nanoclay composites and the film's potential for food products with high moisture content was investigated [69]. It was observed that the permeability of oxygen got increased when humidity crossed 60%. A reversible effect was also observed without any decline of the oxygen barrier property.

3.2   Rheological control agent

In the field of petroleum and pharmacy rheology of polymer composite plays an important role [70]. There is a use of surfactant for reducing the surface energy of clay layers because many petroleum industries depended are not suited with nanoclay materials because of its surface energy difference [3]. Different new applications of nanoclay composite have been explored such as in asphalt mixture rutting and fatigue resistance is being boosted and increased its storage stability [71]. Guo et al. [72] used surfactant with nanoclay as a stabilizer for CO_2 foam and he observed the result that nanoclay has improved balance and formability of CO_2 foam which provides the better gateway for recovery of oil from the porous homogenous medium in a microfluid device. The use of surfactant/nanoclays composite proved to be a great success as a stabilizer of foam for the petroleum industry and increase oil restoration.

3.3   Wastewater treatment

In the current world, water contamination is a situation of concern and many important steps have been initiated to overcome this problem. The main reason for water contamination is the presence of toxic substances like dyes, heavy metals, aromatic molecules, etc., leading to many health problems. To overcome these problems, many techniques are being explored and adsorption is one of them. Nowadays, a lot of novel adsorbents have been devolved for the treatment of contaminated water. For water treatment, polymer clay composites are also used due to their high adsorption capacity, better life cycle, ease of the process, low toxicity and high surface area. Currently, many excellent and better polymer clay composite adsorbents are developed and effectively explored for water remediation. By using the in-suit polymerization method, fabricated poly(1-vinyl-1H-imidazole) (PVI)/sepiolite nanoclay composites were developed and proved as an excellent productive adsorbent for Hg(II) in wastewater. At pH 6, the adsorption of Hg(II) was optimum but with the increase of temperature capacity of adsorption, the value also increased. Various polymer-nanoclay composites can separate, many contaminants from aqueous solutions and it is being proven fruitful in water treatment methods. The PANI/bentonite nanoclay compound was synthesized using plasma-induced polymerization for the removal of radioactive U(VI) ions from the aqueous solution. The adsorption of U(VI) on the surface of bentonite/PANI was significantly affected by pH, ionic strength and temperature [74]. The combination of chitosan and nanoclay adsorbent has become a novel type of pollutant removal technology because of its relatively minimum cost, natural, non-toxicity and natural biodegradability. Regarding the possibility of nanoparticles being disposed into the environment and causing damage through biological variation, chemical conversion, physical variation or gathering, and the interaction between the particles and naturally occurring organic substances or biomolecules, they still exist as a security issue.

3.4   Biomedical applications and drug delivery

The mechanical properties of clay-polymer composites make them useable in the field of biomedical. The mechanical properties of different nanoclays are illustrated in table 2. Nanoclays are being investigated to explore their use in numerous biomedical applications, together with, tissue engineering, wound healing, education of scaffolds, drug delivery, bone cement, most cancers therapy, and immobilization of enzymes. The biomedical applications of the nanoclays are shown in Figure 3.

Drug delivery

Tissue engineering

Biomedical applications of nanoclays

Wound healing

Bone cement

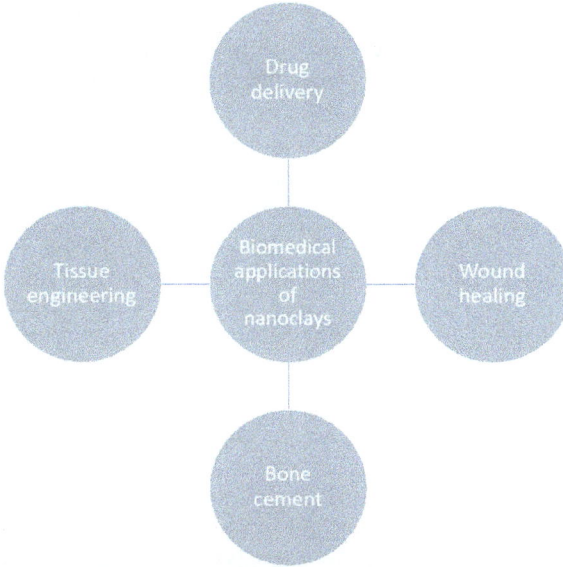

Fig. 3 Biomedical applications of nanoclays

Table 2. clay-polymer composite mechanical characteristics

Nanoclay	Polymer matrix	Improvement in modulus (tension/ compression) (%)	Content of nanoclays (wt.%)	Advancement in flexural strength (%)	Reference
Organic rectorite	Chitosan	-	2	7	[75]
Hectorite	Poly(lactic acid)	75	10	-2.85	[76]
Laponite	P(MEO2MA-co-OEGMA)	1,406	15	1,346	[77]

3.4.1 Bone cement

Bone is a composite material that contains mainly calcium phosphate that provides strength and flexibility, it has organic collagen, hence it is regarded as a composite material. The classification of bones may be done based on cancellous or cortical and around 80% of cortical bones are present in our skeleton. The mechanical properties like flexibility and strength differ in these bones. For example, if modules of elasticity and strength for cortical bone its ranges in between 3 GPa and 30 GPa and 70 MPa and 200 MPa respectively but in the case of cancellous bones these values were much lower as compared to cortical bone. For replacement, implants of the knee into the adjacent bone and hip fixing Polymethylmethacrylate (PMMA) is rapidly taken into use as a bone cement but it has some demerit because PMMA has poor fatigue and poor mechanical power for load taking purpose. Hence the requirement of an alternative for it, and nano-composite of clay material have been investigated for bone cement. The layered PMMA/silicate bone cement nanomixes were prepared by using MMT-based nanoclays having concentrations of 0.5, 1.0, 1.5 and 2.0% by weight [78]. The results showed that the exothermic polymerization temperature was lowered by 12 °C (from 84 °C to 72 °C), which can reduce cell necrosis. Nanoclay fillers help to prevent crack propagation, enhance Young's modulus and toughness, and improve biocompatibility.

3.4.2 Wound healing

The application of nanoclay in wound healing has also been widely explored to get rid of the infection, scar formation, and reduce pain [79–82]. For this purpose, swelling capacity and flexibility are very important. A biodegradable ethanol composite was prepared by using carboxymethyl chitosan (CMCh) and MMT [83]. The swelling behavior was increased correlated to acceptable drugs (like penicillin), and demonstrated good antimicrobial efficacy [83]. The incorporation of MMT nanoparticles in collagen/nisopropylamide hydrogels to regulate their response to stimulus as a scaffold with improved healing and regeneration properties. After 6 days, the wound healing area treated with the nanocomposite material doubled the area covered by gauze, and it recovered almost completely by the 13th day [84]. Gellan Gum Methacrylate (GGMA) wound dressing material was blended with laponite to give a curative agent to the wound [85]. Here, compared with the unfilled hydrogel, laponite regulates the puffiness behavior of the hydrogel network and can reduce the release of antibiotics in the first 8 hours.

3.4.3 Drug delivery

Drug delivery applications are one area where nanoclays have gotten a lot of attention. [86,87], including antibiotics, antihistamines, anti-inflammatory drugs, antibacterial agents, antifungal agents, anti-algae and cancer treatments. Roozbahani et al. studied synthetic laponite nanoplatelets (LAP) [88]. For releasing anionic dexamethasone (DEX) effectively with the help of encapsulation of drug into the interlayer spacing of nanodisk having great efficiency. When the environment is acidic, release occurs more quickly, and it is pH-dependent. Organically modified rectoride, which was changed by Hexadecyltrimethylammonium bromide to increase its proximity to polymer, was coupled with chitosan to produce nano composite film [75]. The film was filled with Bovine Serum

Albumin (BSA) like a pharmacological template to see how the number and spacing of organic resin layers affect drug delivery. Using the same nanoclay, Zeynabad et al. found that the encapsulation efficiency exceeded 90% [89]. In cancer treatment, pH-dependent drug delivery significantly applicable such as ciprofloxacin, an antibacterial agent and methotrexate drug were encapsulated in the naturally present modified laponite composite. Rao et al. prepared a hydrogel for the treatment of bowel cancer drugs [90], comprising of hyaluronic acid sodium in a Glycol methacrylate matrix and anticancer drugs incorporated in HNT. With the help of in vitro simulated gastric juice and gastrointestinal fluid experiments show that these composites hydrogels are pH-dependent sustained release, below than 10% drug deliver in the gastric region and major and restrained amount in intestinal juice making it effective for colon. Pramanik et al. [91] developed a nano-fibre that shows anticancer activity composing of PANI matrix and reinforced MMT with exfoliated nanosheets. These developed nano-composites show high potential against gram-negative and gram-positive bacteria, antifungal and antialgae activities. The different drug-clay hybrids were synthesized and their mechanism used for the preparation of tablets for oral delivery i.e., more prominent and effective method or can be modified by various approaches for the controlled drug release.

Conclusion

The development and the application of various composite of hectorite/clay materials have been explored in this chapter. The properties of composite mainly depend on the method of synthesis and the type of clay material used. The synthesis of nanoclay composite by applying the in-situ polymerization approach is mainly focused due to its controlled nature that helps nanoclays dispersion and clay interlayer spacing in the polymer matrix. The applications of these nanoclay composites have attracted more and more attention in different areas like wastewater treatment, drug delivery, bone cement, food packaging and as a rheological agent.

References

[1] H. Essabir, M. Raji, R. Bouhfid, A. el K. Qaiss, Nanoclay and Natural Fibers Based Hybrid Composites: Mechanical, Morphological, Thermal and Rheological Properties, in: M. Jawaid, A. el K. Qaiss, R. Bouhfid (Eds.), Nanoclay reinforced polymer composites : Natural fibre/nanoclay hybrid composites, Springer Nature Switzerland, 2016, pp. 29-49. https://doi.org/10.1007/978-981-10-0950-1_2

[2] R.S.H. Al-Maamari, J.S. Buckley, Asphaltene precipitation and alteration of wetting: Can wettability change during oil production? In: SPE/DOE Improved Oil Recovery Symposium, Tulsa, Oklahoma, 2000. https://doi.org/10.2118/59292-MS

[3] K. Majeed, M. Jawaid, A. Hassan, A. Abu Bakar, H.P.S. Abdul Khalil, A.A. Salema, I. Inuwa, Potential materials for food packaging from nanoclay/natural fibres filled hybrid composites, Mater. Des. 46 (2013) 391-410. https://doi.org/10.1016/j.matdes.2012.10.044

[4] F. Yu, H. Deng, H. Bai, Q. Zhang, K. Wang, F. Chen, Q. Fu, Confine clay in an alternating multilayered structure through injection molding: A simple and efficient route to improve rarrier performance of polymeric materials, ACS Appl. Mater. Interfaces. 7 (2015) 10178-10189. https://doi.org/10.1021/acsami.5b00347

[5] V. Mittal, Polymer layered silicate nanocomposites: A review, Materials (Basel). 2 (2009) 992-1057. https://doi.org/10.3390/ma2030992

[6] K. Yusoh, S.V. Kumaran, F.S. Ismail, Surface modification of nanoclay for the synthesis of polycaprolactone (PCL) - Clay Nanocomposite, MATEC Web Conf. 150 (2018) 1-6. https://doi.org/10.1051/matecconf/201815002005

[7] M.R. Irshidat, M.H. Al-Saleh, Thermal performance and fire resistance of nanoclay modified cementitious materials, Constr. Build. Mater. 159 (2018) 213-219. https://doi.org/10.1016/j.conbuildmat.2017.10.127

[8] G. Choudalakis, A.D. Gotsis, Permeability of polymer/clay nanocomposites: A review, Eur. Polym. J. 45 (2009) 967-984. https://doi.org/10.1016/j.eurpolymj.2009.01.027

[9] P. Taylor, S. Ganguly, K. Dana, T.K. Mukhopadhyay, T.K. Parya, S. Ghatak, S. Ganguly, K. Dana, T.K. Mukhopadhyay, T.K. Parya, S. Ghatak, Organophilic Nano clay : A comprehensive review, Trans. Indian Ceram. Soc. (2013) 37-41.

[10] N. Öztürk, A. Tabak, S. Akgöl, A. Denizli, Newly synthesized bentonite-histidine (Bent-His) micro-composite affinity sorbents for IgG adsorption, Colloids Surfaces A Physicochem. Eng. Asp. 301 (2007) 490-497. https://doi.org/10.1016/j.colsurfa.2007.01.026

[11] S. Pavlidou, C.D. Papaspyrides, A review on polymer-layered silicate nanocomposites, Prog. Polym. Sci. 33 (2008) 1119-1198. https://doi.org/10.1016/j.progpolymsci.2008.07.008

[12] P. Liu, Polymer modified clay minerals: A review, Appl. Clay Sci. 38 (2007) 64-76. https://doi.org/10.1016/j.clay.2007.01.004

[13] Y. Lvov, E. Abdullayev, Functional polymer-clay nanotube composites with sustained release of chemical agents, Prog. Polym. Sci. 38 (2013) 1690-1719. https://doi.org/10.1016/j.progpolymsci.2013.05.009

[14] T.S. Gaaz, A.B. Sulong, A.A.H. Kadhum, A.A. Al-Amiery, M.H. Nassir, A.H. Jaaz, The impact of halloysite on the thermo-mechanical properties of polymer composites, Molecules. 22 (2017) 13-15. https://doi.org/10.3390/molecules22050838

[15] G. Lazzara, G. Cavallaro, A. Panchal, R. Fakhrullin, A. Stavitskaya, V. Vinokurov, Y. Lvov, An assembly of organic-inorganic composites using halloysite clay nanotubes, Curr. Opin. Colloid Interface Sci. 35 (2018) 42-50. https://doi.org/10.1016/j.cocis.2018.01.002

[16] E.P. Giannelis, Polymer layered silicate nanocomposites, Adv. Mater. 8 (1996) 29-35. https://doi.org/10.1002/adma.19960080104

[17] A.H. Ambre, K.S. Katti, D.R. Katti, Nanoclay based composite scaffolds for bone tissue engineering applications, J. Nanotechnol. Eng. Med. 1 (2010) 1-9. https://doi.org/10.1115/1.4002149

[18] M.I. Carretero, M. Pozo, Clay and non-clay minerals in the pharmaceutical and cosmetic industries Part II. Active ingredients, Appl. Clay Sci. 47 (2010) 171-181. https://doi.org/10.1016/j.clay.2009.10.016

[19] S. Shahidi, M. Ghoranneviss, Effect of plasma pretreatment followed by nanoclay loading on flame retardant properties of cotton fabric, J. Fusion Energy. 33 (2014) 88-95. https://doi.org/10.1007/s10894-013-9645-6

[20] E. Cudjoe, S. Khani, A.E. Way, M.J.A. Hore, J. Maia, S.J. Rowan, Biomimetic reversible heat-stiffening polymer nanocomposites, ACS Cent. Sci. 3 (2017) 886-894. https://doi.org/10.1021/acscentsci.7b00215

[21] V.K. Thakur, M.R. Kessler, Self-healing polymer nanocomposite materials: A review, Polymer (Guildf). 69 (2015) 369-383. https://doi.org/10.1016/j.polymer.2015.04.086

[22] R. Shah, A. Kausar, B. Muhammad, S. Shah, Progression from graphene and graphene oxide to high performance polymer-based nanocomposite: A review, Polym. - Plast. Technol. Eng. 54 (2015) 173-183. https://doi.org/10.1080/03602559.2014.955202

[23] Y. Zhang, S.-J. Park, In situ shear-induced mercapto group-activated graphite nanoplatelets for fabricating mechanically strong and thermally conductive elastomer

composites for thermal management applications, Compos. Part A Appl. Sci. Manuf. 112 (2018) 40-48. https://doi.org/10.1016/j.compositesa.2018.06.004

[24] A. Abdelraheem, A.H. El-Shazly, M.F. Elkady, Synthesis and characterization of intercalated polyaniline-clay nanocomposite using supercritical CO2, AIP Conf. Proc. 1968 (2018). https://doi.org/10.1063/1.5039186

[25] L.A. Utracki, M.R. Kamal, Clay-containing polymeric nanocomposites, Arab. J. Sci. Eng. 27 (2002).

[26] U.U. Ozkose, C. Altinkok, O. Yilmaz, O. Alpturk, M.A. Tasdelen, In-situ preparation of poly(2-ethyl-2-oxazoline)/clay nanocomposites via living cationic ring-opening polymerization, Eur. Polym. J. 88 (2017) 586-593. https://doi.org/10.1016/j.eurpolymj.2016.07.004

[27] A. Saad, K. Jlassi, M. Omastová, M.M. Chehimi, Clay/conductive polymer nanocomposites, in: K. Jlassi, M.M. Chehimi, S. Thomas (Eds.), Clay-polymer nanocomposites, Elsevier, 2017: pp. 199-237. https://doi.org/10.1016/B978-0-323-46153-5.00006-9

[28] Z. Cherifi, B. Boukoussa, A. Zaoui, M. Belbachir, R. Meghabar, Structural, morphological and thermal properties of nanocomposites poly(GMA)/clay prepared by ultrasound and in-situ polymerization, Ultrason. Sonochem. 48 (2018) 188-198. https://doi.org/10.1016/j.ultsonch.2018.05.027

[29] V.S. Vo, S. Mahouche-Chergui, J. Babinot, V.H. Nguyen, S. Naili, B. Carbonnier, Photo-induced SI-ATRP for the synthesis of photoclickable intercalated clay nanofillers, RSC Adv. 6 (2016) 89322-89327. https://doi.org/10.1039/C6RA14724K

[30] M. Guerrouache, S. Mahouche-Chergui, M.M. Chehimi, B. Carbonnier, Site-specific immobilisation of gold nanoparticles on a porous monolith surface by using a thiol-yne click photopatterning approach, Chem. Commun. 48 (2012) 7486-7488. https://doi.org/10.1039/c2cc33134a

[31] V. Georgiadou, C. Kokotidou, B. Le Droumaguet, B. Carbonnier, T. Choli-Papadopoulou, C. Dendrinou-Samara, Oleylamine as a beneficial agent for the synthesis of CoFe2O4 nanoparticles with potential biomedical uses, J. Chem. Soc. Dalt. Trans. 43 (2014) 6377-6388. https://doi.org/10.1039/C3DT53179A

[32] A.K. Nikolaidis, D.S. Achilias, G.P. Karayannidis, Synthesis and characterization of PMMA/organomodified montmorillonite nanocomposites prepared by in situ bulk polymerization, Ind. Eng. Chem. Res. 50 (2011) 571-579. https://doi.org/10.1021/ie100186a

[33] S. Dadashi-Silab, M. Atilla Tasdelen, Y. Yagci, Photoinitiated atom transfer radical polymerization: Current status and future perspectives, J. Polym. Sci. Part A Polym. Chem. 52 (2014) 2878-2888. https://doi.org/10.1002/pola.27327

[34] A. Chakrabarty, L. Zhang, K.A. Cavicchi, R.A. Weiss, N.K. Singha, Tailor-made fluorinated copolymer/clay nanocomposite by cationic RAFT assisted pickering miniemulsion polymerization, Langmuir. 31 (2015) 12472-12480. https://doi.org/10.1021/acs.langmuir.5b01799

[35] M.A. Tasdelen, J. Kreutzer, Y. Yagci, In situ synthesis of polymer/clay nanocomposites by living and controlled/ living polymerization, Macromol. Chem. Phys. 211 (2010) 279-285. https://doi.org/10.1002/macp.200900590

[36] D. Schmidt, D. Shah, E.P. Giannelis, New advances in polymer/layered silicate nanocomposites, Curr. Opin. Solid State Mater. Sci. 6 (2002) 205-212. https://doi.org/10.1016/S1359-0286(02)00049-9

[37] S. Sinha Ray, M. Okamoto, Polymer/layered silicate nanocomposites: a review from preparation to processing, Prog. Polym. Sci. 28 (2003) 1539-1641. https://doi.org/10.1016/j.progpolymsci.2003.08.002

[38] D. Senbet, Measuring the impact and International transmission of monetary policy: A factor-augmented vector autoregressive (favar) approach, Eur. J. Econ. Financ. Adm. Sci. 7 (2008) 121-143.

[39] V.-S. Vo, S. Mahouche-Chergui, V.-H. Nguyen, S. Naili, N.K. Singha, B. Carbonnier, Chemical and photochemical routes toward tailor-made polymer-clay nanocomposites: Recent progress and future prospects, in: K. Jlassi, M.M. Chehimi, S. Thomas (Eds.), Clay-Polymer Nanocomposites, Elsevier, 2017: pp. 145-197. https://doi.org/10.1016/B978-0-323-46153-5.00005-7

[40] Y. Huang, K. Yang, J.Y. Dong, Copolymerization of ethylene and 10-undecen-1-ol using a montmorillonite-intercalated metallocene catalyst: Synthesis of polyethylene/montmorillonite nanocomposites with enhanced structural stability, Macromol. Rapid Commun. 27 (2006) 1278-1283. https://doi.org/10.1002/marc.200600131

[41] S. Abedi, M. Abdouss, A review of clay-supported Ziegler-Natta catalysts for production of polyolefin/clay nanocomposites through in situ polymerization, Appl. Catal. A Gen. 475 (2014) 386-409. https://doi.org/10.1016/j.apcata.2014.01.028

[42] M. Asensio, M. Herrero, K. Núñez, R. Gallego, J.C. Merino, J.M. Pastor, In situ polymerization of isotactic polypropylene sepiolite nanocomposites and its copolymers

by metallocene catalysis, Eur. Polym. J. 100 (2018) 278-289.
https://doi.org/10.1016/j.eurpolymj.2018.01.034

[43] Z. Salmi, K. Benzarti, M.M. Chehimi, Diazonium cation-exchanged clay: An efficient, unfrequented route for making clay/polymer nanocomposites, Langmuir. 29 (2013) 13323-13328. https://doi.org/10.1021/la402710r

[44] Z. Yang, H. Peng, W. Wang, T. Liu, Crystallization behavior of poly(ε-caprolactone)/layered double hydroxide nanocomposites, J. Appl. Polym. Sci. 116 (2010) 2658-2667. https://doi.org/10.1002/app.31787

[45] L. Zang, J. Qiu, C. Yang, E. Sakai, Preparation and application of conducting polymer/Ag/clay composite nanoparticles formed by in situ UV-induced dispersion polymerization, Sci. Rep. 6 (2016) 1-12. https://doi.org/10.1038/s41598-016-0001-8

[46] M. Herrero, K. Núñez, R. Gallego, J.C. Merino, J.M. Pastor, Control of molecular weight and polydispersity in polyethylene/needle-like shaped clay nanocomposites obtained by in situ polymerization with metallocene catalysts, Eur. Polym. J. 75 (2016) 125-141. https://doi.org/10.1016/j.eurpolymj.2015.12.005

[47] D.E. Kherroub, M. Belbachir, S. Lamouri, Synthesis of poly(furfuryl alcohol)/montmorillonite nanocomposites by direct in-situ polymerization, Bull. Mater. Sci. 38 (2015) 57-63. https://doi.org/10.1007/s12034-014-0818-3

[48] B.R. Prado, J.R. Bartoli, Synthesis and characterization of PMMA and organic modified montmorilonites nanocomposites via in situ polymerization assisted by sonication, Appl. Clay Sci. 160 (2018) 132-143.
https://doi.org/10.1016/j.clay.2018.02.035

[49] S. Sharma, M. Kumar Poddar, V.S. Moholkar, Enhancement of thermal and mechanical properties of poly(MMA-co-BA)/Cloisite 30B nanocomposites by ultrasound-assisted in-situ emulsion polymerization, Ultrason. Sonochem. 36 (2017) 212-225. https://doi.org/10.1016/j.ultsonch.2016.11.029

[50] R.S. Cardoso, V.O. Aguiar, M. de Fátima V Marques, Masterbatches of polypropylene/clay obtained by in situ polymerization and melt-blended with commercial polypropylene, J. Compos. Mater. 51 (2017) 3547-3556.
https://doi.org/10.1177/0021998317690444

[51] J. Hua, J. Liu, X. Wang, Z. Yue, H. Yang, J. Geng, A. Ding, Structure and properties of a cis-1,4-polybutadiene/organic montmorillonite nanocomposite prepared via in situ polymerization, J. Macromol. Sci. Part B Phys. 56 (2017) 451-461.
https://doi.org/10.1080/00222348.2017.1327318

[52] C. Dietlin, S. Schweizer, P. Xiao, J. Zhang, F. Morlet-Savary, B. Graff, J.P. Fouassier, J. Lalevée, Photopolymerization upon LEDs: New photoinitiating systems and strategies, Polym. Chem. 6 (2015) 3895-3912. https://doi.org/10.1039/C5PY00258C

[53] M. Chen, M. Zhong, J.A. Johnson, Light-controlled radical polymerization: mechanisms, methods, and applications, Chem. Rev. 116 (2016) 10167-10211. https://doi.org/10.1021/acs.chemrev.5b00671

[54] K. Jlassi, S. Chandran, M. Mičušik, M. Benna-Zayani, Y. Yagci, S. Thomas, M.M. Chehimi, Poly(glycidyl methacrylate)-grafted clay nanofiller for highly transparent and mechanically robust epoxy composites, Eur. Polym. J. 72 (2015) 89-101. https://doi.org/10.1016/j.eurpolymj.2015.09.004

[55] S. Shanmugam, C. Boyer, Stereo-, Temporal and chemical control through photoactivation of living radical polymerization: synthesis of block and gradient copolymers, J. Am. Chem. Soc. 137 (2015) 9988-9999. https://doi.org/10.1021/jacs.5b05903

[56] M. Chen, J.A. Johnson, Improving photo-controlled living radical polymerization from trithiocarbonates through the use of continuous-flow techniques, Chem. Commun. 51 (2015) 6742-6745. https://doi.org/10.1039/C5CC01562F

[57] M. Arslan, M.A. Tasdelen, Polymer nanocomposites via click chemistry reactions, Polymers (Basel). 9 (2017). https://doi.org/10.3390/polym9100499

[58] Z. Zhang, P. Zhang, Y. Wang, W. Zhang, Recent advances in organic-inorganic well-defined hybrid polymers using controlled living radical polymerization techniques, Polym. Chem. 7 (2016) 3950-3976. https://doi.org/10.1039/C6PY00675B

[59] P. Yadav, S. Chacko, G. Kumar, R. Ramapanicker, V. Verma, Click chemistry route to covalently link cellulose and clay, Cellulose. 22 (2015) 1615-1624. https://doi.org/10.1007/s10570-015-0594-2

[60] Y. Zou, L. Zhang, L. Yang, F. Zhu, M. Ding, F. Lin, Z. Wang, Y. Li, "Click" chemistry in polymeric scaffolds: Bioactive materials for tissue engineering, J. Control. Release. 273 (2018) 160-179. https://doi.org/10.1016/j.jconrel.2018.01.023

[61] P.B. Zetterlund, S.C. Thickett, S. Perrier, E. Bourgeat-Lami, M. Lansalot, Controlled/living radical polymerization in dispersed systems: an update, Chem. Rev. 115 (2015) 9745-9800. https://doi.org/10.1021/cr500625k

[62] N. Ballard, M. Salsamendi, P. Carretero, J.M. Asua, An investigation into the nature and potential of in-situ surfactants for low energy miniemulsification, J. Colloid Interface Sci. 458 (2015) 69-78. https://doi.org/10.1016/j.jcis.2015.07.041

[63] Y. Wang, S. Dadashi-Silab, K. Matyjaszewski, Photoinduced miniemulsion atom transfer radical polymerization, ACS Macro Lett. 7 (2018) 720-725. https://doi.org/10.1021/acsmacrolett.8b00371

[64] F. Zhang, S. Li, G.D. Electrolysis, L. Yan, Y. Mengqi, Y. Reyes, Synthesis of polymer hybrid latex poly (methyl methacrylate-co-butyl acrylate) with organo montmorillonite via miniemulsion polymerization method for barrier paper, J. Phys.: Conf. Ser. 985 (2017) 012029. https://doi.org/10.1088/1742-6596/910/1/012029

[65] K. Buruga, J.T. Kalathi, Fabrication of γ-MPS-modified HNT-PMMA nanocomposites by ultrasound-assisted miniemulsion polymerization, JOM. 70 (2018) 1307-1312. https://doi.org/10.1007/s11837-018-2829-9

[66] A. Gurses, Introduction to polymer-Clay nanocomposites, Jenny Stanford Publishing, 2015. https://doi.org/10.1201/b18716

[67] S. Merritt, C. Wan, B. Shollock, S. Patole, D.M. Haddleton, Polymer/Graphene Nanocomposites for Food Packaging, in: G. Cirillo, M.A. Kozlowski, U.G. Spizzirri (Eds.), Compos. Mater. Food Packag., John Wiley & Sons, 2018, pp. 251-267. https://doi.org/10.1002/9781119160243.ch8

[68] F.Y. Fayc, Development of antimicrobial PCL/nanoclay nanocomposite films with enhanced mechanical and water vapor barrier properties for packaging applications, Polym. Bull. 72 (2015) 235-254. https://doi.org/10.1007/s00289-014-1269-0

[69] J.M. Kim, M.H. Lee, J.A. Ko, D.H. Kang, H. Bae, H.J. Park, Influence of food with high moisture content on oxygen barrier property of polyvinyl alcohol (PVA)/vermiculite nanocomposite coated multilayer packaging film, J. Food Sci. 83 (2018) 349-357. https://doi.org/10.1111/1750-3841.14012

[70] A.Y. Malkin, A. Isayev, Applications of rheology, in: A.Y. Malkin, A. Isayev (Eds.), Rheology, ChemTec Publishing, 2017, pp. 377-432. https://doi.org/10.1016/B978-1-927885-21-5.50012-6

[71] J. Yang, S. Tighe, A review of advances of nanotechnology in asphalt mixtures, Procedia-Soc. Behav. Sci. 96 (2013) 1269-1276. https://doi.org/10.1016/j.sbspro.2013.08.144

[72] F. Guo, S. Aryana, An experimental investigation of nanoparticle-stabilized CO2 foam used in enhanced oil recovery, Fuel. 186 (2016) 430-442. https://doi.org/10.1016/j.fuel.2016.08.058

[73] A. Kara, N. Tekin, A. Alan, A. Şafaklı, Physicochemical parameters of Hg(II) ions adsorption from aqueous solution by sepiolite/poly(vinylimidazole), J. Environ. Chem. Eng. 4 (2016) 1642-1652. https://doi.org/10.1016/j.jece.2016.02.028

[74] X. Liu, C. Cheng, C. Xiao, D. Shao, Z. Xu, J. Wang, S. Hu, X. Li, W. Wang, Polyaniline (PANI) modified bentonite by plasma technique for U(VI) removal from aqueous solution, Appl. Surf. Sci. 411 (2017) 331-337. https://doi.org/10.1016/j.apsusc.2017.03.095

[75] X. Wang, Y. Du, J. Luo, B. Lin, J.F. Kennedy, Chitosan/organic rectorite nanocomposite films: Structure, characteristic and drug delivery behaviour, Carbohydr. Polym. 69 (2007) 41-49. https://doi.org/10.1016/j.carbpol.2006.08.025

[76] K. Fukushima, D. Tabuani, M. Arena, M. Gennari, G. Camino, Effect of clay type and loading on thermal, mechanical properties and biodegradation of poly(lactic acid) nanocomposites, React. Funct. Polym. 73 (2013) 540-549. https://doi.org/10.1016/j.reactfunctpolym.2013.01.003

[77] H. Xiang, M. Xia, A. Cunningham, W. Chen, B. Sun, M. Zhu, Mechanical properties of biocompatible clay/P(MEO2MA-co-OEGMA) nanocomposite hydrogels, J. Mech. Behav. Biomed. Mater. 72 (2017) 74-81. https://doi.org/10.1016/j.jmbbm.2017.04.026

[78] G. Kapusetti, N. Misra, V. Singh, S. Srivastava, P. Roy, K. Dana, P. Maiti, Bone cement based nanohybrid as a super biomaterial for bone healing, J. Mater. Chem. B. 2 (2014) 3984-3997. https://doi.org/10.1039/C4TB00501E

[79] S. Noori, M. Kokabi, Z.M. Hassan, Nanoclay enhanced the mechanical properties of poly(vinyl alcohol)/chitosan/montmorillonite nanocomposite hydrogel as wound dressing, Procedia Mater. Sci. 11 (2015) 152-156. https://doi.org/10.1016/j.mspro.2015.11.023

[80] G. Sandri, C. Aguzzi, S. Rossi, M.C. Bonferoni, G. Bruni, C. Boselli, A.I. Cornaglia, F. Riva, C. Viseras, C. Caramella, F. Ferrari, Halloysite and chitosan oligosaccharide nanocomposite for wound healing, Acta Biomater. 57 (2017) 216-224. https://doi.org/10.1016/j.actbio.2017.05.032

[81] M. Liu, Y. Shen, P. Ao, L. Dai, Z. Liu, C. Zhou, The improvement of hemostatic and wound healing property of chitosan by halloysite nanotubes, RSC Adv. 4 (2014) 23540-23553. https://doi.org/10.1039/C4RA02189D

[82] H.A. Heydary, E. Karamian, E. Poorazizi, A. Khandan, J. Heydaripour, A novel nano-fiber of iranian gum tragacanth-polyvinyl alcohol/nanoclay composite for wound healing applications, Procedia Mater. Sci. 11 (2015) 176-182. https://doi.org/10.1016/j.mspro.2015.11.079

[83] M.W. Sabaa, H.M. Abdallah, N.A. Mohamed, R.R. Mohamed, Synthesis, characterization and application of biodegradable crosslinked carboxymethyl chitosan/poly(vinyl alcohol) clay nanocomposites, Mater. Sci. Eng. C. 56 (2015) 363-373. https://doi.org/10.1016/j.msec.2015.06.043

[84] T. Ge, J.T. Kalathi, J.D. Halverson, G.S. Grest, M. Rubinstein, Nanoparticle motion in entangled melts of linear and nonconcatenated ring polymers, Macromolecules. 50 (2017) 1749-1754. https://doi.org/10.1021/acs.macromol.6b02632

[85] S. Pacelli, P. Paolicelli, G. Moretti, S. Petralito, S. Di Giacomo, A. Vitalone, M.A. Casadei, Gellan gum methacrylate and laponite as an innovative nanocomposite hydrogel for biomedical applications, Eur. Polym. J. 77 (2016) 114-123. https://doi.org/10.1016/j.eurpolymj.2016.02.007

[86] S. Karnik, U.M. Jammalamadaka, K.K. Tappa, R. Giorno, D.K. Mills, Performance evaluation of nanoclay enriched anti-microbial hydrogels for biomedical applications, Heliyon. 2 (2016) e00072. https://doi.org/10.1016/j.heliyon.2016.e00072

[87] J.-H. Yang, J.-H. Lee, H.-J. Ryu, A.A. Elzatahry, Z.A. Alothman, J.-H. Choy, Drug-clay nanohybrids as sustained delivery systems, Appl. Clay Sci. 130 (2016) 20-32. https://doi.org/10.1016/j.clay.2016.01.021

[88] M. Roozbahani, M. Kharaziha, R. Emadi, pH sensitive dexamethasone encapsulated laponite nanoplatelets: Release mechanism and cytotoxicity, Int. J. Pharm. 518 (2017) 312-319. https://doi.org/10.1016/j.ijpharm.2017.01.001

[89] A. Carrero, R. van Grieken, I. Suarez, B. Paredes, Development of a new synthetic method based on in situ strategies for polyethylene/clay composites, J. Appl. Polym. Sci. 126 (2012) 987-997. https://doi.org/10.1002/app.36830

[90] F.B. Zeynabad, R. Salehi, M. Mahkam, Design of pH-responsive antimicrobial nanocomposite as dual drug delivery system for tumor therapy, Appl. Clay Sci. 141 (2017) 23-35. https://doi.org/10.1016/j.clay.2017.02.015

[91] K.M. Rao, S. Nagappan, D.J. Seo, C.-S. Ha, pH sensitive halloysite-sodium hyaluronate/poly(hydroxyethyl methacrylate) nanocomposites for colon cancer drug delivery, Appl. Clay Sci. 97-98 (2014) 33-42. https://doi.org/10.1016/j.clay.2014.06.002

Chapter 2

Micro and Nano Clay-Biopolymer Composites for Drug Delivery

Asma Jabeen, Haq Nawaz Bhatti* and Amina Khan

Department of Chemistry, University of Agriculture Faisalabad, Pakistan

* hnbhatti2005@yahoo.com

Abstract

Clay-based biopolymers are among the most efficient and cost-effective composite materials that find applications in a variety of biomedical applications such as drug delivery. This chapter elaborates the different physical, chemical, and structural properties of clay biopolymeric composites, which makes them most reliable to use for the applications of drug delivery. The main focus of writing this chapter is to explore the composition, structural properties, and route of action of clay-based nanomaterials, along with elaborating the mechanism of the drug delivery system. The synthesis of clay-encapsulated drug formulation and their route of administration has been investigated in this chapter. Data for the synthesis of composites and in vitro study has also been included from the existing literature.

Keywords

Biopolymers, Drug-Delivery, Administration, In Vitro Study, Synthesis

Contents

1.   Introduction

During the development of new medicines/drugs, delivery processes and carriers are key concerns of the researchers. Many technologies, particularly micro and nanoencapsulation of drugs are key interests of scientists [1, 2]. Mostly a controlled delivery mechanism is considered as most efficient for delivering medicine in a controlled manner for hours, days, or weeks. The controlled mechanism of drugs-delivery is a procedure in which newly developed micro and nanostructured materials can proficiently encapsulate the high concentration of drugs, and releases it at the targeted site by crossing the cell membrane within a prescribed period of time [3, 4]. Such a mechanism for drug delivery can reduce patient expenses as well as the risks of toxicity. It is also beneficial due to drug-releasing efficacy, specificity of the targeted area, therapeutic indices, and increasing tolerability of corresponding drugs [5]. Therefore, building such a responsive controlled drug delivery system is of crucial significance in the field of clinical sciences. Researchers are also working for the possible applications of targeted drug delivery systems, by producing tiny drugs particles with virtuous blood circulation [6, 7].

A number of excipients are available for drug delivery systems, where the use of polymers and polymeric compounds in a conventional pharmaceutical composition as coatings, binders, suspending agents, and surfactants are trending. A composite is a composition that consists of two or more dispersion components having optimized properties as compared to pure materials [8-10]. These composites might have micro or nanometric sizes in at least one dimension, so we consider this system as micro or nanocomposite. Such kinds of hybrid materials can be obtained by incorporating different polymeric materials into clay (layers, channels) or by dispersing platelets, with enhanced thermal stability and swelling modulation of polymeric composites in aqueous media [11, 12].

In the last decade, different micro and nano clay-based composites materials, biopolymers, and minerals have the great attention of scientists for pharmaceutical applications. Different groups of drug delivery systems have recently been employing drug delivery applications are based on inorganic clay minerals, have the ability to absorb and release drugs in a controlled manner [13-16]. These groups include minerals such as; mesoporous silica, layered double hydroxides, halloysite, and montmorillonite (MMT). Micro silica gels lyophilized to form a dry xerogel with optimal properties were used by Kurczewska *et al.* [17] for wound therapeutic and drug delivery applications.

Montmorillonite materials have a higher surface area, absorption capacity excellent mucoadhesive properties, and higher ion-exchange capacity (cation exchange, 80 mEq/100 mg) as compared to other clay minerals. Kaolin clay minerals with other members of this group such as; kaolinite, nacrite, halloysite, and dickite, have unique structures that are

Adv. App. of Micro and Nano Clay II – Synthetic Polymer Composites   Materials Research Forum LLC
Materials Research Foundations 129 (2022) 24-52   https://doi.org/10.21741/9781644902035-2

composed of tetrahedral-octahedral double layers at ratios 1:1 [18-20]. The tetrahedral-octahedral layers contain Si and Al, O, OH groups at their structures. Most of the clay minerals are pseudo-hexagonal in shape with 300 and 50 nm in diameter and thickness respectively. Based on these arrangements, these tetrahedral-octahedral films categorize the clay minerals named; hectorite, kaolinite, montmorillonite, and halloysite. Sometimes these pharmaceutical-grade kaolinites minerals got contaminated due to the presence of other minerals [21-23].

Since ancient's times, these clay minerals have been extensively used for biomedical applications such as anti-inflammatory agents, antidiarrhea, reducing infections, blood purification, and healing of stomach ulcers. Due to extraordinary surface properties such as; high porosity and large surface area, low density, chemical-thermal stabilities, and controlled drug release process. In addition, excellent biocompatibility makes them able to use as oral antioxidants [24-26].

The interest in clay-minerals-based drug delivery composites amongst the scientific community has dramatically increased so far due to their composition and modification ability which make them able to serve for different purposes. Due to extensive structural flexibility and small particle size, these minerals have been broadly modified to make them suitable for a particular application [27-29]. The natural clay minerals are limited to use due to various factors such as, presence of impurities that affect the clarity of products and instability while the other factors such as the basic structure of a collection of platelets are maintained [30, 31]. On the other hand, synthetic has a well-purified structure that's why its use is continuously expanding from cosmetics industries and personal care product (PCP's) industry to the pharmaceutical industrial sector, where it is being used as anti-diarrheal, anti-bacterial, drug deliveries, acid scavenging applications, and tissue engineering [32-34].

Due to demanding applications of clay-based minerals in the pharmaceutical sector, the principal objective of this article is to provide an outlook of possibilities of the use of micro and nano clay biopolymers and composites for drug delivery applications in controlled and targeted mode.

2.   Medicinal properties of clay biopolymeric composites

Some of the important structural and functional properties make the clay particles unique and preferable to work as drug delivery careers.

2.1 Structural features and physical properties

Clays are fine-grounded crystalline layered based structural materials and are proven to arrange themselves in such a manner to look like well-arranged pages of a book [35, 36]. Each layer is composed of tetrahedral and octahedral layers, this unique layers arrangement makes these clay minerals show distinguishing properties. The tetrahedral layer consists of the silicon-oxygen tetrahedra which shared three corners to link neighboring tetrahedra and the fourth corner of an individual tetrahedron attached as an adjacent part of the octahedral sheet [37]. Similarly, the octahedral layers usually consist of aluminum or and coordinate with oxygen and hydroxyl in a six-fold to from a tetrahedral sheet [18].

2.2 Structural properties

Clay-based nanomaterials divide themselves into cationic and anionic layers reliant on interlayer ions that make the layer charge. The clay nanomaterials consisting of an anionic layer have been typically composed of layered-double hydroxides (LDH's) with moveable anions at interlayer spaces [38]. LDH's include a variety of chemical compounds produce various poly-types of layered structures, such as; MMT's $AlNal_2SiO_3$ nanoparticles (cationic) have highly layered internal-surface. The core structure consists of an octagonal-twisted sheet composed of quadrilateral layers [39, 40].

2.3 Toxicity

Naturally occurring clay-minerals work appropriately to modulate drug delivery. High binding sites, as well as good adsorbing capability, inertness, minimum or least toxicity, and rheological properties, make the clay-minerals revolutionary carriers for drug delivery [41]. The advantageous effects of drug delivery carriers for human health are depending on the selection of targeting sites of a specific quantity of drug for the desired time. The therapeutic application should be achieved at a specific amount without exceeding the permissible toxicity level or tumbling the lowest effective level [42]. Some experiments on mice were conducted [43] to check out the significant effect of dose of LDH nano-materials on the behavior in an oral administration, various factors such as; change in body weight, persistency, and organo-somatic-index were investigated using an amount of 2000mg.kg-1 in a 14-days period of time. The concentration of LDH in plasma was found to be decreased rapidly within 30min reliant on exposure doses, while the excretion of LDH nanomaterials was observed after 24h in urine and feces. Another study was carried out by Kevadiya and Bajaj [45] to check out the toxicity of naturally occurring montmorillonite clay-based nano-silicate platelets (NSP) by feeding it to rats in an acute oral administration, low toxic lethal dose (LD50) was found in Sprague–Dawley rats (both male and female) was greater than 5700 mg kg-1 of the body weight. But the overall study confirmed the safety for the use of NSP in biomedical areas [46].

2.4   Entrapment efficiency

Entrapment efficiency is an important property of clay composites in pharmaceutical applications, high loading of drug could be achieved by encapsulating agents using retardant polymers within the inner spaces of micro/nanotubules, cationic coating helped to restrained delivery rate, and by intercalated with the swiped -water if present [47, 19]. The low encapsulation ability of drugs is the main area of interest of scientists for developing a polymer-based drug delivery system. The major disadvantage of most drug delivery systems is the outflow of drug particles during the encapsulation or cross-linking process. Another disadvantage of some hydrophilic-biodegradable polymers such as; alginate and poly-L-lactide is their high porosity and fast disintegration which lead to a fast drug release. An alternative approach introduced for improved drug release and efficient entrapment is based on the incorporation of the use of water-insoluble materials, such as clay minerals [20]. The entrapment methods of different clay minerals and biopolymers could differ according to the compositions of composites [21].

2.5   Thermal stability

The combination of clay minerals with organic species enhances the thermostability of composites. The thermostability of clay- mineral-based organic species might be up to 750°C indicates the available interactions in the clay sheet (oxygen planes) and aromatic ring, also due to the presence of shielding effect between the aluminum and silicate layers. Udayakumar *et al.* [11] found bentonite useful to increase the temperature bearing capacity and used the bentonite-based composites for naproxen release in a controlled manner. Different analytical techniques found useful for the determination of mechanical-thermal properties of different clay-based composite and nanofibers such techniques are; dynamic mechanical-thermal-analysis (DMTA) (for tensile strength determination) and TGA-DSC analysis [48, 49].

2.6   Biodegradability

Clay (MMT) particles are extensively used for the formation of novel systems suitable for drug-delivery applications. Recently polyurethanes hybrid clay composites were prepared [50], and an organized enhanced structure was formed having long chains, hydrogen bonding expands the chain into a long chain aliphatic structure. A large cluster of aliphatic chains is enhanced by the presence of nano-clay structures. During in vitro studies, it was examined that the release of drugs from MMT-microparticles was carried out by following the partial diffusion, in which swollen matrix layers act as drug carriers (structures). Enzymes media of the body controlled the biodegradation of nano-hybrids (with nano-clay) polyurethanes structures, these structures provide a sustained and controlled delivery of drug through a long aliphatic chain structure which forms an indirect path for delayed diffusion [5, 51].

2.7   Biocompatibility

Clays in the form of tubules and halloysite are mostly found as biocompatible as well as environment friendly. It has been reported [29] that, montmorillonite and polyurethane-based nanocomposites can be used for the control-release of dexamethasone acetate, and their biocompatibility studies of drug release mechanism verified the suitable interfacial interaction between subcutaneous tissue and polyurethane by showing a minute inflammatory response that was later observed to be completely resolved in a 14-days period of time [52]. In another study, montmorillonite-chitosan-based nanocomposite also showed good biocompatibility and mucoadhesive properties in addition to their low solubility in acidic media. Biocompatibility of the above-mentioned composites has extensively been substantiated by aggregation, adhesion, and hemolysis experimentation [53, 54]

Table 1. Clay minerals used for medicinal purposes

Clay based composites	Releasing site	Used for drugs release	Mechanism of delivery	References
HNT based- Folic Acid magnetic formulation	HeLa	Doxorubicin	Targeted	[59]
MMT-	MCF-7	Pacalitaxel	Targeted	[59]
Halloysite-PLGA	Intestine	Tetracycline hydrochloride	Prolonged release	[59]
Chitosan-MMT	clone	Diclofenac-Na	Controlled release	[48]
Chitosan-MMT	Gastro-intestinal trac	Vitaminn-B12	Pulsatile	[48]
Alginate-MMT composition	colon	Mesalazine	Targeted	[60]
MMT-Hybrid Composites	gastrointestinal tract	7-azaindole	Controlled	[39]
HNT-Micro composites	gastrointestinal tract	metoclopramide hydrochloride	Sustained	[21]
PEG-CMC-clay composites	Intestinal Fluid	Ibuprofen	Controlled	[42]
MMT-Hydrogels	Simulated buffer solution	STH	Direct release	[54]

Adv. App. of Micro and Nano Clay II – Synthetic Polymer Composites   Materials Research Forum LLC
Materials Research Foundations 129 (2022) 24-52   https://doi.org/10.21741/9781644902035-2

3.   Drug delivery mechanism

3.1   Mechanisms of clay- drug interactions

Naturally occurring clay-minerals are efficient inorganic cationic exchangers, mostly basic drugs experience ion exchange in solution [55]. The use of different types of clay minerals such as montmorillonite (MMT), halloysite, and saponite are trending in pharmaceutical composites-drug formulations due to their higher cationic exchange properties.

Drug-clay (montmorillonite) minerals interaction mechanisms followed the three major steps:

- Adsorption of the drug onto free sites of the montmorillonite
- Adsorption occurred after the replacement of sodium from the interlayers of montmorillonite
- The -OH groups from montmorillonite interlayers are replaced by drug and makes ionic-bonds with Mg^{2+} and Al^{3+}

The first two-step for drug adsorption mechanisms are considered as major steps for the absorption of the drug onto the interlayers of montmorillonite- clay interlayers by a concentration gradient [56, 57]. After the adjustment of the concentration gradient between the inner and outer sides of the montmorillonite, the diffusion mechanism would stop. In biological media, counter ions show the capability for the displacement and delivery of drug molecules from the substrate to the body by elimination of exchanger [58]. Fig. 1 shows the general mechanism for drug interaction with clay minerals [59].

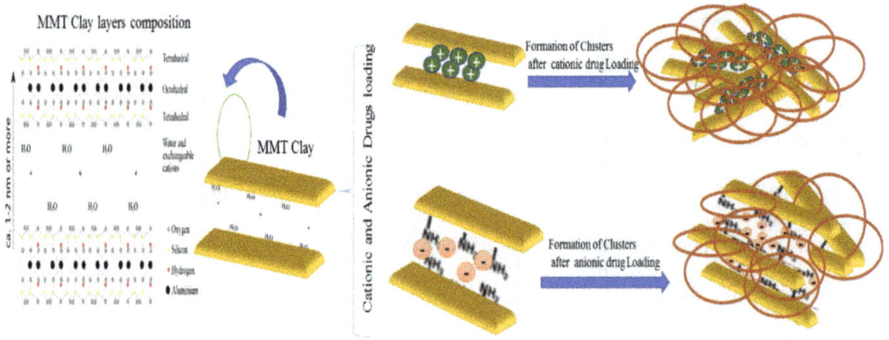

Figure 1. Cationic and anionic drug loading and release by montmorillonite (MMT) clay formulations

3.2   Clay composites targeted drug delivery pathway

The inimitable encrusted structure of montmorillonite (clay) enhances the stability of bioactive agents by electrostatic-interaction which helps in controlled delivering by ion-exchanging of drug molecules with other ionic interactions facilitated by the biological system [8, 60]. Montmorillonite (MMT) clay minerals are mostly reported for the drug-delivering through eye transdermal, colon, and gastro retentive targeted drug delivery. The complex ratio of montmorillonite-drug 20:1 with hydrated MMT boosts the adsorption of the drug in gastric fluid and at intestinal pH. These polyamide-based montmorillonite composites were used by Williams *et al.* [6] for sustained delivery of Metformin in oral administration. It is an anti-diabetic drug that is introduced by ion exchange with silicate layers on the surface of montmorillonite. Montmorillonite clay-based biodegradable nanoparticles are considered efficient for oral chemotherapeutic applications. Different techniques such as solvent extraction followed by evaporation method are being used for the synthesis of drug-loaded montmorillonite nanoparticles formulations and also being testing for in-vitro drug release efficiency of these composites [59] Fig. 2 shows the general mechanism for targeted drug delivery of Montmorillonite encapsulated drug.

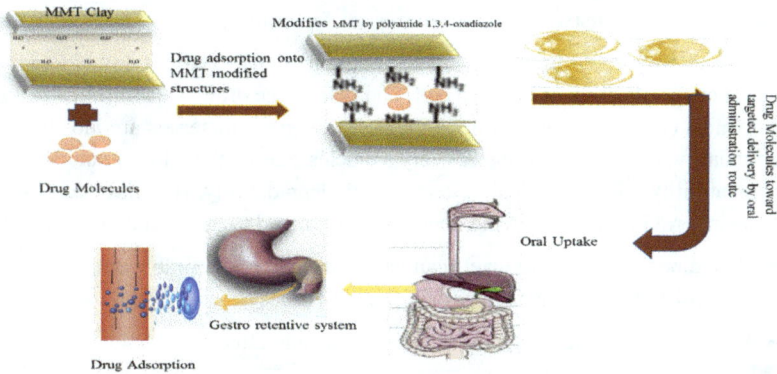

Figure 2. MMT-Clay composites encapsulated drugs delivery into gestor-retentive system

Examples of targeted release

Natural halloysite based composites nano-tubes were considered as more efficient to load drug molecules and showed a sustained delivery of drug up to 10 h. This long-term encapsulation of drugs inside nano-tubes is sported by tube-end stoppers. Cavallaro *et al.*

[61] prepared poly (Niso-propyl-acrylamide) using halloysite nanotube for targeted delivery of curcumin. In-vitro experiments for the loading and releasing of curcumin by halloysite nanotube into the gastrointestinal transit system. It was concluded that the curcumin was released into the targeted sites of the intestine. Fig. 3 presents the general mechanism of targeted delivery of curcumin drug by halloysites.

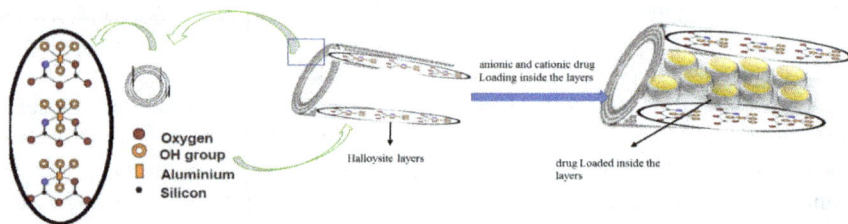

Figure 3. Targeted release of curcumin drug by halloysites nano tubes encapsulation

3.3   Controlled release by nano-composites for of drugs

An ideal and preferred drug delivery system has an ability to carry out the drug molecules to the exact site of action or targeted cells and deliver the drug in a controlled manner [62]. Biopolymers composites of natural inorganic and organic materials are mostly considered as efficient and promising drug delivery vehicles due to their low manufacturing cost, biocompatibility with biological system and biodegradability. A controlled drug delivery mechanism provides the following benefits as compared to targeted delivery [63].

- Maintained plasma drug levels can be achieved at the desirable range in a controlled administration mechanism with improved efficacy
- Also helpful for the reduction of detrimental side effects
- It is possible to deliver the drug in low medical supervision areas by using a controlled drug releasing pathway
- A short half-life could be possible to achieve during in vivo drug administration
- Regular but a small quantity of drug is very effective in this administration instead of several excessive doses

Overall this route has the capability to deliver drugs on action sites as well as negligible side effects. Release of the drug is possible to examine over a long duration of time [64].

In short, the mechanism of controlled drug delivery mainly aims to provide an improved and effective mechanism of drug administration Fig. 4 presents the general mechanism for the attachment of drug molecules by the replacement of Na$^+$ ion and delivery of drug to the action site.

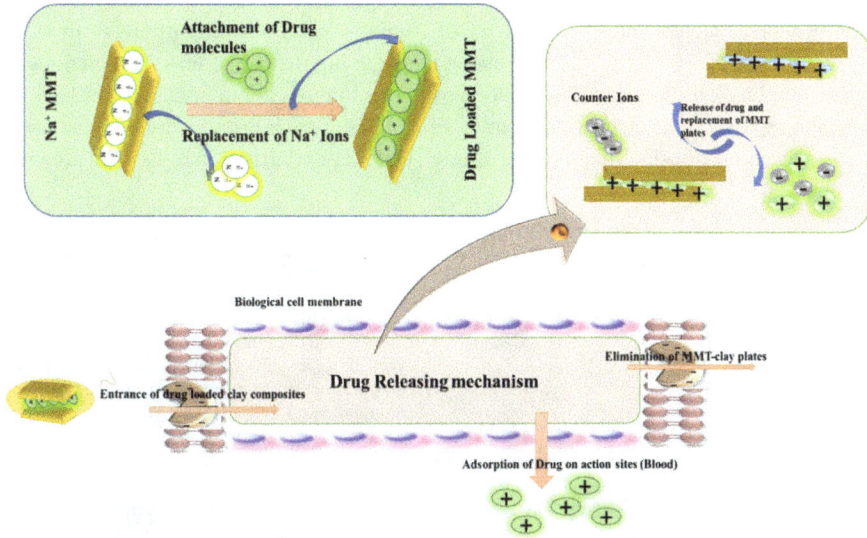

Figure 4. Controlled drug delivery mechanism into biological cell membrane

Examples of controlled released

Hong *et al*. [65] intercalated mesalazine with clay mineral, montmorillonite (MMT), and encapsulated it inside a pH sensitive polymeric alginate-hydrogel bead (product called as MCA-beads), MMT intercalation found as helpful to the prevent the mesalazine dissolution during encapsulation inside the polymers. results declared that the in vitro release of MCA beads showed >5% and ~20% release of mesalazine into gastric and intestinal conditions over 7h period of time.

4.   Administration routes of clay bio-composites for drug delivery

Several administration routes followed by the clays-based nano and micro composites have been reported by different researchers depending on the anticipated therapeutic action of physio-chemical properties of the drug. For example, systemic illnesses in the internal organ are mainly treated by intravenous injections or oral administration, while focal skin lacerations are usually treated topically [14]. The drug dosage depends on disease severity, targeted population, and kinetics of drugs. Similarly, for veterinary applications, a good administration pathway is in the form of food additives for animals, but for human applications, the HNT-clays formulations are more preferred for oral administration in the form of tablets and capsules [4]. Because of the non-biodegradability of HNTs in blood, its use is limited in the form of intravenous injection but the use of clay formulations for topical applications is promising, and most commonly used in the form of creams, bandages, sprays, bone cement and dentist fillers. There are a few examples of the use of clay-based composites for drug release through various administration routes that are given in Table 2. [17].

Table 2 Properties of various Drug Administration Routes

Route	Properties	Draw backs	Ref.
Oral route (in the form of capsules and tablets)	reliable, non-toxic, and convenient	Slow action mechanism, drugs Cause irritation cannot recommend, some Drugs might demolish due to Enzymatic action, e.g., insulin or penicillin etc., absorption of drugs might be ambiguous, cannot allowed for patient showed insentient and un-supportive behavior,	[34]
Except oral route use-full for intravenous and intramuscular injections	A definite beneficial for an immediate action, useable for insentient, noncooperative patients, and also for diarrhea and vomiting patient	May cause pain or sepsis, intravenous injection may react Or cause some kinds of complications, costly, for dosage some aseptic measure and practical skill essentially required	[17]
Transdermal administration: Surface application, in the form of powder, creams or sprays etc.,	The transfer of Active ingredients is direct to the targeted area, safe for liver or Gastro-intestinal trac	By using this route, a local effect might be observed instead of systematic	[68, 73]
Sprays: direct Inhalation route	Direct release of drug to the Lungs	Some kinds of delivery devices for spraying the drug are mandatory	[68]
Implantable administration route	Excess drug bio-availability, smallest dosing frequency	Requires surgery for drug administration, noxiousness of accompanying chemicals, Cause pain, may cause postoperative effect	[49]
In the form of Bone Cement	Increased Toughness and stiffness	Provides low mechanical strength	[66]
Powder application	Finer particles and batter dispersion	Low rate of formation, glomeration	[67]

4.1   Oral administration

Flat sheets of kaolin clay are being rolled into empty halloysite tubes having a diameter of about 50nm, mostly 10 to 15 layers of alumino-silicate rolled into cylindrical halloysite tubes. The assembling of halloysite cylindrical structure and drug could be either by encapsulation of drug molecules inside the halloysite-tube or might be chemically bound the drug onto the outer surface of tubes [4]. The encapsulations offer the improved drug-carrying ability, and the distribution of the drug arises equally in the inner lumen and outer surfaces. Besides that, the chemically bounded drug might be easily washed out from the outer surfaces of carriers. Dissolution of pH-sensitive polymeric composites such as methyl methacrylate could be used 'enteric coated' halloysite particles for drug loading that dissolve at pH 5–8 (pH of the intestine) [24]. The clay nanotubes formulation with carboxymethyl chitosan micro-particle mostly uses oral administration for the protection of albumin from gastrointestinal enzymatic degradation. Such formulations have been used for the removal of various mycotoxins (i.e. zearalenone and de-deoxynivalenol) from the stomach of chicken and piglets [68].

4.2   Intravenous injections

The halloysite-clay-based composites for drug loading followed the 2 major approaches; (I) loading in the inner lumen and (II) adsorption onto the surface by a covalent bond. These two approaches can be used in combination as a first release of drug by bust from the upper surface than slow from inner lumens of the halloysite-tubes. [69]. Sooner applications of kaolin, halloysite nano-clays, and montmorillonite formulations for cosmetic industrial products allow the loading efficiency of up to 10%, and up to 20h constant release of hexuronic acid, glycerol, vitamins, and some other bioactive elements. The loading of halloysite tubes for proteins and DNA is a very fascinating phenomenon, these nano-tube formulations worked as a container for the loading of DNA/proteins which have sizes of 3–8nm. Besides that, the internal-loading of negative proteins with the help of electrostatic interaction into the tubes is also possible by avoiding their external adsorption [70]. Halloysite clay nano-tubes have been reported as a promising and high loading intra-cellular drug-delivery vehicle, the major contests in this technique are; (I) to satisfy the operative uptake and successful delivery (II) to improve the releasing efficiency by various stimuli, cell-triggered and externals [27].

4.3   Topical uses

Topical use of the drug is usually applicable locally and externally, it diffuses by absorbing into the skin/epidermal layer to the action site. This release is mostly attained by making the clay biopolymeric composites-based creams, lotion, sprays, ointment, or patches of bandage or varnish from aqueous-dispersions [68].

4.4   Bone cement/ dentist fillers

Antibiotics loaded on HS-clay-based nano/micro containers and bone cement composed of calcium phosphate or $(C_5O_2H_8)n$ (PMMA) are trending applications of clay composites [71]. Different combinations of antibiotics mixed into the clay minerals formed efficient materials that are being used in a targeted drug delivery application. Halloysite clay doping bone-cement increased the strength and adhesiveness of bone by decreasing the PMMA's polymerization temperature from 66 to 50°C. Another application of halloysite formulation for antimicrobial bone reparents is in dentistry, the halloysite-doped resins of teeth colors are being used for dental caries. Halloysite based adhesives increased the strength of the composites up to 50% [72].

4.5   Creams

The use of clay-based composite in cosmetics is trending to provide good-looking and healthy skin [73]. Clay-based nanotubes carriers act as an active-agents in a controlled released pathway, consulting a slow penetration of loaded emollient agents such as; nutrients and vitamins into the skin [74] carried out the glycerin loading into the clay-nanotubes and its controlled release into the skin via absorption in an intervening period of 4-18h.

4.6   Sprays

Due to high zeta-potential halloysite clay nanotubes and the stability of their colloids in aqueous media, they have been used as drug loading antiseptics for antibacterial sprays. Brilliant green-loaded halloysite-nanotube sprays provide prolonged antimicrobial action. The usual releasing period of brilliant green is 3h for up to 70%, but this duration was increased up to 6 h using halloysite-nanotubes end-stoppers [73].

4.7   Bandages

The antibacterial agents loaded clay nanotubes loaded are being used in modern wound dressing. By shielding the injured area from pathogens, these agents provide a suitable environment for wound healing. Mostly alginate and gelatin/alginate-based HNT agents are reported for the formation of such kinds of bandages [17].

4.8   Pour powder

The use of HNTs and chitosan oligosaccharides in powder formation for wound healing is another innocuous application of clay-based composites [70]. For this purpose, HNTs based powder was prepared by mixing it with chitosan oligosaccharide in an aqueous media resulting in spontaneous electrostatic interactions between positively charged chitosan complex and negatively charged clay-nanotubes.

5.   icro clay biopolymers for drug delivery

5.1   Doxorubicin- montmorillonite (DOX-MMT) formulations

Synthesis: Kohay *et al.* [37] prepared Doxorubicin- montmorillonite (DOX/MMT) high-formulations by incorporating ≈30µ.M of DOX/MMT into an excessive amount of polyethylene glycol-phosphatidyl-ethanol-amine (PEG-PE) (0.096mM) at equilibrium; therefore, montmorillonite (MMT) (0.6g/L) was added to the micelle's (M) mixture. Similarly, low-formulations were prepared by adsorbing PEG-PE to montmorillonite (0.06g/g of clay) and DOX (≈30µM). Two formulations (I) M-DOX and (II) DOX-MMT were prepared and an almost full dose of DOX (about 99%) was adsorbed in both formulations. On the other hand, composites were synthesized by following the same procedure by excluding the addition of DOX.

In-vitro Release: The formulation; DOX (free) also called M-DOX and DOX-MMT applied and investigated after 2 and 6h. The association of M-DOX was found as the highest, as compared to DOX-MMT formulations. The association of DOX with cell-line (MCF-7), released from formulations was reduced. The penetration of M-DOX inside the nucleus was observed completely within 120min. An effective amount of DOX from DOX-MMT low formulation was observed at the outer membranes of the cell, while the high formulation of DOX-MMT was found to be confined inside the nucleus. After 6h, DOX in DOX-MMT formulation is prominently positioned in the nucleus among all other formulations. The structures of MMT showed a prominent effect on stability and DOX releasing rate. This releasing rate is associated with the degree of interaction of DOX and MMT formulations in a sequence; M-DOX > H-MMT > L-MMT > DOX-MMT. Regardless of the slow release of DOX (from high-formulation), its cytotoxicity effect was found as high on sensitive cells as compared to M-DOX. Unexpectedly, the L-formulation showed high cytotoxicity in ADR cells [37].

Conclusion: It was concluded from this research that Internalization of the formulations mechanism could be suggested for increased DOX efficiency, predominantly in the Adriamycin resistant (ADR) cell line. So, the utilization of organic-inorganic hybrid composites as a carrier for drug delivery systems, although not reached its full potential but are functionality efficient for tunable demonstrated system [37].

5.2   Microparticle-montmorillonite (MP/MMT) composite

Synthesis: García-Guzmán *et al.* [39] synthesized MP-MMT composites by an exfoliation followed by adsorption. For synthesis, 50mg of MMT was mixed with distilled water at pH=3.0±0.1 and sonicated (at frequency 43 kHz) for about 30 min at ambient temperature. A fresh dispersion of microparticle (MP) (about 5mL at pH aprox.3.0) was poured dropwise in the MMT mixture, this mixture was shaken for 60min at 760rpm at an ambient temperature. The MMT-microparticles dried for 12h in the open air at about 25.0±1°C.

Drug release: these micro composites were used in oral administration, as passes through various intestinal tracks, these composites tolerate various pH ranges. By following this path, release Atorvastatin calcium (AC) salt trihydrate was investigated during 48h. Experiments were performed using dialysis-bag-method; under various conditions (gastric, colonic, and intestinal) for fasted (FAS) and fed (FED) states. Instant delivery of AC in intestinal as well as gastric conditions was observed as 14.9% and 40.60% at FAS and about 5.3% and 17.84% at FED-state respectively within 2h. At FAS and FED-states, the release of AC was delayed, but after that, it was continuous and controlled for up to 48h. AC release in the formulations of MMT-modified composites was through two mechanisms. One was the creation of a tortuous path by the enhancement of barrier properties that resists the diffusion of AC molecules released from the surface of microparticles. On the other hand, the decrease in the swelling efficiency of composites was observed due to the presence of electrostatic attraction. Both systems MP and MP-MMT composite formulations worked simultaneously for the continuous delivery of AC during 48h bearing extreme pH conditions [39].

Conclusion: It was concluded from the results that, both formulations (MP and MP-MMT) showed the release of AC through a controlled release mechanism for up to 48h. The MP-MMT compositions were found as a decrease cost-effective and efficient drug-releasing pathway envisioned for controlled release in the intestinal tract [39].

5.3   Mesalazine-clay based alginate (MCA) bead

Synthesis: Hong *et al.* [60] prepared Mesalazine- clay composites (MCC) by mixing a specific amount of MMT in D. ionized water after swelling of MMT interlayer, solution of mesalazine was prepared and its pH 4 was adjusted. Then loading of mesalazine was performed by the vigorous stirring of mixture for 5 days, the resultant mixture was filtered and washed with D. ionized water and oven-dried at 60°C. Mesalazine- clay alginate (MCA) beads were synthesized by dissolving mesalazine powder (previously prepared) in water. After that, a fixed amount of sodium alginic acid was added to the mesalazine mixture. Various concentration of MCC (0.5-1g) and 10g of mesalazine-alginic acid solution was mixed together, after that 30ml solution of $CaCl_2$ was added dropwise to this mixture. After 15min, the prepared beads were washed with distilled water and dried in an oven at 60°C for 24h.

In vitro release: MCA and cross-linker ($CaCl_2$) used in targeted drug release applications. Due to mesalazine intercalations with clay, MCA beads showed extensive loading efficiency of mesalazine (about 9%) and release in the intestinal track, this efficiency was due to its pH-sensitive and swelling property of alginate in alkaline media. In particular, increased concentration of MCC in MCA-beads compositions showed improved loading of mesalazine and prohibited instant devastation of MCA-beads due to over-swelling in intestinal pH. Mesalazine is a colon targeted-drug extensively used for Crohn's (disease) treatment. To avoid the side effects and insufficient absorption of Mesalazine (drug), clone and targeted release at the action site should be preferred [60].

Conclusion: particularly, the increasing amount of MCC during the formation of MCA-bead enhanced the entrapping efficiency of mesalazine and prevent the certain destruction that results from over-swelling caused by MCA-bead at intestinal pH conditions. The increased amount of $CaCl_2$ decreased the drug swelling and drug entrapment efficiency by MCA-beads, but the side effects were comparatively negligible [60].

6.   Nano-clay biopolymers for drug delivery

6.1   Clay/(PEG-CMC) bio-composites

Synthesis: Belmessaoud *et al.* [42] prepared clay-based composite materials using CMC, PEG, and CMC-PEG mixture. A homogeneous blend was obtained by dissolving these materials in a 1:1 weight ratio in distilled water and the mixture was stirred for 120min. Later on, 0.4g of Montmorillonite- Ibuprofen (MtIb) (0.4g) hybrid mixture was stirred with 0.1g mixture of each polymeric blended solution, a homogeneous suspension was obtained after 3h stirring at 60°C into an ultrasonic bath. Mixtures of each composite were poured into Patri-plates for evaporation at 50°C in a vacuum chamber at 50°C. The films were obtained and milled into 6mm tablets. Now, these synthesized composites were called MtIb-PEG, MtIb-CMC, and MtIb-PEG-CMC.

In vitro study: In experimental work, a sustained release of Ib at pH 7.4 in phosphate buffer solution was investigated, The MtIb-hybrid composites showed a burst effect by releasing about 80% of Ib in 60min. this fast release was due to the repulsive interactions between negatively charged ionic species. The negative charges on the surface of clay increase at pH 7.4 medium provide by buffer solution, which speeds up the mechanism of Ib release from the surface of the clay. The rate of Ib release was greatly Interrupted in the presence of CMC, PEG, and PEG-CMC combinations in composites mixture, and sustained release Ib from MtIb-PEG combination was observed 80% after 6.2h. Similarly, in the case of MtIb-CMC combinations, the release period for Ib was observed more prolonged as compared to the above cases and only 50% release was observed in this case. The release of Ib from MtIb-PEG-CMC composite system observed was 80% after 7h, this might be due to the interactions of carboxylic groups with CMC and PEG groups [42].

Conclusion: Overall it was concluded from the data different Mt-Ib compositions with PEG, CMC, and PEG-CMC composites resulted in the efficient delivery of Ibuprofen. In these formulations, the presence of polymer-clay, as well as polymer-polymer interactions, favored the controlled delivery of Ibuprofen in a biological system [42].

6.2   Silver-doped polyacrylamide-dextran hybrid nano composites

Synthesis: Patra and Swain [33] prepared polyacrylamide by polymerization (in situ) of Acryl-amide using MBS (N, N-Methylene bis acrylamide) in the presence of a crosslinker, dextran, and an initiator ammonium per-sulfate. Initially, acrylamide (4ml of 1.4M), dextran (at percentages 1-10%), and 2ml of APS (0.1M) mixture using a three-neck vessel on constant stirring at 55°C. After 20min, 2ml of MBS (0.1M) was added as a cross-linker into the vessel, the mixture was constantly stirred at 55°C till the formation of gel-like product. Various ratios of nano hydrogel were prepared. 50mg of dried hydrogels dissolved for 24h into distilled water, then swollen discs were shifted into 0.1mM of silver nitrate solution for the next 24h. The silver ions were penetrated through the free space into the gel's crosslinked networks. After 24h hydrogel mixture was treated with 0.1mM Sodium borohydride (NaBH4) for 3h at 5°C for the reduction of silver ions into nano-particles. After that, swollen discs of silver-nanohydrogels were dried in an oven at 50°C.

In vitro experiments: The release of ornidazole (OD) drug from polyacrylamide-dextran (PAM-D) hydrogel and PAM-D-Ag (silver) nano hydrogel was examined in a buffer capsule (pH 7.0) at 32°C. The quantification release of OD was investigated by UV–Vis. Spectrophotometer. The loading of OD related to the swelling ability in each case was observed 40mg OD in 5ml water for both compositions PAM-D hydrogel (5%) and PAM-D-Ag nano hydrogel (1.5%). However, the OD release varied from PAM-D hydrogel formulations up to 68.07% but it was observed a little bit higher up to 98.96% OD release from PAM-D/-Ag nano hydrogel formulations. It was due to the structural compositions of both formulations, that releasing efficiency also depends on the swelling property of nanohydrogels which provide a large surface for the release of OD (drug) from the interior to the outer environment from the nano-hydrogel surface. It was observed that the presence of MBS structure in PAM-D-Ag nano hydrogel formulations enhances the hydrophilicity which favored the release of OD in a more sustainable way [33].

Conclusion: It was observed from the results that the prepared nanohybrid hydrogels formulations showed excellent biocompatibility towards various bacterial types, also have high confrontation for Bacillus c. The released rate of OD (drug) was found about 98.5% in a 6h period of time with nanohydrogels at 1.5% of silver-NPs concentrations. Hence, the PAM-D-silver nanohydrogels formulations in drug delivery applications are considered as a potential vehicle [33].

6.3　Salecan-clay polymer nanocomposites

Synthesis: Florian *et al.* [27] synthesized clay-based nanocomposites by methacrylic acids polymerization in the presence of clay-Cl 93A and Salecan by keeping the clay amount and methacrylic acid concentrations constant while using different biopolymer concentrations. In order to assure a good swell-ability together with improved mechanical properties and based on the previously published results [15–17], A constant amount of clay (10% w/v or 0.6g in monomer approximately 1%) was selected from a reported work for all combinations of hydrogel nanocomposites, the various concentrations of Salecan 0, 0.06, 0.24, 0.48g were used for the synthesis of, PCS-1, PCS-2, and PCS-3 formulations. Briefly, 0.6g of Cl-93A clay was dispersed in 42mL distilled water and magnetically stirred at room temperature following the use of an ultrasonic probe. After that, Salecan powder (at the above-mentioned amounts) was added to the mixture and again ultrasonicated. Then, 6mL methacrylic acid (MAA), and 6mL solution of (1% w/v) BIS (N, N'-methylene bis acrylamide) were added into the mixture at constant stirring and nitrogenous atmosphere. Further, the mixture was again sonicated by keeping it in an ice bath, then 6mL of solution of ammonium persulfate (APS) (1.2% w/v) was added dropwise into the dispersion. After further stirring for 6h, the resultant hydrogel nanocomposite samples were washed and stored. Hydrogels dried by lyophilization and places in a vial containing 1.5mL of doxorubicin (DOX) solution (100µg/mL) for adsorption into a swollen hydrogel at ambient temperature for 24h.

In vitro experiments: Releasing efficiency of DOX from hydrogel formulations (PCS-1, PSC-2, and PCS-3) were investigated on normal and tumoral (HT29 and Colo-205) cells at time intervals of 30min and 180min. It was observed from the initial investigation the biopolymeric clay-based composites have a non-toxic effect, and also found no effect on morphological structure and propagation of normal (MD-BK) and HT29 (adenocarcinoma cells) regardless of incubation time, so the overall compositions do not interfere with the biocompatibility of the nanocomposites system. Nanocomposites systems should be efficient during drug (DOX) release and up-take by tumoral cells. The Ki-67 protein remains available during all the active phases (during the cell cycle) but unavailable at G0 -phase (resting cells phase), this is favorable for analyzing the cell population. The working mechanism of drug release was visualized using fluorescent microscopy. Observations were recorded after 30min intervals and an enhancement of fluorescence intensity was observed at the intranuclear level after 180min. Moreover, the DOX (drug) loading efficiency was found as increasing on increasing the concentration of Salecan due to decreasing the size of pores which provides a favorable condition for the entrapment of DOX molecules. At acidic conditions, the releasing amount of DOX was correlated positively with the amount of Salecan in the composites, while at basic pH (7.4) the releasing amount of DOX was highest due to the presence of the lowest amount of Salecan in composites. Toxicity assays verified the negligible toxicity in the blank sample (PC) and normal cells, on the other hand, DOX-loaded Salecan-clay based nano-composites (PCS-1, PCS-2, and PCS-3) were cytotoxic after 180min. It was also found from data that both tumor cell behaves differently cytotoxic effect, HT29 cells were found as more resistant to internalization of DOX, conversely, Colo-205 cells showed an antiproliferative effect in the presence of the high amount of Salecan in nanocomposites formulations [27].

Conclusions: The results show the role-played presence of hydrogen bonds in the formulations of Salecan-clay nano-composites hydrogels, each component influences the structure of composites that demonstrate achievement of a highly progressive crosslinked network of the nanocomposite. Substantiating all the results it can be recommended that Salecan-clay-based nanocomposites formulations could be considered as efficient candidates for the entrapment of chemotherapeutic agents [27].

6.4   MMT-ciprofloxacin-TiO$_2$ nano-structure

Synthesis: Gulen and Demircivi [34] prepared montmorillonite-based TiO$_2$ nano-materials for the delivery of ciprofloxacin. The materials were prepared by dissolving a specific amount of MMT in 50mL D. water (de-ionized water) at shaking for 24h. After that, 0.25g of CPx (ciprofloxacin) was added to the experimental solution, the suspension was filtered after stirring for 24h and dried at room temperature. A series of MMT-CPx formulations were prepared by coating different TiO$_2$ concentrations. For the preparation of MMT-CPx-TiO$_2$ formulations, different amounts of MMT [0.001, 0.01, 0.03, 0.07, 0.1, 0.2 and 0.3g (X1-X7 respectively)] were stirred for 24h at 360rpm with 50mL volume of n-hexane. In the 2nd formulation, the calculated amount of TiO2 at different percentage amounts (10 (Y1), 30 (Y2), 50 (Y3), 80 (Y4), and 100 (Y5) wt%) were dissolved in n-hexane solution (5mL) and further stirred for 24h at 700rpm. The TiO$_2$ solution was added dropwise into a previously prepared mixture of MMT-CPx solution and further stirred for the next 2h.

In vitro experimental: To determine the drug-releasing efficiency of nanocomposites experiments were performed at different buffer solutions at temperature $37\pm5°C$. The effect of TiO_2 amounts was investigated on releasing time of a drug from a controlled drug release mechanism by MMT-CPx-TiO_2, MMT-CPx, and CPx (commercial) in simulated gastric and blood medium at pH 1.2 and 7.5 respectively. From the results, it was concluded that the TiO_2 coated formulations were unable to deliver a huge amount of CPx as compared to uncoated nanoformulations. The release from CPx from Y1-profile in the gastric fluid was higher as compared to the other formulations (Y2, Y3, and Y5) but Y4-profile showed a close releasing time as of Y1 formulation. The releasing profile of Y1 reached up to 99.99% in 24h and the release of CPx from X5 formulation to 99.99% in 15h. As porous structures of nanoformulations were due to the presence of TiO_2 so the longer time was taken by the TiO_2 coated nano-particles as compared to uncoated formulations. Furthermore, the controlled release of CPx (drug) in simulated blood (at pH 7.4) showed best (100% in 69h) from as Y2 formulations, while the other formulations (Y1, Y3, Y4, and Y5) showed approximately similar releasing profiles [34].

Conclusion: It was concluded from the results that, the effect of pH on the releasing efficiency of CPx at pH 1.2 was comparatively faster than pH range 5.3 and 7.4 due to less and slow solubility of the drug (CPX) at acidic pH. The Korsmeyer-Peppas kinetics studies best explain the CPx releasing profile from various formulations. The results also exhibited that the structure of clay promises for increasing the bioavailability of releasing time of CPx [34].

Conclusion

Nano and micro clay bio-polymer composites have conventionally been used in several drug delivery applications, such as for the treatment of diseases, skin chemotherapy, and in the form of a number of medicines for human health. Encapsulation of drugs in the unique nano, microstructures of clay composites is trending due to their, low cost, low toxicity, biocompatibility, thermal stability, and efficient drug delivery mechanism. Recently, these clay-based formulations have been used for the development of tablets, additives, injections, dental medication, cosmetics, sprays, etc. Drugs encapsulated into these formulations entered into the body through different administration routes by following controlled delivery or targeted delivery mechanism. It can be concluded from the literature study that clay-based drug formulations are efficient carriers for taking and safely delivering the drugs into the body. In most cases, it has been observed that the extensive doses of clay composites damage the cells which is a serious concern for biomedical applications.

Adv. App. of Micro and Nano Clay II – Synthetic Polymer Composites   Materials Research Forum LLC
Materials Research Foundations 129 (2022) 24-52   https://doi.org/10.21741/9781644902035-2

References

[1] S.M. Mousavi, S.A. Hashemi, S. Salahi, M. Hosseini, A.M. Amani, A. Babapoor, Development of clay nanoparticles toward bio and medical applications. IntechOpen, U.K., 2018, pp. 167-191. https://doi.org/10.5772/intechopen.77341

[2] S. Hua, H. Yang, W. Wang, A. Wang, Controlled release of ofloxacin from chitosan-montmorillonite hydrogel, Appl. Clay Sci. 50 (2010) 112-117. https://doi.org/10.1016/j.clay.2010.07.012

[3] L. Chen, C.H. Zhou, S. Fiore, D.S. Tong, H. Zhang, C.S. Li, W.H. Yu, Functional magnetic nanoparticle/clay mineral nanocomposites: preparation, magnetism and versatile applications, Appl. Clay Sci. 127 (2016) 143-163. https://doi.org/10.1016/j.clay.2016.04.009

[4] R. Yendluri, Y. Lvov, M.M. de Villiers, V. Vinokurov, E. Naumenko, E. Tarasova, R. Fakhrullin, Paclitaxel encapsulated in halloysite clay nanotubes for intestinal and intracellular delivery, J. Pharm. Sci. 106 (2017) 3131-3139. https://doi.org/10.1016/j.xphs.2017.05.034

[5] H. Mahdavi, H. Mirzadeh, M.J. Zohuriaan-Mehr, F. Talebnezhad, Poly (vinyl alcohol)/chitosan/clay nano-composite films, J. Am. Sci. 9 (2013), 203-214.

[6] L.B. Williams, S.E. Haydel, Evaluation of the medicinal use of clay minerals as antibacterial agents, Int. Geol. Rev. 52 (2010) 745-770. https://doi.org/10.1080/00206811003679737

[7] D. Tan, P. Yuan, F. Annabi-Bergaya, D. Liu, H. He, High-capacity loading of 5-fluorouracil on the methoxy-modified kaolinite, Appl. Clay Sci. 100 (2014) 60-65. https://doi.org/10.1016/j.clay.2014.02.022

[8] H. Kaurav, S. Manchanda, K. Dua, D.N. Kapoor, Nanocomposites in Controlled & Targeted Drug Delivery Systems. In nano hybrids and composites, Trans Tech Publ. Ltd. 20 (2018) 27-45. https://doi.org/10.4028/www.scientific.net/NHC.20.27

[9] E. Ruiz-Hitzky, P. Aranda, M. Darder, G. Rytwo, Hybrid materials based on clays for environmental and biomedical applications, J. Mater. Chem. 20 (2010) 9306-9321. https://doi.org/10.1039/c0jm00432d

[10] J.M. Oh, T.T Biswick, J.H. Choy, Layered nanomaterials for green materials, J. Mater. Chem. 19 (2009) 2553-2563. https://doi.org/10.1039/b819094a

[11] G.P. Udayakumar, S. Muthusamy, B. Selvaganesh, N. Sivarajasekar, K. Rambabu, F. Banat, P.L. Show, Biopolymers and composites: Properties, characterization and

their applications in food, medical and pharmaceutical industries, J. Environ. Chem. Eng. 9 (2021) 105322. https://doi.org/10.1016/j.jece.2021.105322

[12] M. Abdollahi, M. Rezaei, G. Farzi, A novel active bionanocomposite film incorporating rosemary essential oil and nanoclay into chitosan, J. Food Eng. 111 (2012) 343-350. https://doi.org/10.1016/j.jfoodeng.2012.02.012

[13] Y. Zhao, J. Li, X. Han, Q. Tao, S. Liu, G. Jiang, D. Hou, Dual controlled release effect of montmorillonite loaded polymer nanoparticles for ophthalmic drug delivery, Appl. Clay Sci. 180 (2019) 105167. https://doi.org/10.1016/j.clay.2019.105167

[14] W. Li, D. Liu, H. Zhang, A. Correia, E. Mäkilä, J. Salonen, H.A. Santos, Microfluidic assembly of a nano-in-micro dual drug delivery platform composed of halloysite nanotubes and a pH-responsive polymer for colon cancer therapy, Acta Biomater. 48 (2017) 238-246. https://doi.org/10.1016/j.actbio.2016.10.042

[15] R. Surya, M.D. Mullassery, N.B. Fernandez, D. Thomas, Synthesis and characterization of a clay-alginate nanocomposite for the controlled release of 5-Flurouracil, J. Sci. Adv. Mater. Dev. 4 (2019), 432-441. https://doi.org/10.1016/j.jsamd.2019.08.001

[16] P. García-Guzmán, L. Medina-Torres, F. Calderas, M.J. Bernad-Bernad, J. Gracia-Mora, B. Mena, O. Manero, Characterization of hybrid microparticles/Montmorillonite composite with raspberry-like morphology for Atorvastatin controlled release, Col. Surf. B: Biointer. 167 (2018) 397-406. https://doi.org/10.1016/j.colsurfb.2018.04.020

[17] J. Kurczewska, P. Pecyna, M. Ratajczak, M. Gajęcka, G. Schroeder, Halloysite nanotubes as carriers of vancomycin in alginate-based wound dressing, Saudi Pharm. J. 25 (2017) 911-920. https://doi.org/10.1016/j.jsps.2017.02.007

[18] J. Xue, Y. Niu, M. Gong, R. Shi, D. Chen, L. Zhang, Y. Lvov, Electrospun Microfiber Membranes Embedded with Drug-Loaded Clay Nanotubes for Sustained Antimicrobial Protection, ACS Nano. 9 (2015) 1600-1612. https://doi.org/10.1021/nn506255e

[19] Y. Zhang, R. Gao, M. Liu, B. Shi, A. Shan, B. Cheng, Use of modified halloysite nanotubes in the feed reduces the toxic effects of zearalenone on sow reproduction and piglet development, Theriogenol. 83 (2015), 932-941. https://doi.org/10.1016/j.theriogenology.2014.11.027

[20] M. Liu, C. Wu, Y. Jiao, S. Xiong, C. Zhou, Chitosan-halloysite nanotubes nanocomposite scaffolds for tissue engineering, J. Mater. Chem. B, 1(2013) 2078-2089. https://doi.org/10.1039/c3tb20084a

[21] S. Sharif, G. Abbas, M. Hanif, A. Bernkop-Schnürch, A. Jalil, M. Yaqoob, Mucoadhesive micro-composites: Chitosan coated halloysite nanotubes for sustained drug delivery, Col. Surf. B: Biointer. 184 (2019) 110527. https://doi.org/10.1016/j.colsurfb.2019.110527

[22] V. Bertolino, G. Cavallaro, G. Lazzara, M. Merli, S. Milioto, F. Parisi, Effect of the biopolymer charge and the nanoclay morphology on nanocomposite materials, Ind. Eng. Chem. Res. 55 (2016) 7373-7380. https://doi.org/10.1021/acs.iecr.6b01816

[23] C.D. Nunes, P.D. Vaz, A.C. Fernandes, P. Ferreira, C.C. Romao, M.J. Calhorda, Loading and delivery of sertraline using inorganic micro and mesoporous materials. Eur. J Pharm. Biopharm. 66 (2007), 357-365. https://doi.org/10.1016/j.ejpb.2006.11.023

[24] A. Vikulina, D. Voronin, R. Fakhrullin, V. Vinokurov, D. Volodkin, Naturally derived nano-and micro-drug delivery vehicles: halloysite, vaterite and nanocellulose, New J. Chem. 44(2020), 5638-5655. https://doi.org/10.1039/C9NJ06470B

[25] V. Vinokurov, A. Novikov, V. Rodnova, B. Anikushin, M. Kotelev, E. Ivanov, Y. Lvov, Cellulose nanofibrils and tubular halloysite as enhanced strength gelation agents. Polymers, 11(2019), 919. https://doi.org/10.3390/polym11050919

[26] C. Aguzzi, G. Sandri, P. Cerezo, E. Carazo, C. Viseras, Nanosized Tubular Clay Minerals: Health and Medical Applications of Tubular Clay Minerals, Elsevier, 2016, pp. 708-725. https://doi.org/10.1016/B978-0-08-100293-3.00026-1

[27] K. Florian, S.K. Swain, (2018). Nano silver decorated polyacrylamide/dextran nanohydrogels hybrid composites for drug delivery applications, Mater. Sci. Eng. C, 85 (2018) 130-141. https://doi.org/10.1016/j.msec.2017.11.028

[28] L. Bounabi, N.B. Mokhnachi, N. Haddadine, F. Ouazib, R. Barille, Development of poly (2-hydroxyethyl methacrylate)/clay composites as drug delivery systems of paracetamol, J. Drug Deliv. Sci. Technol. 33 (2016) 58-65. https://doi.org/10.1016/j.jddst.2016.03.010

[29] M. Jafarbeglou, M. Abdouss, A.M. Shoushtari, M. Jafarbeglou, Clay nanocomposites as engineered drug delivery systems, RSC Adv. 6 (2016) 50002-50016. https://doi.org/10.1039/C6RA03942A

[30] S.K. Swain, B. Shur, S.K. Patra, Poly (acrylamide-co-vinyl alcohol)-Superabsorbent materials reinforced by modified clay, Poly. Comp. 34(2013) 1794-1800. https://doi.org/10.1002/pc.22583

[31] M. Rahim, M.R.H.M. Haris, N.U. Saqib, An overview of polymeric nano-biocomposites as targeted and controlled-release devices. Biophy. Rev. (2020)1-9. https://doi.org/10.1007/s12551-020-00750-0

[32] C. Viseras, C. Aguzzi, P. Cerezo, M.C. Bedmar, Biopolymer-clay nanocomposites for controlled drug delivery, Mater. Sci. Tech. 24 (2008) 1020-1026. https://doi.org/10.1179/174328408X341708

[33] S.K. Patra, S. K. Swain, Swelling study of superabsorbent PAA-co-PAM/clay nanohydrogel, J. Appl. Poly. Sci. 120 (2011), 1533-1538. https://doi.org/10.1002/app.33381

[34] B. Gulen, P. Demircivi, Synthesis and characterization of montmorillonite/ciprofloxacin/tio2 porous structure for controlled drug release of ciprofloxacin tablet with oral administration, Appl. Clay Sci. 197 (2020) 105768. https://doi.org/10.1016/j.clay.2020.105768

[35] F. García-Villén, E. Carazo, A. Borrego-Sánchez, R. Sánchez-Espejo, P. Cerezo, C. Viseras, C. Aguzzi, Clay minerals in drug delivery systems. In Modified clay and zeolite nanocomposite materials, Elsevier. (2019) 129-166. https://doi.org/10.1016/B978-0-12-814617-0.00010-4

[36] C. Aguzzi, G. Sandri, P. Cerezo, E. Carazo, C. Viseras, Health and medical applications of tubular clay minerals. In Developments in clay3 science, Elsevier. 7 (2016) 708-725. https://doi.org/10.1016/B978-0-08-100293-3.00026-1

[37] H. Kohay, C. Sarisozen, R. Sawant, A. Jhaveri, V.P. Torchilin, Y.G. Mishael, PEG-PE/clay composite carriers for doxorubicin: Effect of composite structure on release, cell interaction and cytotoxicity, Acta. Biomater. 55 (2017) 443-454. https://doi.org/10.1016/j.actbio.2017.04.008

[38] K.M. Rao, A. Kumar, M. Suneetha, S.S. Han, pH and near-infrared active; chitosan-oated halloysite nanotubes loaded with curcumin-au hybrid nanoparticles for cancer drug delivery, Int. J. Biol. Macromol. 112 (2018) 119-125. https://doi.org/10.1016/j.ijbiomac.2018.01.163

[39] R. García-Vázquez, E.P. Rebitski, L. Viejo, C. de los Rios, M. Darder, E.M. Garcia-Frutos, Clay-based hybrids fo r controlled release of 7-azaindole derivatives as neuroprotective drugs in the treatment of Alzheimer's disease, Appl. Clay Sci. 189 (2020) 105541. https://doi.org/10.1016/j.clay.2020.105541

[40] E.P. Rebitski, P. Aranda, M. Darder, R. Carraro, E. Ruiz-Hitzky, Intercalation of metformin into montmorillonite. Dalton Trans. 47 (2018) 3185-3192. https://doi.org/10.1039/C7DT04197G

[41] V. Trivedi, U. Nandi, M. Maniruzzaman, N.J. Coleman, Intercalated theophylline-smectite hybrid for pH-mediated delivery, Drug Deliv. Transl. Res. 8 (2018) 1781-1789. https://doi.org/10.1007/s13346-018-0478-8

[42] N.B. Belmessaoud, N. Bouslah, N. Haddadine, Clay/(PEG-CMC) biocomposites as a novel delivery system for ibuprofen, J. Polym. Eng., 40 (2020) 350-359. https://doi.org/10.1515/polyeng-2019-0390

[43] W. Chrzanowski, S.Y. Kim, E.A. Abou Neel, Biomedical applications of clay, Aus. J. Chem., 66 (2013) 1315-1322. https://doi.org/10.1071/CH13361

[44] B.D. Kevadiya, H.C. Bajaj, The layered silicate, montmorillonite (MMT) as a drug delivery carrier. In Key Engineering Materials, Trans Tech Pub. Ltd. 571 (2013) 111-132. https://doi.org/10.4028/www.scientific.net/KEM.571.111

[45] K. Chen, B. Guo, J. Luo, Quaternized carboxymethyl chitosan/organic montmorillonite nanocomposite as a novel cosmetic ingredient against skin aging, Carbohydr. polym., 173 (2017) 100-106. https://doi.org/10.1016/j.carbpol.2017.05.088

[46] F. Guo, S. Aryana, Y. Han, Y. Jiao, A review of the synthesis and applications of polymer-nanoclay composites, Appl. Sci., 8 (2018) 1696. https://doi.org/10.3390/app8091696

[47] Y.M. Lvov, M.M. DeVilliers, R.F. Fakhrullin, The application of halloysite tubule nanoclay in drug delivery, Expert Opin. Drug Deliv. 13(2016) 977-986. https://doi.org/10.1517/17425247.2016.1169271

[48] L.M.A. Meirelles, F.N. Raffin, Clay and polymer-based composites applied to drug release: A scientific and technological prospectio, J. of Pharm. Pharm. Sci. 20 (2017) 115-134. https://doi.org/10.18433/J3R617

[49] N. Selvasudha, U.M. Dhanalekshmi, S. Krishnaraj, Y.H. Sundar, N.S.D. Devi, I. Sarathchandiran, Multifunctional Clay in Pharmaceuticals. In Clay Sci. Techn. Intech Open, U.K., 2020, pp. 99-119 https://doi.org/10.5772/intechopen.92408

[50] X. Wang, Y. Du, J. Luo, Biopolymer/montmorillonite nanocomposite: preparation, drug-controlled release property and cytotoxicity, Nanotech. 19 (2008) 065707. https://doi.org/10.1088/0957-4484/19/6/065707

[51] Y. Dong, S.S. Feng, Poly (d, l-lactide-co-glycolide)/montmorillonite nanoparticles for oral delivery of anticancer drugs, Biomater. 26(2005), 6068-6076. https://doi.org/10.1016/j.biomaterials.2005.03.021

[52] C.N. Cheaburu-Yilmaz, R.P. Dumitriu, M.T. Nistor, C. Lupusoru, M.I. Popa, L. Profire, C. Vasile, Biocompatible and biodegradable chitosan/clay nanocomposites as new carriers for theophylline controlled release, J. Pharm. Res. Int. (2015) 228-254. https://doi.org/10.9734/BJPR/2015/16525

[53] S. Leporatti, Polymer Clay Nano-composites, Polym.11 (2019) 1445. https://doi.org/10.3390/polym11091445

[54] F. Kianfar, N. Dempster, E. Gaskell, M. Roberts, G. Hutcheon, Lyophilised Biopolymer-Clay Hydrogels for Drug Delivery, MJNDR.1 (2017) 1-9. https://doi.org/10.18689/mjndr-1000101

[55] P.V.K. Kumari, Y.S. Rao, S. Akhila, Role of nanocomposites in drug delivery, GSC Biol. Pharm. Sci. 8(2019) 094-103. https://doi.org/10.30574/gscbps.2019.8.3.0150

[56] R.I. Iliescu, E. Andronescu, G. Voicu, A. Ficai, C.I. Covaliu, Hybrid materials based on montmorillonite and citostatic drugs: Preparation and characterization, Appl. Clay Sci. 52 (2011) 62-68. https://doi.org/10.1016/j.clay.2011.01.031

[57] L.A. De Sousa Rodrigues, A. Figueiras, F. Veiga, R.M. de Freitas, L.C.C. Nunes, E.C. da Silva Filho, C.M. da Silva Leite, The systems containing clays and clay minerals from modified drug release: a review, Colloids Surf. B. 103 (2013) 642-651. https://doi.org/10.1016/j.colsurfb.2012.10.068

[58] R. Suresh, S.N. Borkar, V.A. Sawant, V.S. Shende, S.K. Dimble, Nanoclay drug delivery system, IJPSN. 3(2010), 901-905

[59] N. Khatoon, M.Q. Chu, C.H. Zhou, Nanoclay-based drug delivery systems and their therapeutic potentials, J. Mater. Chem. B. 8(2020), 7335-7351. https://doi.org/10.1039/D0TB01031F

[60] H.J. Hong, H.S. Jeong, K.M. Roh, I. Kang, "Preparation of Mesalazine-Clay Composite Encapsulated Alginate (MCA) Bead for Targeted Drug Delivery: Effect of Composite Content and CaCl2 Concentration", Macromol. Res. 26 (2018) 1019-1025. https://doi.org/10.1007/s13233-019-7033-4

[61] G. Cavallaro, G. Lazzara, M. Massaro, S. Milioto, R. Noto, F. Parisi, S. Riela, Biocompatible poly (N-isopropyl-acryl-amide)-halloysite nanotubes for thermo responsive curcumin release, J. Phys. Chem. C. 119(2015) 8944-8951. https://doi.org/10.1021/acs.jpcc.5b00991

[62] M. Baek, J.H. Choy, S.J. Choi, Montmorillonite intercalated with glutathione for antioxidant delivery: Synthesis, characterization, and bioavailability evaluation, Int. J. Pharm. 425 (2012) 29-34. https://doi.org/10.1016/j.ijpharm.2012.01.015

[63] S. ul Haque, A. Nasar, Montmorillonite clay nanocomposites for drug delivery. In Applications of Nanocomposite Materials in Drug Delivery, Woodhead Publishing, Sawstone U.K., 2018, pp. 633-648. https://doi.org/10.1016/B978-0-12-813741-3.00028-5

[64] G.V. Joshi, B.D. Kevadiya, H.M. Mody, H.C. Bajaj, "Confinement and controlled release of quinine on chitosan-montmorillonite bionanocomposites", J. Polym. Sci. A: Polym. Chem. 50 (2012) 423-430. https://doi.org/10.1002/pola.25046

[65] H.J. Hong, J. Kim, Y.J. Suh, D. Kim, K.M. Roh, I. Kang, "pH-sensitive mesalazine carrier for colon-targeted drug delivery: A two-fold composition of mesalazine with a clay and alginate", Macromol. Res. 25 (2015) 1145-1152 https://doi.org/10.1007/s13233-017-5150-5

[66] W. Hur, M. Park, J.Y. Lee, M.H. Kim, S.H. Lee, C.G. Park, S.N. Kim, H.S. Min, H.J. Min, J.H. Chai, S.J. Lee, "Bioabsorbable bone plates enabled with local, sustained delivery of alendronate for bone regeneration", J. Control. Release. 222 (2016), 97-106 https://doi.org/10.1016/j.jconrel.2015.12.007

[67] E. Kolanthai, P. Abinaya Sindu, K. Thanigai Arul, V. Sarath Chandra, E. Manikandan, S. Narayana Kalkura, "Agarose encapsulated mesoporous carbonated hydroxyapatite nanocomposites powder for drug delivery", J. Photochem. Photobiol. B Biol. 166 (2017) 220-231. https://doi.org/10.1016/j.jphotobiol.2016.12.005

[68] A.C. Santos, C. Ferreira, F. Veiga, A.J. Ribeiro, A. Panchal, Y. Lvov, A. Agarwal, Halloysite clay nanotubes for life sciences applications: From drug encapsulation to bioscaffold, Adv. Colloid Interf. Sci. 257 (2018) 58-70. https://doi.org/10.1016/j.cis.2018.05.007

[69] R. Onnainty, G. Granero, Chitosan-clays based nanocomposites: promising materials for drug delivery applications. Nanomed, Nanotechnol. J. 1 (2017) 114.

[70] M. Delyanee, A. Solouk, S. Akbari, M.D. Joupari, "Hemostatic Electrospun Nanocomposite Containing Poly (lactic acid)/Halloysite Nanotube Functionalized by Poly (amidoamine) Dendrimer for Wound Healing Application: in Vitro and in Vivo Assays", Macromol. Biosci. (2021) 2100313. https://doi.org/10.1002/mabi.202100313

[71] R.K. Matharu, L. Ciric, M. Edirisinghe, Nanocomposites: Suitable alternatives as antimicrobial agents, Nanotechnol. 29 (2018) 282001. https://doi.org/10.1088/1361-6528/aabbff

[72] S.K.S. Kushwaha, N. Kushwaha, P. Pandey, B. Fatma, "Halloysite Nanotubes for Nanomedicine: Prospects, Challenges and Applications", Bio Nano Sci.. 11 (2021) 200-208. https://doi.org/10.1007/s12668-020-00801-6

[73] U. Jammalamadaka, K. Tappa, D. K. Mills, Calcium phosphate/clay nanotube bone cement with enhanced mechanical properties and sustained drug release, IntechOpen, U.K., 2018, pp. 123-146 https://doi.org/10.5772/intechopen.74341

[74] R.D. Sagare, F.S. Dasankoppa, H.N. Sholapur, K. and Burga, "Halloysite nanotubes: design, characterization and applications. A review". Farmacia, 69 (2021) 208-214. https://doi.org/10.31925/farmacia.2021.2.3

Chapter 3

Nanoclay-based Polymer Nanocomposites for Electromagnetic Interference Shielding

M.R. Nigil[1], B.T.S. Ramanujam[2,*] and R. Thiruvengadathan[1*]

[1]Department of Electronics and Communication Engineering,

[2]Department of Sciences, Amrita School of Engineering, Coimbatore, Amrita Vishwa Vidyapeetham, INDIA.

* t_rajagopalan@cb.amrita.edu; bts_ramanujam@cb.amrita.edu

Abstract

In the past decade, there has been rapid growth in the miniaturization of electronic devices without compromising their performance for a particular application. Due to electromagnetic interference, the performance of those devices can be affected very seriously and hence there is a requirement to develop electromagnetic interference shielding materials. Compared to metals, conducting polymer nanocomposites can be developed as an efficient EMI shielding material. Thus, this chapter presents a detailed review of the research developments on nanocomposites formed by meticulous mixing of conducting polymer and nanoclay in appropriate proportions, specifically applied as microwave absorption materials. Employing nanoclay into the conducting polymers is found to improve the overall EMI shielding material's mechanical strength, dielectric properties, thermal stability, barrier properties and most importantly the shielding efficiency.

Keywords

Conducting Polymers, Nanoclay, Shielding Efficiency, Polymer Nanocomposites, Electrical Conductivity

Contents

1.   Introduction

There is a significant increase in the number of electronic devices used in both military and civilian applications. These devices operate at different frequency ranges and are susceptible to electromagnetic interference (EMI) [1-5]. It is required that their performance should not be impacted due to EMI. The EMI not only affects the performance of the electronic gadgets, but also has adverse effects on human health.

The need to prevent the EMI of electronics devices has led to the development of novel shielding material with high shielding effectiveness (SE). The three mechanisms of shielding are reflection, absorption and multiple reflection [2, 5, 6]. It is important to develop an EMI shielding material with high absorption properties because reflections of EM waves have undesirable effects on the performance and the durability of the electronic devices surrounding the primary device. At the beginning, the conventional metals have been used as EMI shielding materials. But, owing to high density, cost and corrosion problems associated with metals, alternative EMI shielding materials with desirable characteristics such as high absorption, superior mechanical strength, light weight and easy processability are being developed and reported in literature [7-11].

In this chapter, our focus is to summarize recent developments and the current status on EMI shielding materials that exhibit high absorption mechanism. The absorption mechanism can be enhanced by choosing materials with the following properties like high dielectric loss, high magnetic loss, excellent electrical conductivity, and high permeability. The thickness of the EMI shield also affects the absorption mechanism. The potential candidates and alternate to conventional metals as EM shields are *conducting polymer-based composites* (CPCs) [3, 5, 6, 12-14]. The conducting polymers have low density, lightweight, cost effective and can be used to prepare ultra-thin flexible EMI shield. The physical and chemical properties of the conducting polymer composites can be enhanced by the inclusion of suitable nano additives. There are many types of nanoparticles that are used as reinforcement in the conducting polymer to enhance its shielding properties also namely carbon nanotubes (CNTs) [15], graphene [16], metal-based nanoparticles [17], magnetic nanoparticles [18], and also nanoclays [7, 9, 11].

Specifically, this chapter presents a review and summary of recent research and development on the effects of incorporating nanoclays in conducting polymers, with focus towards enhancing the mechanical, thermal, electrical and dielectric properties of the nanocomposites. The CPC-based shield materials are expected to absorb microwaves in X Band (12-18 GHz) and Ku Band (18-27 GHz) part of the EM spectrum. In the first section of the chapter, the EM shielding theory and the basic characteristic that an EMI shielding material should possess are discussed in detail. In the later section, the methods of preparing polymer nanocomposites with nanoclay particles are summarized along with physical and chemical improvements provided by the nanoclay particles to the polymer-nanocomposite.

2.   Electromagnetic interference shielding theory

Electromagnetic shielding material reduces the radiated EM energy through suitable mechanism. Based on the wavelength and distance between the source and the EMI shield, three regions namely, near field ($r < \lambda/2\pi$), transition region ($r = \lambda/2\pi$) and the far field region ($r > \lambda/2\pi$) (radiation zone) can be defined. The far fielding approximation is valid because of the constant *intrinsic impedance* (E/H) and is equal to 377 ohms in free space but, where as in near field the intrinsic impedance is not constant because depending on the current and voltage variation, the dominance of electric and magnetic field varies continuously.

The plane EM wave striking the material with different intrinsic impedance is separated into reflected and transmitted part. The EM waves amplitude is determined by the shield's intrinsic impedance. The SE provides information about the EMI shield's ability to reduce EM energy or power. The SE also depends on the frequency of the EM wave, shield's thickness and the shield to the source distance. The upcoming sections will summarize in detail about the mechanism of shielding efficiency and what are the properties/mechanism that the material manufactured from polymer nanocomposites should possess in them for good microwave absorption and what factor contributes to the shielding efficiency and how these parameters can be calculated using the scattering matrix (S-parameter).

2.1 Shielding effectiveness and mechanisms of shielding

The shielding effectiveness can be defined as the ratio of incident EM energy to the transmitted energy through the EMI shield, The shielding efficiency is usually expressed in decibel (dB).

$$SE(dB) = 20 \log_{10} \left(\frac{E_T}{E_I}\right) \tag{1}$$

$$SE(dB) = 20 \log_{10} \left(\frac{H_T}{H_I}\right) \tag{2}$$

$$SE(dB) = 20 \log_{10} \left(\frac{P_T}{P_I}\right) \tag{3}$$

Where E, H and P represent the electric field, magnetic field intensity and power of the EM radiation respectively the subscripts 'T' represent the transmitted wave and the subscript 'I' represent incident wave. Higher the SE, lower will be the transmitted energy through the shield. The SE of -20 dB is desirable for attenuation of almost 99% of EM energy [8]. Note that the ratio is taken as incident wave to the transmitted wave the SE will be positive. Hence, the total EMI SE_T is

$$SE_T = SE_R + SE_A + SE_M \tag{4}$$

where SE_R, SE_A and SE_M corresponds to the shielding effectiveness attributed to reflection, absorption and multiple reflection of EM waves respectively. Nanoclays with high interfacial area and porosity can provide high multiple reflection attenuation contributing to increase in SE [8].

The reflection, absorption and multiple reflection shielding efficiency can be expressed as,

$$SE_R = 39.5 + 10 \log \left(\frac{\sigma}{2f\mu\pi}\right) \propto \frac{\sigma}{\mu} \tag{5}$$

$$SE_A = 8.7d(f\mu\pi\sigma)^{\frac{1}{2}} \tag{6}$$

$$SE_M = 20 \log \left(1 - e^{-\frac{2d}{\delta}}\right) \tag{7}$$

where the symbols d, f, σ and δ represent the shield thickness, the frequency, the bulk conductivity and the skin depth respective. Skin depth is given as $\delta = (\pi \mu \sigma f)^{-1/2}$.

It is evident that SE_R will dictate the total SE in highly conducting and non-magnetic materials. When interacting with EM waves, the material's ability to create electric and magnetic dipoles gives rise to internal electric and magnetic fields, and when the incident and internal fields interact, attenuation occurs. This phenomenon can be achieved by using materials with high magnetic properties and electrical properties. As mentioned previously the multiple reflection dominates when there is porosity present in the shielding material.

2.2   Absorption and multiple reflection

The microwave absorption is aided by dipole formation, absorption mechanism, and multiple reflection mechanisms, in addition to impedance matching. Multiple reflections can be ignored in the calculation if the SE_A is larger than 10 dB. Also, the multiple reflection can be ignored for plane waves.

Minimum reflection happens when the EM wave and shielding material impedances are matched. The fabrication of polymer nanocomposites with the homogeneous dispersion of nanoclays enhance the synergistic effect between the particles, which provides good impedance matching and improves absorption. When the impedance of air and the EMI shielding material match, good EM wave absorption can be obtained. The impedance matching of a material can be calculated using the normal impedance matching formula [8].

$$Z_{in} = Z_0 (\mu_r/\varepsilon_r)^{1/2} \tanh \left[j \left(\frac{2\pi fd}{c}\right) (\mu_r \epsilon_r)^{1/2}\right] \tag{8}$$

$$|Z| = \frac{Z_{in}}{Z_0} = (\mu_r/\varepsilon_r)^{1/2} \tanh \left[j \left(\frac{2\pi fd}{c}\right) (\mu_r \epsilon_r)^{1/2}\right] \tag{9}$$

Here, Z_{in} is the absorber's input impedance, Z_0 is the free space characteristic impendence, d is the sample thickness, c is the light velocity and Z is the normalised impedance. The range of Z is '0' to '1'. Closer the Z value to 1 higher is the impedance matching and higher is the absorption. High dielectric loss ($\tan \delta_E = \varepsilon''/\varepsilon'$) and magnetic loss ($\tan \delta_M = \mu''/\mu'$) of the material can also contribute to almost zero reflection or high absorption. For zero reflection of EM waves by the shield, the ideal condition would be both magnetic loss and dielectric loss being equal to each other ($\tan \delta_E = \tan \delta_M$), which is highly unlikely in practical conditions.

2.3 Dependence on thickness

Though the thin film material prepared are focus on absorption of the EM radiation there will be some amount of reflection which can be reduced by preparing material of appropriated thickness as stated in **Eq 10**. The reflections can be further decreased by preparing material with thickness defined by *quarter-wavelength cancellation model*. According to quarter-wavelength cancellation model [8], the air-absorber reflected wave and air-metal reflected wave should be in 180° out of phase at appropriate thickness and frequency then there will be destructive interference.

$$T_m = \frac{n\lambda}{4} = \frac{nc}{4F_m\sqrt{|\mu_r||\varepsilon_r|}} \quad (10)$$

This can reduce the reflection waves.

2.4 Shielding efficiency calculation

The four experimental methods to determine the SE of a sample are open field method, shield box method, shielded room method and coaxial transmission line method. The coaxial transmission method is preferred because it could identify the components due to reflection, absorption and transmission.

Both scalar network analyser (SNA) and vector network analyser (VNA) are being used to measure the EMI SE of the samples. VNA is preferred over SNA due to the former's ability to measure both amplitude and phase of the signal. SNA method enables measurement of only the amplitude of the signal.

The incident and transmitted waves in two port VNA are represented by complex scattering parameters known as S parameters. The subscript numbering convention for S-parameter usually, the first number represents the output or the port from where the signal emerges, and the second number represents the input or the port through which the signal is applied. Measurement of reflection and transmission coefficient requires four S-parameters S_{11}, S_{21}, S_{12}, S_{22} or a pair (S_{11} and S_{21}) of S parameters of the chosen material. The SE of different mechanisms can be obtained using the following expression,

$$SE_R = 10 \log_{10}(1 - |S_{11}|^2) \quad (11)$$

$$SE_A = 10 \log_{10}\left(1 - \frac{|S_{11}|^2}{|S_{21}|^2}\right) \quad (12)$$

$$SE_T = SE_R + SE_A \quad (13)$$

The reflectance (R), absorbance (A) and transmittance (T) are related to the S-parameter by the following equations,

$$R = \left|\frac{E_R}{E_I}\right|^2 = |S_{11}|^2 = |S_{22}|^2 \tag{14}$$

$$T = \left|\frac{E_T}{E_I}\right|^2 = |S_{12}|^2 = |S_{21}|^2 \tag{15}$$

$$A = 1 - R - T = 1 - |S_{11}|^2 - |S_{12}|^2 \tag{16}$$

Thus, it is possible to obtain the SE_T of unknown samples using the VNA. In the forthcoming sections, the methods of preparing polymer nanocomposites with nanoclay particles are summarized along with physical and chemical improvements provided by the nanoclay particles to the polymer-nanocomposite.

3. Clays and nanoclays

Before discussing the nanoclays, it is appropriate to understand the structure and properties of clays and also about what makes them a potential nanoparticle filler. Clays are in general inorganic solids-made primarily of silica, alumina and magnesia. Particle size of naturally occurring clays is typically of few microns. Clays are layered materials with each layer having high surface to volume ratio, high adsorption rate and negative charge. The solid colloids can be divided into four types of crystalline silicate clays, non-crystalline silicate clays, Fe and Al oxides and organic (humus) [19]. In EMI shielding application, the crystalline silicate clays will be used because of their high negative charge and layer by layer (LbL) structuring [19]. The crystalline silicate clays are further divided into 6 types based on their physical structure. They are nesosilicates, sorosilicates, inosilicates, cyclosilicates, phyllosilicates and tectosilicates [20].

Nesosilicates are single tetrahedral compound made of one silicon atom and 4 oxygen atoms. *Sorosilicates* are compounds made of two or more linked tetrahedrals. *Inosilicates* are made of single or double chain of tetrahedrals. *Cyclosilicates* are made of closed rings or double rings of tetrahedrals. *Tectosilicates* are made of framework of tetrahedrals. Finally, *Phyllosilicates* $(SiO_5)^{2-}$ are made of sheets of tetrahedrals. This extra oxygen atom connects with other tetrahedrals forming a sheet network which makes phyllosilicates appropriate for our application. The sheet like structure of phyllosilicates have high aspect ratio which promotes surface interaction between the conducting polymer, nanoparticle filler and the insulating matrix [21].

3.1   Sheet-structure and cation exchange

Clays are relatively bad conductors when compared to copper or metals but based on our application, EMI shielding materials requires conductivity of order 10^{-2} S/m, at this range clays do not affect the electrical conductivity due to low filler loading but they significantly improve other properties due to the high reactiveness which is attributed to *isomorphous substitution* between different layer of clays [19].

The *sheet-structure* of crystalline silicate are made of tetrahedral with a silicon at centre surrounded by oxygen atoms and octahedral with either magnesium or aluminium at the centre surrounded by hydroxide molecules held together by an *apical oxygen* atom [22]. Based on the number of cation atoms, the octahedral is classified into dioctahedral which is aluminium Al^{3+} dominant and trioctahedral which is magnesium Mg^{2+} dominant. Based on the stacking of layers and type of octahedral, the crystalline clays are further classified into 1:1 type clay (tetrahedral-octahedral) and 2:1 type clay (tetrahedral-octahedral-tetrahedral) as shown in **Figure 1** [23]. Where the first number contributes to the number of tetrahedrals, and the second number contributes to the number of octahedral.

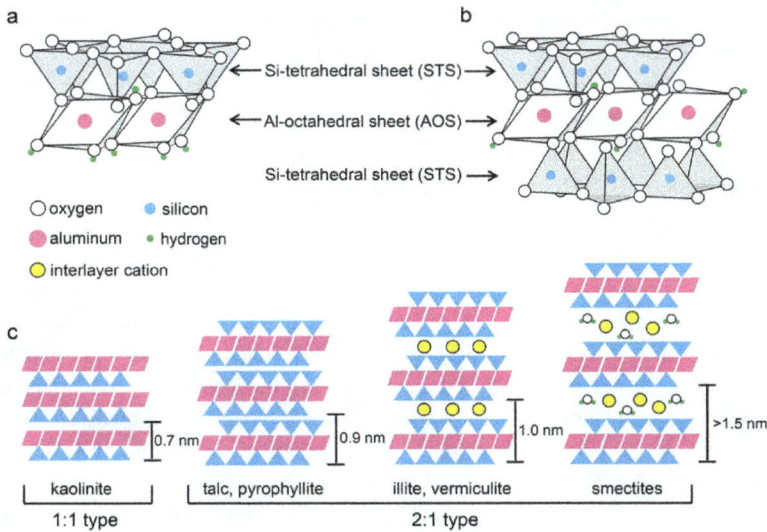

Figure 1 A Schematic representation of unit cell (a) 1:1 type and (b) 2:1 type clay. c) Stacking arrangements of distinct structures. Reprinted with Permission from Luyi Sun et al. Advanced Functional Materials, Volume: 29, Issue: 16, 1807611, (2019). [23]. Copyright 2019, Wiley-VCH.

Before looking into why we prefer 2:1 type of clays over 1:1 types of clays let us look about what isomorphous substitution means. In the $(SiO_5)^{2-}$ one of the oxygens is replaced with then same size cation Al^{3+} making it SiO_4^- (negatively charged). Similarly in the octahedral $(OH)_2Al_2O_2$ one of the aluminium is replaced with Mg^{3+} and making it $(OH)_2AlMgO_2^-$ (negatively charged). This is called *isomorphous substitution* and this makes the clay particles highly reactive and negative, this attracts the incoming cations to the variable spacing and thus promotes surface interactions.

In case of the 1:1 clay there is negligible isomorphous substitution and there is no swelling action due to the formation of hydrogen bond between adjacent layers. The 2:1 clay are divided into two types expanding and non-expanding. The expanding clays are further divided into smectite and vermiculite, and the non-expanding clays are divided into grained mica and Chlorite. In our application we will be using the 2:1 layered expanding smectite clays owing to their high isomorphous substitution, negative charge, high shrinking and swelling property due to the loosely bound oxygen atom and high plasticity. Some examples of smectite clays are montmorillonite and bentonite clay.

The nanoclays (<100 nm) are preferred over micro clays and macro clays due to their high surface area to volume ratio. As evident from above, higher surface area of the nanoclays, higher is the interaction with the other particles in the mixture. In the nano meter regime, the high surface area improves the mechanical, thermal properties and the interfacial layers that aids in multiple reflection [11].

4.   Insulating matrix and conducting polymer

EMI shielding material requires conduction of the order 10^{-2} S/m [11]. The conventionally used insulating polymers cannot be used directly in the EMI shielding application owing to their low mechanical strength and low thermal stability. Also, the insulating polymer matrix should be made conducting, to achieve the electrical conductivity of the order 10^{-2} S/m. The nanoclay fillers are essential for increasing mechanical strength and thermal stability of the insulating matrix. The inclusion of the conducting polymers aids in achieving the required electrical conductivity. The role played by conducting polymers in achieving the desired electrical conductivity is explained in **Section 5.4**. The hybrid mixture of the conducting polymer and nanoclay is added as filler into an *insulating matrix* to enhance its strength, modulus, conductivity and operating temperature. The insulating matrix can be divided into three categories thermoplastics, thermosetting plastics and elastomers. Based on our application some of the most commonly used insulating polymers are ethylene-vinyl-acetate (EVA) [11], polyetherimide (PEI) [7], polyvinylchloride (PVC) [10] and polystyrene.

Adv. App. of Micro and Nano Clay II – Synthetic Polymer Composites   Materials Research Forum LLC
Materials Research Foundations 129 (2022) 53-78   https://doi.org/10.21741/9781644902035-3

The organic polymers which contain highly delocalised π-electron that conduct electricity are called *conducting polymers*. The conducting polymers are classified into two types intrinsic conducting polymers and extrinsic conducting polymers. Intrinsically Conducting Polymers (ICPs) have tuneable conductivity that varies between metal conductivity to semiconductor conductivity. Doped conducting polymers are categorised as n-doping and p-doping depending on how they are doped. The oxidation process is used to achieve p-doping, while the reduction process is used to achieve n-doping. Depending on the doping extent, conductivity of the polymers can increase significantly [4]. Some of the conducting polymers with their conductivity and doping material are listed below in **Table 1** [24].

Table 1 List of Conducting Polymers with Doping Materials and Conductivity[24]

Polymers	Doping Materials	Conductivity (S/cm)
Polyacetylene	I_2, Br, Li, Na, AsF_5	10^4
Polypyrrole	BF_4^-, ClO_4^-, Tosylate	$500\text{-}7.5 \times 10^3$
Polythiophene	BF_4^-, ClO_4^-, Tosylate, $FeCl_4^-$	10^3
Poly(3-alkylthiophene)	BF_4^-, ClO_4^-, $FeCl_4^-$	$10^3\text{-}10^4$
Polyphenylene-vinylene	AsF_5	10^4
Polythienylene-vinylene	AsF_5	2.7×10^3
Polyisothi-anaphthene	BF_4^-, ClO_4^-	50
Polyfuran	BF_4^-, ClO_4^-	100
Polyazulene	BF_4^-, ClO_4^-	1
Polyphenylenesulphide	AsF_5	500
Polyphenylene	AsF_5, Li, K	10^3
Polyaniline	HCl	200

Thus, in the preceding sections, the properties of main constituent materials that make up the EMI shielding material have been addressed. The following section summarizes the synthesis routes to realize EMI shielding film using the constituent materials.

Adv. App. of Micro and Nano Clay II – Synthetic Polymer Composites   Materials Research Forum LLC
Materials Research Foundations 129 (2022) 53-78   https://doi.org/10.21741/9781644902035-3

5.   Polymer nanocomposite preparation

The most important factor is the mixing of the nanoclay particles with the conducting polymers because depending on the level of mixing the surface interactions are increased. Various methods are used in the dispersion of conduction polymers and nanoclays. Methods like solution casting, melt blending, and in-situ polymerization have been employed to form the composite as explained below [25, 26]. The clay particles are hydrophilic in nature and the conducting polymer is hydrophobic in nature, therefore both these constituent materials go against each other making the dispersion harder, thus creating phase separation. In order to avoid this, before the mixing of nanoclays with conducting polymers is carried out, the surface of the clay is modified accordingly such that high intercalation and exfoliation is favoured.

In solution casting or solution blending, the conducting polymers are added to solution like epoxy resins and are subjected to *ultrasonication* to create cavities and the nanoclay are reinforced with these cavities and later, the prepared nanocomposite mixture is added into the insulating matrix. Later after mixing the solution evaporates leaving thin film. The epoxy resin solution helps in the mobility of polymeric chains and this helps in intercalation of polymeric chains with layered nanoclay particles [26, 27]. In case of *melt blending*, the polymer is mixed with nanoclay using an extruder and after intense mixing, a die press machine is used to apply pressure and obtain the polymer nanocomposite thin film. In this method, the polymeric mobility which helps to form intercalation or exfoliation is largely due to the thermal energy [26, 27].

Though the above-mentioned methods can be also used in the preparation, the *in-situ polymerization* provides high intercalation and exfoliation. In this method the layered silicate materials are *swollen* using ultra sonification and then the monomer of the conducting polymer is added into the interlayer gaps and then polymerization of the material is initiated to obtain the intercalation [26, 27].

Intercalation is the reversible inclusion or insertion of a molecule into the materials with layered structure. During intercalation the *van der Waals* gap in between the sheets increases by 2-3nm, and the energy required for this process is usually supplied by charge transfer between the guest and the host solids. *Exfoliation* is the extreme case of dispersions where the complete separation of the layers is achieved by aggressive separation method(s). The above-mentioned different type of polymer-clay composites systems are shown in **Figure 2** [28]. Following the addition of CPC as a filler to the insulating polymer matrix, the resultant composite material is pressed and made into a film of appropriate thickness, that would act as EMI shield.

| Intercalated | Intercalated and flocculated | Exfoliated |

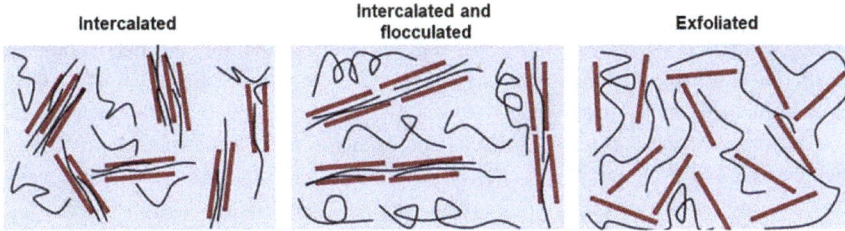

Figure 2 Schematic representation of the dispersions of clay in polymer systems that includes intercalation and exfoliation processes. Reprinted with Permission from Suprakas Sinha Ray et al. Progress in Polymer Science, Volume: 28, Issue: 11, 1539 2003.[28] Copyright 2003, Elsevier.

6. Characterization methods and characteristics of nanoclay-based conducting composites

Inherently poor mechanical and thermal properties of the insulating polymer matrix are enhanced by the addition of optimum filler loading. Here, the filler loading is typically made of conducting polymer reinforced with nanoclays. The presence of nanoclays in this matrix is found to provide enhancements to the properties of the EMI shield films [7, 11]. These are usually studied using the following tests that are shown in **Table 2**.

Table 2 Summary of Characterization Techniques employed to Determine the Properties of Nanocomposite Films.

Testing	Properties Verified
X Ray Diffraction (XRD)	Basal spacing or interlayer spacing
Scanning Electron Microscope (SEM), Transmission Electron Microscope (TEM)	Provide high resolution images of the polymer nanocomposite structures
Four-Point Probe Device	Electrical Conductivity of the Polymer Nanocomposite (σ_{dc})
Dynamic Loading Test	Provides information about the complex modulus (G)
Thermogravimetric Analysis (TGA), Differential Thermal Analysis (DTA) and Differential Scanning Calorimetry (DSC)	Provides details on mass or weight changes with temperature changes

Adv. App. of Micro and Nano Clay II – Synthetic Polymer Composites   Materials Research Forum LLC
Materials Research Foundations 129 (2022) 53-78   https://doi.org/10.21741/9781644902035-3

Besides, Fourier transform infrared spectroscopy and ultraviolet-Visible spectroscopy have been used to study and verify nature of the reinforcement of the polymer nanocomposite prepared. Depending on the absorption pattern, the vibration and stretching between different molecules can be known and thus helping us to know the intercalation and exfoliation range of the nanoclay particles. The details of these experimental techniques are beyond the scope of this chapter. They are tabulated to give the reader some basic information on how the material improvement with nanoclays inclusion are studied.

There are six primary characteristic improvements provided by the nanoclay based conducting polymer composites when incorporated into an insulating matrix. The first three **Sections 6.1 to 6.3** summarizes the contribution of the nanoclay based conducting polymer in the improvement of the physical properties of the EMI shield thin film prepared for the encapsulation of electronic devices. These improvements in the mechanical integrity, thermal resistance and barrier properties achieved by the inclusion of nanoclays will increase the quality and lifetime of the EMI shield thin film, thus protecting the device for a long time from EMI with less wear and tear.

The **Sections 6.4 to 6.6** summarizes the improvements provided by the nanoclay-based conducting polymer to the EMI shielding effectiveness.

6.1   Mechanical strength

Considering the structure of clays, it is well known that nanofillers increases the mechanical properties of the composites. Among the mechanical properties of the polymer nanocomposite thin film's, tensile strength and modulus are two key parameters. The enhancement of these properties is attributed to the high aspect ratio of the nanofillers used in the matrix. For example, a composite with an high aspect ratio for 1D and 2D nanomaterials showed a modulus enhancement by a factor of six at 0.01 nanoclay filler [25]. It has been documented in the literature that increasing nanofiller content at lower volume fractions results in a significant increase in mechanical strength [25]. However, at high volume fractions of the fillers, mechanical strength is found to decrease, this is attributed to issues related to the dispersion and agglomeration of the clays. For example, in case of polyaniline-clay nanocomposite prepared using 3-pentadecyl phenol 4-sulphonic acid (PANICN-PDPSA) at 15 and 20 wt. % of the conductive filler in the blend, the conductivity and shielding efficiency increased but the tensile strength started to decrease as the content of the conductive filler was increased [11]. The minimum filler loading where the trade-off occurs cannot be explicitly given, because it purely depends and varies with the constituent materials that are used in making the polymer-nanocomposites. For a thumb-rule 5-15 wt.% loading can be considered as optimal filler loading [9, 11].

The reinforcement of the nanoclays within the insulating polymer matrix plays an important role in increasing the physical property of the EMI shield thin film, so it is suitable to use nanoclays with high surface area for better reinforcement. Note that there is a significant increase in both tensile modulus and interaction between the insulating polymer matrix-nanoclay surfaces when the nanofillers were aligned using melt drawing, and polymer stretching methods, reflecting better reinforcement of matrix with fillers. Also, it is not necessary that all the aligned nanofillers with nanoclay should interact well with the insulating polymer matrix. Depending on the nanofillers prepared, insulating matrix can be chose.

Nanoclays enhance the modulus and tensile strength of the material but decrease the elongation at break. However, when compared to graphene and carbon nanotube-based composites, the relative reinforcement for a given volume percent of filler in the composite (with nanoclay as a filler) is substantially lower [25, 26].

This is one of the most significant benefits of employing nanoclays, when nanocomposites are created with either immaculate or organically altered clays, the tensile modulus of a polymeric compound is improved. This improved mechanical property does not improve EMI shielding effectiveness, but it does potentially improve the mechanical integrity of the EMI thin film developed.

6.2　Thermal stability

As mentioned in **Section 5** the prepared polymer nanocomposite with nanoclay is added to an insulating matrix and made into a thin film of appropriate thickness using die pressing machine. For a more comparative study, thin films were made with polymer nanocomposites with no nanoclay inclusion. These samples were tested for their thermal stability using Thermogravimetric Analysis (TGA).

Depending on the conducting polymer, insulating polymer-matrix and nanoclay used, the EMI thin film showed different degradation temperature ranges. To summarize the significance of inclusion of nanoclays and its influence on the thermal properties of the materials an example have been provide with polyaniline-clay nanocomposite prepared using 3-pentadecyl phenol 4-sulphonic acid (PANICN-PDPSA), polyaniline-clay nanocomposite prepared using dodecyl benzene sulfonic acid (PANICN-DBSA), polyaniline without clay nanocomposite prepared using 3-pentadecyl phenol 4-sulphonic acid (PANI-PDPSA) and polyaniline without clay nanocomposite prepared using dodecyl benzene sulfonic acid (PANI-PDPSA). The loss of solvent and low molecular weight volatile contaminants caused a minor weight loss in all of these conductive film samples with and without nanoclay below 100 °C [9-11]. As the temperature increased, the composites with nanoclays PANICN-PDPSA and PANICN-DBSA showed better thermal stability than the composite without nanoclays PANI-PDPSA and PANI-DBSA. The samples made of polymer nanocomposite with nanoclays showed two-stage degradation. The first stage-degradation corresponds to a 25 percent weight loss and is only linked with the decomposition of polymers and usually occurs between 225 and 300 °C. The second stage of thermal degradation occurred between 400 and 600 °C, resulting in a 65 percent weight loss due to the degradation of organic compounds. In addition, when compared to PANI-PDPSA and PANI-DBSA, weight loss during the second stage of decomposition was reduced [9-11].

The high aspect ratio of nanoclays, which inhibits degradation *(weight loss in material)* and provides a barrier to prevent evaporation of tiny molecules formed during the thermal decomposition process, is responsible for the nanoclay filled composite's exceptional thermal stability. The barrier effect of clay increases during the volatilization because of the re-assembly of the silicate layers of the polymer surface during thermal decomposition. As the percentage of conductive filler with nanoclay increased in the insulating matrix the second stage of degradation increased due to the residual formation. The residual formation is not seen in conductive films without nanoclays [10, 11]. This shows that, the inclusion of nanoclays increased thermal resistance of the polymer-nanocomposite, therefore increasing the thermal stability.

The thermal stability was further increased by the formation of extended polymeric chain on clay surface, therefore creating more polymer-clay interactions [29]. Furthermore, when compared to metals, polymers have a high thermal expansion coefficient. Because of the interaction with the fillers, fillers like nanoclays and graphene decrease the thermal expansion of polymers by confining the movement of a large volume of polymer chains [25]. The mechanical strength and thermal degradation temperature of some of the polymer-nanocomposite materials with nanoclays are tabulated in **Table 3**.

6.3   Barrier properties

If the absorbed water molecules are constrained and unable to rotate, they cannot dissipate EM energy; but, if the water molecules are unbounded, they can dissipate energy through unrestricted rotation in the existence of an EM field. This energy dissipation is associated with frictional and inertial losses during the reorientation of absorbed dipole water molecules when exposed to an EM field. This reorientation or rotation of water molecules has significant impact on the relative permittivity and loss tangent of the polymer nanocomposite materials [30].

Therefore, it is necessary to make the EMI shielding material less susceptible to moisture diffusion. To highlight the drastic effect of moisture uptake, consider an epoxy system *(primarily used in radome structures)*, immersed in distilled water at 25 °C. The dielectric properties deviated significantly due to the moisture uptake. The relative permittivity improved by 15% and the loss tangent increased by 220% at a moisture content of 5%. The moisture content not only affected the dielectric properties of the insulating polymer-matrix nanocomposite composition, but it also reduced the modulus of elasticity and mechanical strength. The addition of nanoclays has been proved effective in improving the moisture barrier properties. Different weight fraction loading of clays showed different moisture barrier effect due to different dispersion rates. Optimal loading of 2 wt.% showed great extent of dispersion and increased moisture-barrier properties. Several nanoclay loadings revealed a higher moisture uptake rate, which is owing to insufficient clay dispersal in the polymer structure caused by the natural moisture content in the organoclays ahead of usage [30].

Table 3 Mechanical Strength and Thermal Degradation Temperature of Some Key Nanoclay-based Polymer Composites.

S.No	Composite Material with Nanoclay	Filler Loading Percentage (wt.% or vol. %)	Tensile Strength (MPa)	Thermal Degradation Temperature
1	EVA-Polyaniline with Nanoclay [11]	15 vol. %	17.50	400-600 °C *(Second Stage)*
2	La-Doped Barium Hexaferrite-Polyaniline with Sepiolite Clay [8]	60 vol. %	-	-
3	Polyetherimide with Cloisite 30B (MMT) [7]	1 wt. %	95.9	-
4	PVC-Polyaniline with Bentonite Clay [9]	5 vol. %	-	102-123 °C
5	Polystyrene-Polyaniline with Bentonite Clay[10]	10 vol. %	70.6	360-380 °C

The diffusion route of penetrating molecules can be obstructed and altered by polymer composites by including nanofillers with a high aspect ratio. The penetrating molecules can be either oxygen (water vapor) or moisture absorbed from the environment. This approach to prevent moisture uptake and water vapor penetration is primarily used in Radome structures in defence applications [30, 31]. A tortuous pathway for water diffusion through the polymer-matrix nanocomposite is created by well dispersed nanofillers, thereby lowering the water absorption rate and transverse diffusivity [30]. In addition, regardless of the kind of gas molecules, a polymer nanocomposite reinforced with clay reported a decrease in gas permeability. Polymeric nanocomposites with graphene oxide conjugated polymer nanocomposites showed high hinderance to both oxygen and carbon dioxide permeation [25]. Therefore, EMI shields prepared with nanoclay fillers have high moisture and gas barrier properties and provided longevity by delaying significant dielectric property degradation thus, increasing the lifetime and quality of the EMI shield.

6.4 Electrical conductivity

The resistivity of the sample is measured using the four proe electrode methods. Then conductivity is determined using the inverse of Van der Pauw's relation $\sigma_0 = \left(\frac{In2}{\pi d}\right) * \frac{V}{I}$. The electrical conductivity of the conducting polymer with nanoclays are in the same range as conduction polymer without any reinforcement. This is because the conducting polymers inside the clay layers are in a well-expanded configuration, and the conducting polymer chains operate as a bridge, aiding electron movement. When this conductive filler is mixed with the insulating matrix, it aggregates and breaks down into minute particles, allowing continuous conducting channels to form and the material transitions from insulating to conductive phase at a specific concentration of conductive filler called percolation threshold [11]. For example, EVA, an insulating matrix with conductivity in the range of 10^{-11} S/m, A significant change in electrical conductivity was obtained with the addition of small amount conductive filler with nanoclay. With 5 wt. % fraction of polyaniline-clay nanocomposite prepared using 3-pentadecyl phenol 4-sulphonic acid (PANICN-PDPSA) and polyaniline-clay nanocomposite prepared using dodecyl benzene sulfonic acid (PANICN-DBSA) conductivity increased to 8.03 x 10^{-3} S/m and 8.72 x 10^{-3} S/m respectively, similar trends can be seen in papers [7, 32]. SEM images shown in **Figure 3** demonstrate the formation of interconnected conductive networks in the presence of nanoclay in such nanocomposites at a very low filler concentration and the desired value of electrical conductivity in the range of 10^{-1} S/m is shown to be achieved with 5 wt. % filler besides realizing excellent tensile strength of 67 MPa [10]. This is in accordance with percolation theory, which posits that conductive networks can be effectively produced with optimal filler loading if one of the dimensions of the conductive filler is in the nanometre size and has a high aspect ratio. With the inclusion of nanoclays in CPCs, the percolation threshold concentration is achieved at a reduced concentration of conducting fillers. This is likely due to the high polymer mobility provided by the nanoclays and also, the extended polymeric chains with high π-electron conjugation with nanoclay can enhance conduction by transfer of electron between different layers of clays [11] and this type of electron movement within the backbone polymer chain is called as variable range hopping (VRH) [33].

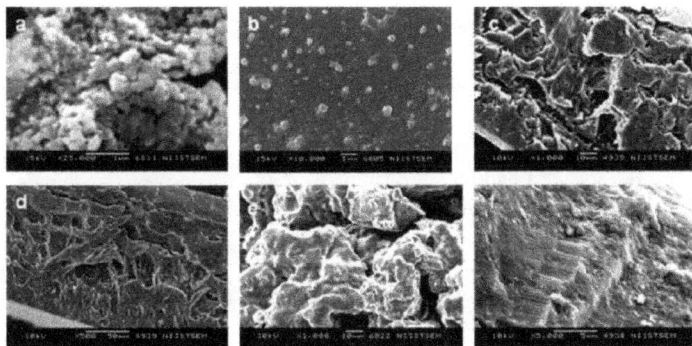

Figure 3. SEM images of different nanocomposites that reveals the extent of dispersions and formation of conductive networks. Bright contrast denotes regions of conductive filler, and the black portion denotes regions of matrix polymer. Composites are (a) PANI-PDPSA, (b) PANI/PS(1.25% filler) (c) PANI/PS(5% filler), (d) PANICN/PS 1(2.5% filler), (e) PANICN/PS 1(5% filler), and (f) PANICN/PS 1(7.5% filler). Reprinted with Permission from Sudha et al. Composites: Part A Volume 41 1647, (2010).[10] Copyright (2010) Elsevier.

There is a significant amount of non-covalent interactions present between the clays, conducting polymers and the insulating matrix. These interactions are responsible for the formation of porous templates. Through edge-to-edge association, the porous walls were produced by parallel stacking of silicate layers and their aggregates [10]. The creation of a continuous and uniform conductive film is thus due to the many covalent and non-covalent interactions among the various entities present in clay, conductive polymers, and the insulating matrix. The multiple covalent and non-covalent interactions among the clay tactoids, insulating matrix, and conducting polymer get strengthened as polymerization progresses, causing the porous wall to collapse and form the electrical percolated network [10].

6.5 Dielectric properties

The complex permittivity ($\varepsilon = \varepsilon' - j\varepsilon''$) and complex permeability ($\mu = \mu' - u''$) and their related loss tangents (tan $\delta\varepsilon$ and tan $\delta\mu$) define the dielectric characteristics of the polymer-nanocomposite sample, such as dielectric loss and magnetic loss. The real parts ε' and μ' are related with the EM wave's storage capacity, in this instance microwave energy, whereas the imaginary parts ε'' and μ'' indicate microwave energy's dissipation capability. The electric loss tangent (tan $\delta\varepsilon = \varepsilon''/\varepsilon$) and the magnetic loss tangent (tan $\delta\mu = \mu''/\mu$') are commonly employed to calculate a shielding material's attenuation capabilities [8].

Table 4 Dielectric Constant, Conductivity, Operation Frequency, Reflection Loss and Effective Absorption Bandwidth of Some Key Nanoclay-based Polymer Composites.

S No	Composite Material with Nanoclay	Filler Loading Percentage (wt.% or vol. %)	Thickness (mm)	Frequency	Reflection Loss RL$_{min}$ (dB)	Effective Absorption Bandwidth (EAB)
1	EVA-Polyaniline with Nanoclay [11]	15 vol. %	10 mm	8 GHz	75.7 dB	-
2	La-Doped Barium Hexaferrite-Polyaniline with Sepiolite Clay [8]	60 vol. %	2.4 mm	10.80 GHz	71.98 dB	3.54 GHz
3	Polyetherimide with Cloisite 30B (MMT) [7]	1 wt. %	0.2 mm	8-12 GHz	0.5 dB	4 GHz
4	PVC-Polyaniline with Bentonite Clay [9]	5 vol. %	10 mm	10 GHz	55.2 dB	-
5	Polystyrene-Polyaniline with Bentonite Clay [10]	10 vol. %	10 mm	8 GHz	16.2 dB	-

After adding nanoclays to the composite, the complex permeability improves. This is due to dipole relaxation and polarisation. According to the free electron hypothesis, adding high conductive polymers to a composite can increase its conductivity, contributing to an increase in ε'' [8]. Also, the addition of conducting polymers can increase the resulting interface effects between the nanoclays and the conducting polymers, which in turn, will cause improved interfacial polarization, thus increasing the value of ε'. Note that as the frequency increases, both ε' and ε'' drops, suggesting frequency dispersion behaviour, which Debye theory attributes to polarisation relaxation lagging in high frequencies [8]. For a good absorption mechanism and impedance matched sample the electric loss tangent should be equal to the magnetic loss tangent, which is highly unlikely to happen or be achieved in practical situations. So, it is preferred to have materials with either high dielectric loss or magnetic loss along with good conduction loss. In most practical cases materials with high dielectric losses are preferred.

As per Maxwell's equation, the movement of charges generates an AC electric field under the applied EM field, which produces an internal magnetic field and releases some magnetic energy. However, when a polymer with a relatively high conductivity is added, the intrinsic magnetic loss cannot counterbalance the radiation of magnetic energy produced by the internal magnetic field, especially at high frequencies, and the redundant energy is radiated out, resulting in a decrease in magnetic loss tangent. So, in this chapter we will be focusing on materials with high dielectric loss and conductive loss. As well know the inclusion of nanoclay contributed to increase of the dielectric loss which mainly comes from polarization relaxation and conductive losses. Polarization relaxation is involved with ionic, electronic, dipole orientation and interfacial polarization. However, the ionic and electronic polarization are always detected in the higher frequency region 10^3-10^6 GHz theory [8, 31]. Thus, the dipole and interfacial polarization are dominant in the GHz region.

The exfoliated samples had a greater dielectric constant and a larger loss factor than the neat polymers, which is remarkable. The dielectric constant and loss factor of intercalated polymers were only slightly higher than those of pure polymers. Furthermore, for exfoliated nanocomposite with higher nanoclay concentration, the real and imaginary parts of dielectric constants are larger. The distinct morphologies created in these films can be used to provide more in-depth reasoning. The platelets are totally delaminated, exfoliated, and disseminated in the conduction polymer, resulting in a broad interfacial area between the nanoclay and the polymer matrix in such nanocomposites. In an applied electric field, the different conductivities of the two phases results in substantial interfacial area or space charge polarisation of trapped charges at the interface. This effect of polarization is also termed as Maxwell Wagner polarization [34]. It is worth noting that conductive loss is another important factor that influences the dielectric loss of polymer nanocomposite. The creation of a highly organised conductive network is responsible for this increased conductive loss.

6.6   Shielding efficiency

Though the incorporation of nanoclay has not contributed significantly to the increase of electrical conductivity, it has decreased the percolation threshold and increased the shielding efficiency with increased loading of nanoclay filled conducting polymer into the insulating matrix. The increased attenuation supported by the numerous reflections caused by the nanoclay accounts for the improved shielding efficiency obtained for conductive nanocomposite films. Note that absorption loss is due to the heat loss under the action of magnetic and electric dipoles observed in the shielding material and the EM field, as stated in detail in **Section 2.2**.

Multiple reflection arises as a result of the shielding material's porosity and increased number of interfaces, which provides the clays with a large surface area. The lower the energy that passes through the sample, the better the shielding effectiveness [10, 11, 29]. Mixing clay with ferrite materials like La-doped barium hexaferrite contributed to increase in magnetization. The increased magnetization promotes dipole formation, improving conductive losses and dielectric losses. The dielectric losses are mainly due to different polarization but, the materials with ferrite content can also increase the dielectric losses [8].

There are factors other than multiple reflection that contributes to the loss or attenuation of the EM wave and increase the shielding efficiency. These are transmission loss and reflection loss. Both transmission loss and reflection loss depend on various factor like loss tangent, dielectric constant of the whole material, thickness of the material, refraction, Bore's sight error etc. [7]. The formula to calculate the transmission and reflection loss are given in **Section 2.4**. The electrical conductivity measurements also showed that the conductivity of the material remained almost of the same order after the addition of nanoclays [9]. This is due to the formation of more ordered conductive network in the presence of nanoclays, which contributed to conductive losses [9]. Briefly saying, the main factor for all the enhancement is the high aspect ratio that increases surface interaction. This increased surface interaction is achieved as a result of good reinforcement of nanofillers in the polymer with high degree of polymerization.

7.   Summary and perspectives

The nanoclays may not have many advantages as carbon nanotubes or graphene nanofillers in terms of increasing the electrical conductivity or contributing to the shielding effectiveness. But the insulating polymer matrix with the inclusion of nanoclay and conductive polymer is also one of the suitable candidates for the preparation of EMI shield thin films with increased life expectancy due to the increased mechanical strength, thermal resistance and moisture-gas barrier properties and protecting the device encapsulated for a long time from EMI [1].

The factors that provide challenge in the incorporation of nanoclay fillers into the polymer matrix are exfoliation, orientation, compatibility and reaggregation. The hydrophilic nature of clays and the hydrophobic nature of the conducting polymers makes these materials highly incompetent for mixing, both these constituent materials go against each other making the dispersion harder, creating phase separation. To avoid this, the surface of the clay is required to be chemically modified. The morphology of clays allows easy surface modification and nanometre dispersion in polymer matrix [35]. New methods of surface modification and nanometre dispersion that increase the interaction between the nanoclay surface and the polymer matrix and providing high exfoliation, orientation, compatibility and reduced reaggregation can be considered as a prospective research idea.

Focusing on the shielding application, the possibility of achieving high magnetic loss can be considered as an important area of research. The magnetic losses can be increased by using conducting polymers with high magnetic susceptibility, leading to increased magnetic dipole formation and thus increasing the absorption of EM waves. High magnetic loss can also be achieved by incorporating ferrite materials into nanoclays and creating a hybrid ferrite-nanoclay filler [8]. Also, various conducting polymers can be mixed to create hybrid conducting polymers and their electrical conductivity can be studied.

References

[1] S. Geetha, K.K.S. Kumar, C.R.K. Rao, M. Vijayan, D.C. Trivedi, EMI shielding: Methods and materials - A review, J. Appl. Polym. Sci. 112 (2009) 2073-2086. https://doi.org/10.1002/app.29812

[2] L. Sevgi, Electromagnetic screening and shielding-effectiveness (SE) modeling, IEEE Antennas Propag. Mag. 51 (2009) 211-216. https://doi.org/10.1109/MAP.2009.4939074

[3] V. Shukla, Review of electromagnetic interference shielding materials fabricated by iron ingredients, Nanoscale Adv. 1 (2019) 1640-1671. https://doi.org/10.1039/C9NA00108E

[4] V. Shukla, Advances in Hybrid Conducting Polymer Technology for EMI Shielding Materials, in: S. Shahabuddin, A.K. Pandey, M. Khalid, P. Jagadish (Eds.), Advances in Hybrid Conducting Polymer Technology, Springer International Publishing, Cham, 2021, pp. 201-247. https://doi.org/10.1007/978-3-030-62090-5_9

[5] M. Sidhaarth, R. Suriyanarayanan, G. Srigovindan, R. Thiruvengadathan, A holistic approach to evaluate EMI shielding characteristics of carbon nanotube-based polymer composites, 12th IEEE Nanotechnology Materials and Devices Conference, NMDC 2017, Institute of Electrical and Electronics Engineers Inc., 2018, pp. 131-132. https://doi.org/10.1109/NMDC.2017.8350530

[6] S.S. Prabhu, R. Bhavani, G.K. Thonnuthodi, R. Thiruvengadathan, A computational approach to determine shielding effectiveness of carbon nanotube-based nanocomposites for EMC application, Comput. Mater. Sci. 126 (2017) 400-406. https://doi.org/10.1016/j.commatsci.2016.10.006

[7] M. Dwivedi, A. Dixit, S. Alam, A.K. Ghosh, Dielectric and tensile behavior of nanoclay reinforced polyetherimide nanocomposites, Journal of Applied Polymer Science 122 (2011) 1040-1046. https://doi.org/10.1002/app.34196

[8] Q. Shang, H. Feng, K. Pan, N. Chen, L. Tan, J. Qiu, Modified sepiolite/BaLa0.5Fe11.5O19@polyaniline composites with superior microwave absorption properties, Journal of Materials Science: Materials in Electronics 31 (2020) 8523-8535. https://doi.org/10.1007/s10854-020-03388-6

[9] J.D. Sudha, S. Sivakala, C.K. Chandrakanth, K.S. Neethu, K.N. Rohini, R. Ramakrishnan, Percolated conductive polyaniline-clay nanocomposite in polyvinyl chloride through the combined approach porous template and self-assembly, Express Polymer Letters 8 (2014) 107-115. https://doi.org/10.3144/expresspolymlett.2014.13

[10] J.D. Sudha, S. Sivakala, K. Patel, P. Radhakrishnan Nair, Development of electromagnetic shielding materials from the conductive blends of polystyrene polyaniline-clay nanocomposite, Compos. - A: Appl. Sci. Manuf. 41 (2010) 1647-1652. https://doi.org/10.1016/j.compositesa.2010.07.015

[11] J.D. Sudha, S. Sivakala, R. Prasanth, V.L. Reena, P. Radhakrishnan Nair, Development of electromagnetic shielding materials from the conductive blends of polyaniline and polyaniline-clay nanocomposite-EVA: Preparation and properties, Composites Science and Technology 69 (2009) 358-364. https://doi.org/10.1016/j.compscitech.2008.10.026

[12] B.T.S. Ramanujam, P.V. Adhyapak, S. Radhakrishnan, R. Marimuthu, Effect of casting solvent on the structure development, electrical, thermal behavior of polyvinylidene fluoride (PVDF)-carbon nanofiber (CNF) conducting binary and hybrid nanocomposites, Polym. Bull. 78 (2021) 1735-1751. https://doi.org/10.1007/s00289-020-03176-6

[13] B.T.S. Ramanujam, C. Gopalakrishnan, Investigations of structure development, electrical and thermal properties of polyvinylidene fluoride-expanded graphite nanocomposites, Bull. Mater. Sci. 44 (2021) 66. https://doi.org/10.1007/s12034-021-02354-0

[14] R. Thiruvengadathan, S. Mahadevan, Experimental and computational aspects of electronic properties of carbon-based polymer nanocomposites, Carbon-Based Polymer Nanocomposites for Environmental and Energy Applications, Elsevier Inc.2018, pp. 175-198. https://doi.org/10.1016/B978-0-12-813574-7.00007-1

[15] M.H. Al-Saleh, U. Sundararaj, Electromagnetic interference shielding mechanisms of CNT/polymer composites, Carbon 47 (2009) 1738-1746. https://doi.org/10.1016/j.carbon.2009.02.030

[16] Z. Chen, C. Xu, C. Ma, W. Ren, H.M. Cheng, Lightweight and flexible graphene foam composites for high-performance electromagnetic interference shielding, Adv. Mater. 25 (2013) 1296-1300. https://doi.org/10.1002/adma.201204196

[17] M. Hu, J. Gao, Y. Dong, K. Li, G. Shan, S. Yang, R.K.Y. Li, Flexible transparent PES/silver nanowires/PET sandwich-structured film for high-efficiency electromagnetic interference shielding, Langmuir 28 (2012) 7101-7106. https://doi.org/10.1021/la300720y

[18] P. Saini, V. Choudhary, N. Vijayan, R.K. Kotnala, Improved electromagnetic interference shielding response of poly(aniline)-coated fabrics containing dielectric and magnetic nanoparticles, J. Phys. Chem. C 116 (2012) 13403-13412. https://doi.org/10.1021/jp302131w

[19] C.D. Barton, A.D. Karathanasis, Clay Minerals. Encyclopedia of Soil Science, 2002.

[20] N.C.W.R.R. Brady, The nature and properties of soils, Pearson Prentice Hall, Upper Saddle River, N.J., 2008.

[21] E. García-Romero, M. Suárez, A structure-based argument for non-classical crystal growth in natural clay minerals, Mineral. Mag. 82 (2018) 171-180. https://doi.org/10.1180/minmag.2017.081.031

[22] M.F. Brigatti, E. Galan, B.K.G. Theng, Chapter 2 Structures and Mineralogy of Clay Minerals, in: F. Bergaya, B.K.G. Theng, G. Lagaly (Eds.), Developments in Clay Science, Elsevier2006, pp. 19-86. https://doi.org/10.1016/S1572-4352(05)01002-0

[23] Y. Zhou, A.M. LaChance, A.T. Smith, H. Cheng, Q. Liu, L. Sun, Strategic Design of Clay-Based Multifunctional Materials: From Natural Minerals to Nanostructured Membranes, Adv. Funct. Mater. 29 (2019). https://doi.org/10.1002/adfm.201807611

[24] D. Kumar, R.C. Sharma, Advances in conductive polymers, Eur. Polym. J. 34(8) (1998) 1053-1060.

[25] M. Bhattacharya, Polymer nanocomposites-A comparison between carbon nanotubes, graphene, and clay as nanofillers, Materials 9 (2016). https://doi.org/10.3390/ma9040262

[26] Z. Spitalsky, D. Tasis, K. Papagelis, C. Galiotis, Carbon nanotube-polymer composites: Chemistry, processing, mechanical and electrical properties, Prog. Polym. Sci. 35 (2010) 357-401. https://doi.org/10.1016/j.progpolymsci.2009.09.003

[27] P. Nguyen-Tri, T.A. Nguyen, P. Carriere, C. Ngo Xuan, Nanocomposite Coatings: Preparation, Characterization, Properties, and Applications, Int. J. Corros. 2018 (2018). https://doi.org/10.1155/2018/4749501

[28] S. Sinha Ray, M. Okamoto, Polymer/layered silicate nanocomposites: A review from preparation to processing, Prog. Polym. Sci. 28 (2003) 1539-1641. https://doi.org/10.1016/j.progpolymsci.2003.08.002

[29] H. Sehaqui, J. Kochumalayil, A. Liu, T. Zimmermann, L.A. Berglund, Multifunctional nanoclay hybrids of high toughness, thermal, and barrier performances, ACS Appl. Mater. Interfaces 5 (2013) 7613-7620. https://doi.org/10.1021/am401928d

[30] C. Garcia, J.L. Cohn, C. Chesley, L. Grace, Effects of moisture absorption on the dielectric properties of nanoclay-reinforced epoxy for radome applications, ASME 2014 International Mechanical Engineering Congress and Exposition, IMECE 2014, American Society of Mechanical Engineers (ASME), 2014. https://doi.org/10.1115/IMECE2014-38815

[31] P.C. Kim, D.G. Lee, I.S. Seo, G.H. Kim, Nanocomposite stealth radomes with frequency selective surfaces, Compos. Struct. 86 (2008) 299-305. https://doi.org/10.1016/j.compstruct.2008.03.045

[32] C.A. Stergiou, A.Z. Stimoniaris, C.G. Delides, Hybrid Nanocomposites With Organoclay and Carbon-Based Fillers for EMI Suppression, IEEE Trans. Electromagn. Compat. 57 (2015) 470-476. https://doi.org/10.1109/TEMC.2014.2384014

[33] M. Vellakkat, A. Kamath, S. Raghu, S. Chapi, D. Hundekal, Dielectric constant and transport mechanism of percolated polyaniline nanoclay composites, Ind. Eng. Chem. Res. 53 (2014) 16873-16882. https://doi.org/10.1021/ie502922b

Adv. App. of Micro and Nano Clay II – Synthetic Polymer Composites   Materials Research Forum LLC
Materials Research Foundations 129 (2022) 53-78   https://doi.org/10.21741/9781644902035-3

[34] N. Anwar, M. Ishtiaq, A. Shakoor, N.A. Niaz, T.Z. Rizvi, M. Qasim, M. Irfan, A. Mahmood, Dielectric properties of polymer/clay nanocomposites, Polym. Polym. Compos. 29 (2021) 807-813. https://doi.org/10.1177/0967391120953250

[35] K.P. Dasan, Nanoclay/polymer composites: Recent developments and future prospects, Adv. Struct. Mater., Springer Verlag, 2015, pp. 561-579. https://doi.org/10.1007/978-81-322-2470-9_19

Chapter 4

Micro Clay/Nano Clay Polymer Composite Flame Retardant Applications

Momina, Haq Nawaz Bhatti* and Amina Khan*

Department of Chemistry, University of Agriculture, Faisalabad, Pakistan

Mominamustafa427@gmail.com: hnbhatti2005@yahoo.com*:aminakhan1649@gmail.com*

Abstract

The discovery of fire is one of the earliest and most significant achievements of man. However, it is a lethal power that has mostly stayed uncontrollable. Big fires occur practically daily and result in significant loss of life and property. The majority of the contents in the houses are flammable. Clothes, blankets, furniture, paper, synthetic polymeric materials, automobile and plane interiors, etc., all burn when the conditions are favorable. The addition of flame retardant to polymeric materials improves their flame retardancy and thermal stability. In this chapter, we will study the addition of micro clay/ nano clay to improve the flame retardancy of polymeric composites to make them feasible to be used in various potential applications.

Keywords

Fire, Flame Retardant, Polymer, Clay Composite, Applications

Contents

Micro Clay/Nano Clay Polymer Composite Flame Retardant Applications79

Adv. App. of Micro and Nano Clay II – Synthetic Polymer Composites   Materials Research Forum LLC
Materials Research Foundations 129 (2022) 79-107   https://doi.org/10.21741/9781644902035-4

1. Introduction

The discovery of fire is one of the earliest and most significant achievements of man. Fire has been a basic necessity of all civilizations and has a far greater impact on mankind's progress than any other innovation. Fire, on the other hand, is a lethal power that has mostly stayed uncontrollable [1]. As some of the following catastrophes demonstrate, fires have wreaked immense damage throughout history.

The wildfire season in California erupted at the start of July 2021 and has been quickly increasing since then. By September 2021, 7,377 fires have burnt over 2.2 million acres, an area larger than the Grand Canyon. According to the National Interagency Fire Center's (NIFC) status report dated September 14-2021, 44,647 wildfires were burning more than 5.6 million acres throughout the country. In the United States, there are 19,761 workers employed on 79 major, ongoing fires, 60 of which are still burning. So far, 6.3 million acres have been burnt by 44,654 fires.

Australia experienced one of the worst forest fire outbreaks in recorded human history from September 2019 to March 2020. The wildfire season began in June 2019, making 2019 Australia's warmest year on record. This resulted in widespread devastation across the country, with flames raging in every state and territory. Mega-fires, like the Currowan wildfire, which seems to be one of several devastating bushfires on the eastern seaboard over the 2019-2020 season, caused enormous devastation. The 2019-2020 fires were the worst in terms of property burnt, wildlife mortality, and environmental damage—some of it everlasting, like the destruction of the residual rainforest.

A series of more than 200 wildfires burned 1700 km² of woodland in Turkey's Mediterranean Area in July - August 2021, making it the country's terrible wildfire season. On July 28, 2021, flames broke out in Manavgat, Antalya Region, with temperatures hovering around 37 °C (99 °F). Two fires were still blazing on August 9, 2021, both in Mula. The flames are part of a wider sequence of wildfires, including some in nearby Greece, caused by a heatwave made more likely by changing climate.

The potential for anything to catch fire in a specified situation is known as a fire risk. When you have any electronic devices, you are putting yourself at risk of catching fire. This fire risk situation could burn the electronic item if the battery shorts out or the power supply fails and causes electrical discharge on the circuit board. Preventing the material from sparking and reducing the rate of temperature rise once it does is one way to reduce fire risks.

In today's world, the chances of a catastrophic fire destroying a whole town during a peaceful period are minimal. Techniques for putting out fires have been developed. The efficacy of firefighting operations has increased due to technological developments such

Adv. App. of Micro and Nano Clay II – Synthetic Polymer Composites   Materials Research Forum LLC
Materials Research Foundations 129 (2022) 79-107   https://doi.org/10.21741/9781644902035-4

as strong water jets and firefighting equipment. Smoke detectors and sprinkler systems keep buildings safe. Attempts are also undertaken to insulate materials against fire by the use of flame retardants. As a result, one may believe that man has conquered fire. Nobody could be farther from the truth, since big fires occur practically daily and result in significant loss of life and property.

A material's potential to contribute meaningfully to a fire that could become out of control and cause destruction is called a fire hazard. This underlying force of nature that allows the fire to exist on this planet also ensures that many of the things we use today may burn. Combustion can occur in any carbon-based material, from wood to plastics, provided that heat and oxygen are present. As oxygen is abundant, combustion is a continual force of nature on our earth. Although some substances burn more easily than others, any carbon-based objects, including thermally stable forms of carbon like graphite, will burn in excess supply of oxygen and heat [2].

Modern polymeric materials are classified as carbon-based combustible material, and they can be significantly more dangerous than organic materials such as timber, cotton, or other cellulosic mass in some instances. Although these materials are more flammable than natural-based materials, they are used because they are considerably superior in other areas, like cost and fabrication efficiency [3]. However, humanity's current transition to a more stable society has prompted all segments of society to review our use of natural assets (energy, water, and so on), toxicity issues, and environmental impact. This has shifted the way the materials sector approaches strategic planning [4].

The life-cycle evaluation compares the effects of organic polymers, green chemical approaches, sustainable manufacturing, and recycling on the production of sustainable goods using standardized (ISO14040) methods. While the combustibility of polymers occurs in nature, like wool and cellulose, is well understood, the flammability of new bio-based polymers with bio-derived monomers, such as those from corn, soy, castor, and other sources, is unknown, and the finest fire retardant strategies for these advanced materials have yet to be developed [5]. If the usage of plastics, whether petroleum-based or bio-based, grows in many parts of daily life, it's important to remember that these substances bring with them a slew of negative consequences.

2.   Flame retardants

Chemicals that are applied to materials to prevent or delay the spread of fire are known as flame retardants. Since the 1970s, they've been employed in a variety of consumer and commercial items to reduce the susceptibility of materials to burn fabric materials [6]. The word "flame retardant" also refers to textiles that do not burn and are self-extinguishing.

Adv. App. of Micro and Nano Clay II – Synthetic Polymer Composites   Materials Research Forum LLC
Materials Research Foundations 129 (2022) 79-107   https://doi.org/10.21741/9781644902035-4

This sort of fabric material will not facilitate the spread of flame in the event of an unintentional fire. Practically all textiles are flammable to some degree; words like flame-resistant, fireproof and flameproof are often useless or misleading. The burning rate varies, from material to material [7].

Flame retardants (FR) are substances that are applied to flammable materials to make them more fire-resistant. They're made to reduce the risk of fire if they come into touch with a heating element, such as a cigarette or a candle, or if there's an electrical failure. The flame retardant will slow or stop ignition and prevent the fire from spreading to other things if the material is ignited. Inorganic acids, hydrates, acid salts, organophosphorus and organ bromine chemicals, antimony salts/halogen complexes, and polymers are used as FR for textile fabrics [8].

Flame retardants have been used in a variety of applications since antiquity. Flame retardants were restricted to use in naturally occurring polymers like wood and cotton. In 360 B.C., a book on defenses advised that timbers be preserved from fire by coating them with vinegar. During the fight for Piraeus in 83 B.C., wooden defensive towers were coated with alum to prevent them from igniting by fire. In that fight, these towers were effectively deployed without burning up. After that, nothing notable happened until Nicolas Sabbatini proposed that clay and gypsum might be utilized as flame retardants for painted canvas used in Parisian theatre in 1638. Obadiah Wyld was awarded English Patent No. 551 in 1735 for imparting fire resistance to papers and fabrics by employing a combination of borax, alum, and ferrous sulfate [9].

Following multiple fires in French theatres in 1786, King Louis XVIII hired Gay-Lussac to investigate ways to safeguard theatrical materials. Gay-Lussac discovered that ammonium salts of hydrochloric acid, phosphoric acid, and sulfuric acid were highly efficient in flame-retarding kemp and linen textiles after 34 years of his systematic investigation of all known chemicals. He also mentioned that a combination of ammonium phosphate and chloride was much more efficient and produced better outcomes. He also noted that although borax alone could not prevent a warm glow, it was an extremely effective Quartermaster Corps at multiple places, resulting in incomplete knowledge of the phenomena, including the dangers of employing iron and zinc chemicals. Until the early 1960s, there were plenty of patents on the subject, indicating that the concept had matured. Lyons has provided an outstanding review of the literature in this area (1970). The three primary findings that established the framework for current flame retardants include Gay- and Lussac Henry Perkin's work, as well as the impact of combining antimony oxide with organic halogen compounds. Although today's technology is considerably more advanced, it nevertheless reflects variants on these previous themes.

They may be categorized into three groups:

(1) Primary FR based on phosphorous and/or halogen—phosphorous derivatives often act in the crystalline or condensed form, whereas halogen (chlorine or bromine) typically acts in the gaseous phase [10].

(2) For phosphorous and halogen-based FRs, synergists such as nitrogen and antimony are used. Synergists are not fired retardants but when are added to phosphorus or halogen-phosphorus-based flame retardant, increase their durability [11].

Alumina trihydrate, silicates, boron compounds, and carbonates are examples of additive or physical FRs. Their activity is mostly physical; however recently there has been some evidence of a chemical impact. These are non-durable and should only be used if washing durability is not a concern [12].

3.   Toxicity of conventional flame retardant

However, due to possible toxicity and environmental issues, several FRs have now been prohibited. Metal hydroxides, like aluminum and magnesium hydroxides, reduce the burning of polymeric materials by chilling the polymeric material and encouraging the development of an insulating ceramic coating. However, due to metal hydroxides' low efficiency, they often require a large loading (commonly >50 wt. %), which harms the physical and functional characteristics of composite materials.

Silica-containing FRs increase the thermal and mechanical properties of polymers by creating a protective coating in the condensed phase by accumulating low-surface-energy silica (SiO_2) clusters, but their poor efficiency makes them unsuitable for actual commercial use [13].

Red phosphorus(inorganic) and organic phosphorus-containing chemicals, for example, function primarily by interacting with reactive H (%) and OH (%) in the gas phase, leading to significant flame-resistant efficacy [14]. But, due to its red hue, plasticity, ability to diffuse to time, and production of poisonous phosphine, red phosphorus is not the best available option [15].

Halogenated chemicals have severe disadvantages in terms of environmental permanence and toxicity [16]. Furthermore, their usage is now restricted by the European Commission REACH, which governs the registration, evaluation, authorization, and limitation of chemicals. As a result, developing environmentally acceptable and incredibly efficient FRs for polymeric materials that incorporate strong flame retardancy and mechanical characteristics has remained a significant problem [15].

Adv. App. of Micro and Nano Clay II – Synthetic Polymer Composites   Materials Research Forum LLC
Materials Research Foundations 129 (2022) 79-107   https://doi.org/10.21741/9781644902035-4

Typical fire-resistant polymer composites are easily turned into char layers or clinker with little self-sustaining capacity after burning, which cannot prevent secondary catastrophes like detonation/radiation exposure triggered by tripping or short circuit after the fire. Conventional fire-retarding polymeric composites have limitations in some applications, like high-speed rail transit, nuclear reactors, and other areas where the self-sustaining potential of substances after combustion is also urgently needed to ensure the typical operation of electrical devices [17].

As previously stated, the most frequent way to enhance a polymer's flame retardancy performance, – i.e., greater ignition resistance, reduced flame spreading, escape time, confinement of flame, and degradation materials, is to incorporate FRs.

4. Combustion (oxidation) of polymers

Polymer ignition is a multi-step procedure that often involves several chemical and physical reactions. To begin and maintain the combustion of polymers, three essential elements, namely heat, blaze, and oxygen that are generally represented as a fire triangle must accompany each other. In general, a polymer's combustion begins with the polymer's thermal breakdown in the presence of an external heating element. The polymer thermally decomposes (pyrolysis) when exposed to an external heat source, generating volatile fragments by bond or chain scissions. These flammable degradation pieces then spread into the atmosphere, forming a flammable mixture of gaseous [18].

When the self-igniting temperature exceeds a particular value with the help of external heating systems, or when interacting with an exterior source of intense energy, such as flame the mixture can be used as fuel and burst into flames to start a fire. The combustion of polymeric materials is sustained by thermal oxidation processes of combustible breakdown products with the help of heat and oxygen. It should be mentioned that in the burning of certain polymers, reactive hydrogen ($H°$) and hydroxyl radicals ($HO°$) are present, which are essential for flame combustion [19].

Aside from the external heat source, polymer combustion produces a significant quantity of combustion heat, which gives heat feedback to the underlying polymer and so accelerates thermal deterioration. Self-sustaining combustion is produced when the combustion heat is sufficient to sustain the polymer's thermal decomposition rate, which keeps the developed combustible fuels below flammability limitations. As a result, the flame does not go out until the polymer has entirely burned out without disruption [20].

5.   Ways to reduce the combustion of polymer

Based on our overall understanding of polymer combustion, physical and/or chemical operations can disrupt and impede the burning process.

Physical activities involve: (1) cooling the underlying composite material at a temperature less than the critical temperature necessary for endothermic ignition to continue.

(2) Forming a protective carbonaceous layer that can function as a heat protector to prevent the transmission of both combustible fuels and O_2 between the combustion zone and the thermal decomposition zone. (3) Diluting the fuel by adding inactive additives and fillers that produce nonflammable inert gases.

Polymer oxidation can also be hampered;

- By introducing fire retardants that remove and terminate active $H°$ and $HO°$ in the condensed and gas phases.

- By forming a char layer via, cyclization, dehydration, and cross-linking interactions of polymer matrix and /or the fire retardants[21].

6.   Synthesis of polymer clay micro/nanocomposites

Various methods can be used to synthesize polymer clay composites. Here are some methods that can be used to synthesize flame retardant polymer clay composites

6.1   *In situ* polymerization

Herein the technique, prepolymers /liquid monomers are successfully incorporated into clay layers and coagulate inside the clay layers, causing the bandgap to increase (d spacing). Heat or an appropriate initiator can be used to start polymerization. This technique produces the majority of exfoliated nanocomposites because it allows for the selection of appropriate chemicals and polymerization pathways, bringing about a strong interaction between polymeric material and clay. For the manufacture of composites based on poly (e-caprolactone), polystyrene (PS), polyamide (PA), polyolefin, epoxy, and polyethylene terephthalate (PET) in situ polymerization can be employed [22].

6.2   Melt intercalation

It is a process in which clay is mixed with a molten polymer matrix. If the polymer has a strong affinity for the layer surfaces, it can invade between the layers of clay and produce exfoliated or an intercalated nanocomposite is produced. Nanocomposites made of polyamides such as nylon and polyethylene terephthalate are made using the melt intercalation method (PET) [23]. This technique is less expensive and easier to use than others. Owing to its incredible potential for application with high computational approaches like twin-screw extrusion and injection molding, this technology has grown in popularity. When treated with supercritical carbon dioxide, this procedure is more economical as creates nanocomposites having enhanced mechanical characteristics [24].

6.3   Solution casting

Polymer clay nanocomposites are made using the solution technique, which involves the use of solvents in which polymeric material, as well as clay, is miscible. The most common solvents are water, chloroform, and acetone [25]. When both solutions (one having the polymeric material and the other having clay) are mixed, the polymer chains replace the solvent molecules and intercalate between the layers of clay. After that, the solvent is removed, while the intercalated polymers will remain in the clay layers. Because of the desorption of solvent molecules entropy increases and this is considered as the driving factor for polymer intercalation from solution [26]. This technique has been used to intercalate water-soluble polymers such as poly (ethylene oxide) and poly (ethylene vinyl alcohol) between the clay layers. This technique has been used to make nanocomposites based on cellulose, high-density polyethylene, polyimide, and other non-aqueous solvents. The benefit of this approach is that it allows for the creation of interlayer composites from polymers with low or no polarity [23].

7.   Structures of polymer clay composites

The flame retardancy of clay/polymer composites is defined by the level of exfoliation rather than the type of SiO_2 and Al_2O_3. The degree to which the clay particle is dispersed in the polymer matrices is determined by two factors: (1) the clay material's capacity to interact physically or chemically with the polymer matrix, and (2) the technique employed to gather polymer clay composites. Three primary structures are listed below:

1. Conventional micro composites: if polymer chains are unable to pass through the silicate layers then micro composites are formed. This is similar to clay particle reinforced polymers, with no improvement in many characteristics.

2. Intercalated nanocomposites: Intercalated nanocomposites are formed when polymer chains penetrate the galleries between silicate layers in a crystallographically regular way within a few nanometers. In many characteristics, this structure leads to a certain degree of improvement [27].

3. Exfoliated nanocomposites: Exfoliated nanocomposites are those in which deep penetration of polymer chains results in constantly distributed silicate layers throughout the matrix. This results in significant improvements in a wide range of characteristics [28].

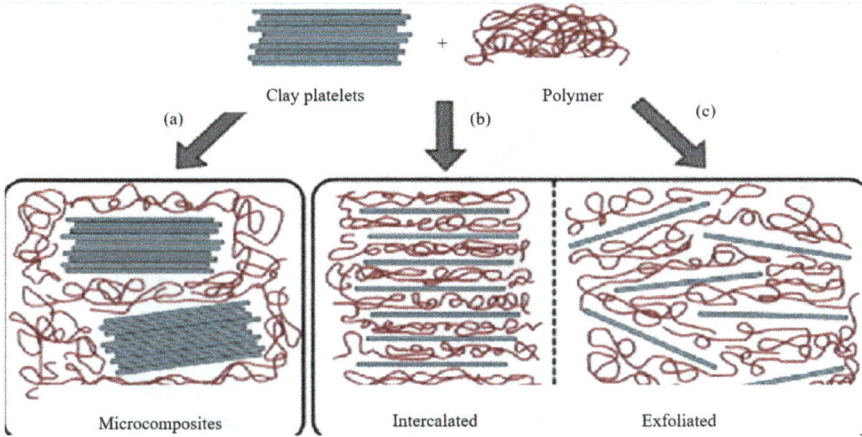

Figure 1. Primary structures of polymer clay composites

8. The flame-retardant mechanism

The mechanism of polymer clay composites in reducing heat and hazardous gases are mostly based on the two effects:

- barrier effect
- charring effect

Polymer clay composites are thought to work by forming a char that acts as a potential barrier to mass and energy transmission (protect the polymer from heat and oxygen attack. A ceramic-like layer is created on the material's surface by cross-linking and cyclization and the efficiency is determined by the forming layer's homogeneity [11]. Alumina and silica are also formed from clay during this process. The development of char layers might potentially be accelerated by flame retardant components. Thus, it was thought that the exfoliated layered materials' barrier effect on volatiles was significant in delaying thermal-oxidat

The mass-loss rate for polymer/layered compound composites is generally slower than for pure polymer, which is owing to the layered compounds' excellent dispersion. The PHRR might be decreased considerably by layer materials when they were effectively distributed in polymer matrixes, according to prior research. Although the PHRR had been considerably reduced in earlier studies, the THR had not been much reduced. These might be described as follows: the addition of layered compounds could slow the burning rate or fire size of these nanocomposites. The nanocomposites, on the other hand, would generally

Adv. App. of Micro and Nano Clay II – Synthetic Polymer Composites   Materials Research Forum LLC
Materials Research Foundations 129 (2022) 79-107   https://doi.org/10.21741/9781644902035-4

burn out utterly and the polymer would consume the same quantity of oxygen, resulting in a THR that was almost equivalent [4].

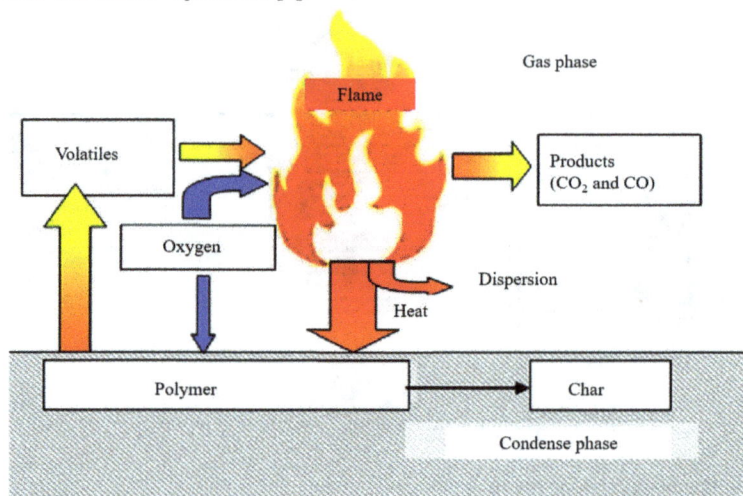

Figure 2. Mechanism of flame retardancy of polymer clay composites

The impact of the char residues can be described as follows: because the boiling point of most volatiles is significantly lower than the temperature at which the polymer matrix decomposes, the volatiles is superheated as they are produced. The bubbles then nucleate and develop under the heated polymer surface, releasing them into the gas phase. These bubbles stir the molten polymer surface, preventing the formation of a carbonaceous char layer and a heat transmission barrier on the polymer surface. The increased char production in the condensed phase clearly illustrates the lower flammability of polymer/layered compound nanocomposites. The presence of layered chemicals in the polymer matrix, on the other hand, slows the bubbling process and makes it easier to form a consistent protective char layer on the substrate surface. The layered compounds on the burning surface can act as insulation for the underneath polymeric substrate, slowing heat and mass transfer between the vapor and condensed phases. The layered clay's barrier effect might increase the polymer's barrier properties, thus the transformation of clay during ignition could also serve a flame retardant role under heat radiation. Clays migrate to the interface of nanocomposites, and that exfoliated platelets move to the surface of molten nanocomposites, obstructing oxygen penetration and diffusion into the melt. They combine with char and increase its efficiency. In this way clay also served as a char promoter. The flammability of polymeric materials depends upon the thermal stability of micro clay/nano clay used. The development of an uninterrupted protective layer, which seemed to function as a heat shield, appeared to be the flame retardant mechanism for polymer/layered compound micro/nanocomposites [29].

9.   Characterization and testing

The flame retardancy research has used a variety of characterization approaches, including temperature stability, mechanical strength, shape and size, and flame resistance characterization. The objective of classification varies depending on the intended result of the experiment, and it usually requires the use of two or more approaches at the same time. The test techniques may also differ based on sample characteristics and test conditions, which may influence how samples are prepared

10.   Morphology and structure analysis

Transmission Emission Microscopy (TEM), Scanning Electron Microscopy (SEM), and X-Ray Diffraction (XRD) is the imaging techniques that have been widely used in the study of polymer clay composites to gain a better understanding of the correlation between the shape and structure of the micro/nano clay and polymer. These pictures aid in the understanding of clay dispersion within the polymer matrix and the detection of potential agglomeration. The functional groups in the composites and chemical constituents, as well as the char generated after burning the clay polymer composite samples, are determined using FTIR and other spectrometry analyses.

10.1 SEM

Scanning electron microscopy (SEM) imaging can reveal shape, topology, composition, and other details, as well as indicate if the clay has been correctly distributed inside the polymer matrix. There will be agglomeration if the additives are not dispersed well, which might lead to blockage of the percolation network and a lower the flame retardancy of the polymer composites [30].

10.2 XRD

To describe the polymer nano clay-composite structure and identify whether it is agglomerated (flocculated), exfoliated, or intercalated, X-Ray diffraction (XRD) is used. Agglomeration happens when separate nano-sheets join together to produce a physically bigger layered structure. Because of the limited dispersion, just a small portion of the Nano filler is visible to the polymer. Flocculation occurs when the layers of silicates in polymer micro/nano clay are in contact with each other due to the hydroxyl group. In an intercalated form, the layered structure of the polymer clay nano-composites is formed that enlarges the spacing in between the layers. The nanoparticles are evenly distributed and appear distinct in the polymer matrix after exfoliation. The nanoclay content of exfoliated structures is often significantly lower than that of intercalated structures.

XRD uses Bragg's equation to detect if flocculation, intercalation, or exfoliation has happened by measuring the d-spacing between the nanosheets.

Bragg's equation is

$$2 d \sin\theta = n\lambda \qquad (1)$$

Where λ = wavelength of the light source, d = spacing between the layers of material, θ is the angle of diffraction, and n is an integer, (its value is 1) [31]. A prominent peak arises for agglomeration when d-spacing is equivalent to the agglomerated d-spacing. The peak for agglomeration will occur at a lesser value of 2 if intercalation has occurred. This value suggests that the spacing between the plates has risen however period spacing has not [32].

10.3 FTIR

The ultimate aim of flame retardancy is char formation, which assures that flammable volatiles do not migrate to the vapor phase and, as a result, the rate of heat release is lowered. Char residue that is formed after calorimeter test can be used to find out its elemental composition via elemental analysis on the char samples produced after the cone calorimeter examination, FTIR may be required to evaluate the distinctive peaks and subsequently identify various char elements. This study can also aid in the discovery of key stages in the polymer breakdown process. A thorough study of the relationship of structure and property of char can help us to understand the function of the char layer and its role in fire resistance. FTIR helps to determine the composition of a molecule, especially to identify the specific functional groups in compounds. Other researchers have used this screening technique to determine the chemical composition of composite materials, and therefore establish a link between the structure and properties of polymer clay composite [32].

10.4 NMR

NIST has developed a quantitative method based on solid-state nuclear magnetic resonance (NMR). The approach is based on the measurement of T1H (proton longitudinal relaxation time). It relies on two mechanisms: (1) the paramagnetic nature of clay, which decreases the spin-lattice relaxation time (T1H) of adjacent protons directly, and (2) spin diffusion, which propagates this locally increased relaxation to distant protons. The evaluation of such effects allows for the extraction of two dispersion-related factors. The proportion of

Adv. App. of Micro and Nano Clay II – Synthetic Polymer Composites   Materials Research Forum LLC
Materials Research Foundations 129 (2022) 79-107   https://doi.org/10.21741/9781644902035-4

the potentially accessible clay surface that has been converted into polymer/clay interfaces is the first parameter, f. The second parameter is a measure of the dispersion homogeneity of these real polymer/clay interactions. In the next sections of this study, these factors will be explored to define the shape of polymer clay composite and measure the extent of mixing [33].

10.5   Thermal analysis

The kind and packing of nanomaterials, the interaction between polymer and clay, structure and stability of char, and other variables all have a role in heat degradation mechanisms in polymer clay composites. Differential scanning calorimetry (DSC) and thermogravimetric analyzer (TGA) are commonly employed to investigate the thermal degradation and thermal stability of polymer-clay composites.

10.6   TGA

Thermodynamic stability, decomposition process, and other factors are measured using TGA. TGA is used to quantify the relationship between mass loss and temperature or time while keeping the reaction temperature constant. The loss of mass at different temperatures brings forth detailed information and the mechanism of polymer disintegration. When the decay pathways of the polymer and its nanocomposites are identical, it indicates the existence of a physical barrier does not affect the degradation process. For catalytic charring, on the other hand, the degradation route for Nano clay composites differs from that of micro clay polymer composites, as seen by the difference in form and size of the nanocomposite's curve in comparison to that of the micro clay polymer. TGA gives critical information about:

- Reaction kinetics, like activation energy and the kinetic rate equation,
- Information about thermal degradation, such as the beginning of thermal degradation temperature, thermal stability, and char residual [1].

Generally, the data obtained from TGA is isothermal or dynamic temperature gradient over time. *Isothermal operation* maintains a steady temperature while measuring the amount of substance in a specific range of time. Isothermal TGA has a significant drawback that a specific temperature must be selected for the test, even though degradation processes often vary dramatically depending on temperature. In dynamic temperature TGA sample is generally weighed while it is heated at a consistent temperature gradient rate, like 10 °C/min, it covers a broad range of temperatures, allowing variations in the disintegration mechanism to be observed with one test only. Moreover, dynamic testing helps to evaluate the kinetic variables like high-temperature residuals and activation energy.

The relationship between the rate of reaction and temperature is determined by the Arrhenius equation which is given as follows

$$\Delta \ln k / \, \delta T = E_a / RT^2 \tag{2}$$

In above equation

k = rate of reaction, t= temperature in kelvin, R= ideal gas constant

By rearranging the above equation

$$\Delta \ln k = \delta T \, E_A / RT^2 \tag{3}$$

By integrating the equation

$$\int \delta \ln k = \int \delta T \, E_A / RT^2$$

$$\ln k = -E_a / RT^2 + \text{integration constant} \tag{4}$$

By taking antilog

$$K = -E_a / Ae^{RT} \tag{5}$$

A= integration constant

By using this equation thermal activation energy of polymer clay composites can be determined. Micro/Nano clay polymer composites had a greater initial degradation temperature than the pure polymer, according to TGA and DTG measurements. This is due to a decrease in composite mobility due to fire-retardant materials. The amount of char yield of the micro/nano clay composites in contrast to the plain polymer can be calculated using the mass loss.

10.7   DSC

Heat transformations of polymer clay composites are evaluated using DSC. The heat flow from a material is measured using DSC as a function of temperature. A DSC may be used to identify the temperature for phase transition for a substance by examining the significant shift in heat flow versus temperature graph [34]. To distinguish between a phase shift and a reaction, Differential scanning data is generally combined with TGA [35].

The glass transition temperatures of micro/nano clay composites can also be determined by using DSC. Glass transition temperature of micro clay polymer composites is less than that of nano clay polymer composites because in micro clay polymer composites agglomeration occurs due to an increase in filler loading. As a result, TGA and DSC may be used to evaluate the degradation method, initiation of disintegration, char production, and reaction kinetics; all of these give understanding about flame retardancy.

11. Test for flammability

There are various approaches to analyzing the fire response behavior of polymeric materials. To understand the processes of fire retardancy of polymeric materials and formulations of FRs, a thorough understanding of various flame testing methodologies is required. Clearly, using one of the techniques alone cannot adequately define fire response behavior and can result in inaccurate inferences [36].

11.1 Cone calorimetry

A cone calorimeter is a type of calorimeter that measures the temperature. In the subject of fire retardancy, cone calorimeters are widely used to assess firefighting aspects during a well-developed fire incident. It is predicated on the idea that oxygen uptake is proportional to the rate of heat losses, and it's been accepted as a reference by the International Standards Organization (ISO-5660) and the American Society for Testing and Materials (ASTM E1354). The cone calorimeter displays a material's fire behavior using a smaller sample, offering it as a more cost-effective test method [37].

Cone calorimeters give acceptable insight into a material's fire performance in a growing fire and have therefore become a common test technique for measuring the fire response characteristics of flame retardant polymeric materials [38].

Cone calorimetry can be used to determine:
- time to ignition
- specific extinction area (SEA)
- fire growth rate index (FIGRA)
- peak heat release rate (PHRR)
- Heat release rate (HRR)
- total heat release (THR)
- fire performance index (FPI)

The HRR determines thermal energy/unit area, whereas the PHRR represents the highest quantity of heat released through burning. We can define THR as "the time integral of HRR". The smoke extinction area (SEA) corresponds to the mass loss and is utilized as an indicator of fumes production [39]

Cone calorimeters have been utilized in different flame retardancy investigations as a bench-scale instrument that evaluate the fire response characteristics of clay polymer composites and give a forecast of a huge fire situation than that of other thermal resistance research methods like TGA. The flammability characteristics of silicate layered polyester resin were investigated using TGA and cone calorimeter. The incorporation of nano clay does not influence the effective heat of combustion. However, when the loading of nano clay is increased, the PHRR, THR, and FIGRA decrease, suggesting enhanced fire retardancy. HRR has a linear relationship with mass loss rate, SEA production, and carbon monoxide yield.

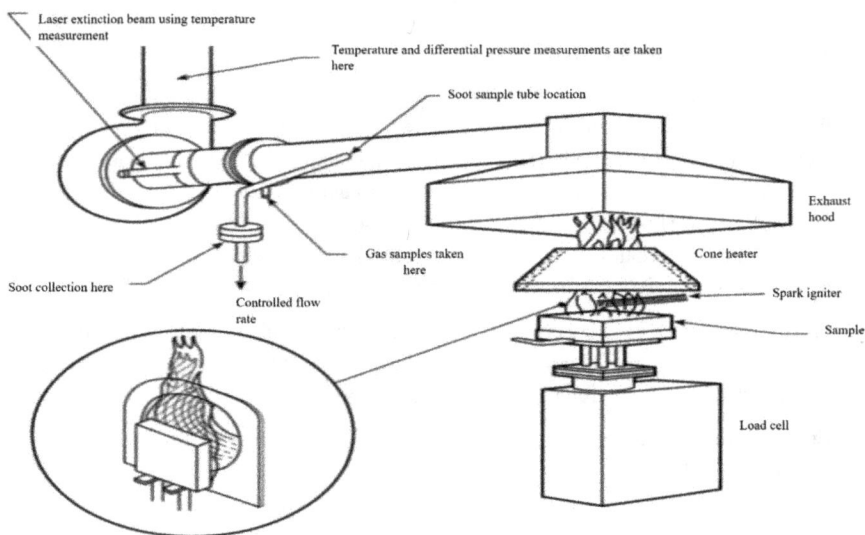

Figure 3. Cone calorimetry

Cone calorimeters may therefore give precise results that can be used to predict a material's preliminary response in high flame, analyze fire response characteristics amongst composites, and calculate fume toxicity factors that can be utilized to design rescue efforts and departure during a fire.

Limitation

It cannot cover fire spreading and only provides data for one-dimensional combustion.

11.2 Pyrolysis Burning Flow Calorimetry (PCFC)

Lyon and Walters developed the PCFC technique in 2004 that was first utilized by the Federal Aviation Administration in the United States of America to evaluate the heat release rate of different polymeric materials. Burning Flow Calorimetry (PCFC) is a way to determine the rate at which a material releases heat during burning in an inactive environment. The heat of combustion generated by burning is measured employing oxygen uptake calorimetry in the course of controlled burning of the composite material.

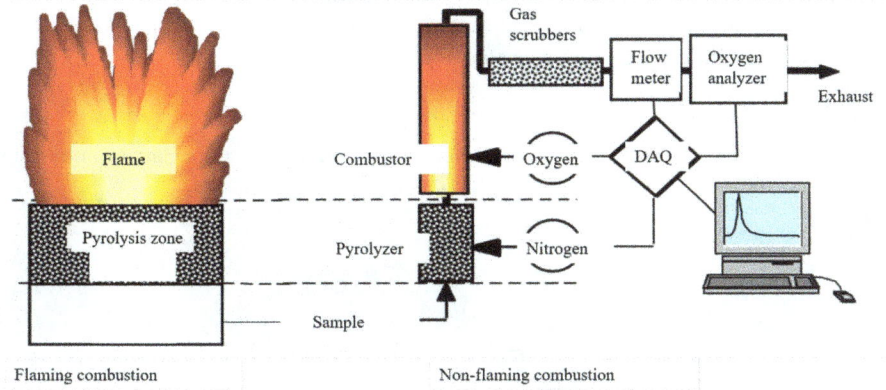

Figure 4. Pyrolysis burning flow calorimetry

Advantages of PCFC

1) PCFC be beneficial, especially if the size of the sample is milligrams.

2) In PCFC tests, the endergonic effect and radical trapping may be detected.

3) PCFC is utilized to determine heat release rate, the heat of solid and combustible gas, and heat release capacity (HRC) which is a derived quantity. HRC is an inherent aspect of the substance that does not depend upon the heat release rate and is regarded as a critical attribute for the determination of fire hazards.

$$HRR = PHRR / \text{heating rate} \tag{6}$$

It's the normal value of heat release rate, which is calculated by dividing the peak heat release rate by the corresponding heating rate

Adv. App. of Micro and Nano Clay II – Synthetic Polymer Composites   Materials Research Forum LLC
Materials Research Foundations 129 (2022) 79-107   https://doi.org/10.21741/9781644902035-4

Limitation of PCFC

As seen in comparatively larger scale tests such as cone calorimeter, PCFC does not detect physical or barrier effects such as dripping, intumescence, and so on.

PCFC analyzes forced combustion, but flame retardancy is linked to incomplete burning. Therefore, establishing a relationship between PCFC and other flame tests can be challenging. Because of the variations in test circumstances and working principles, the cone calorimeter and PCFC are frequently applied in parallel. According to their findings, the cone calorimeter's normalized peak heat release rate was always less than or equal to the PCFC's, and the difference was labeled as the barrier effect [32]

11.3 Limiting oxygen index (LOI)

LOI is used to determine the lowest oxygen concentration necessary for flame combustion to continue. It can be presented in the following way:

$$LOI = (O_2) / (N_2) + (O_2) \tag{7}$$

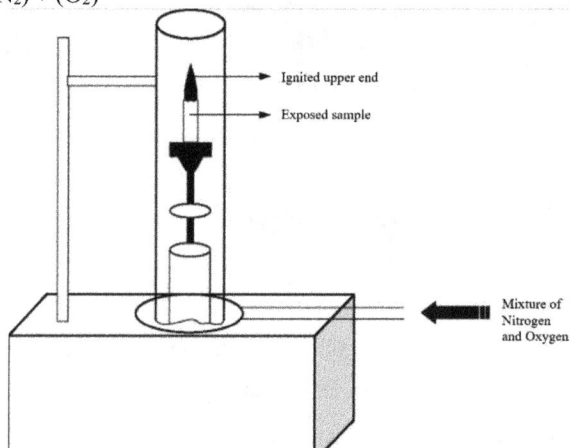

Figure 5. Limiting oxygen index testing

Advantages of LOI

However, this approach is still employed for material testing since it is simple, quick, and inexpensive.

Many researchers have utilized the LOI test to estimate the flame retardancy properties of polymers and their composites. The oxygen index and char residue were seen to be related. The degree of flammability of a substance can be related to its chemical structure. The LOI may also be used to evaluate the performance of pure polymer and fire retardant composites of polymer in a simple way. Polyamide integrated with aluminum di-isobutyl phosphinate (ABPA) and organically modified clay (OMMT) were investigated for their flame retardancy and mechanical characteristics. Pure polyamide has 25.6 % LOI, while 12 wt. % ABPA loading increases the LOI to 35 percent.

Limitation of LOI

LOI is not often considered a valid method of determining flammability characteristics, and there is a little direct link between LOI and other conventional flammability criteria [32].

UL 94 test

UL testing, which was created by UL, is a technique of classification of materials based on their ability to quench a spark when it has been started. It is included in several National and International Standards (ISO 9772 and 9773). In this test, a flame is exposed to a sample in different directions and then taken away from the sample to evaluate the response of the material. The UL classification method favors substances that ignite slowly or self-extinguish and does not spill burning stuff.

11.4 Horizontal burning test (HB94)

A flame is applied to one end of a horizontal piece of plastic until the flame front reaches the required point in the test (usually 30 sec). If the flames continue to burn, the time it takes to reach the second mark is calculated. When the fire ends, the length of the damaged area and the time it took to burn are measured. If it takes more time than that is recommended to reach the second mark or if burning stops before the second mark, the material will be categorized as 94HB.

11.5 Vertical burning test (V94)

Flame is applied to the material at the bottom end of the sample which warms the material in the higher portions of the sample. It is a more persistent test than the horizontal burning test. The bottom end of the test specimen is exposed to a test flame, and the material is categorized using the table below. To pass the UL94V test, materials must be self-extinguishing [40].

Adv. App. of Micro and Nano Clay II – Synthetic Polymer Composites   Materials Research Forum LLC
Materials Research Foundations 129 (2022) 79-107   https://doi.org/10.21741/9781644902035-4

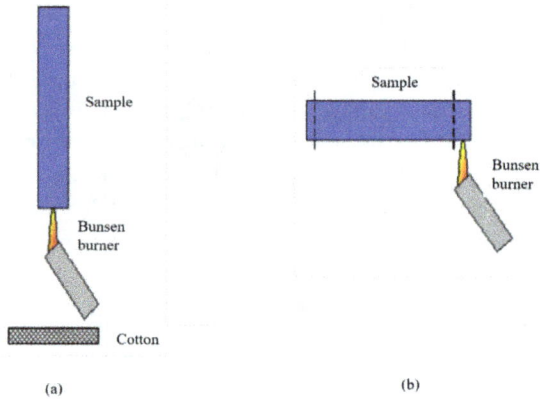

Figure 6. (a) UL94 vertical burning test (b) UL94 horizontal burning test

11.6   Radiation gasification apparatus

This is quite similar to a cone calorimeter in terms of design. The main distinction is that the sample is exposed to light in a sealed cylindrical chamber (which is continually purged with a fixed gas mixture, usually nitrogen). Rather than using calorimetry, the gasification processes of polymeric samples are tested by monitoring the mass loss rate and temperatures of the sample subjected to a fire-like heat flux (no burning). As a result, the results are entirely dependent on condensed phase processes. Due to the lack of heat feedback from a flame, it enables uniform exposure of the sample to the heat flux during the time of the experiment; and (ii) it allows for direct visual observations of the burning activity using a video camera [41].

Adv. App. of Micro and Nano Clay II – Synthetic Polymer Composites   Materials Research Forum LLC
Materials Research Foundations 129 (2022) 79-107   https://doi.org/10.21741/9781644902035-4

Table 1. Advantages and disadvantages of different flame retardants

Types of flame retardants	Advantages	Disadvantages	References
Halogen compounds	▪ Inexpensive ▪ Commonly used ▪ Efficient at low loadings ▪ Easy to process	▪ Emission of smoke and toxic gases ▪ Low heat and light stability ▪ Corrosive emission	[16]
Metal hydroxide	▪ No acidic gas emission ▪ Effective smoke reduction ▪ Non toxic ▪ Low price	▪ Very high loading necessary ▪ Decrease mechanical properties	[12]
NOR compounds	▪ Halogen free ▪ Good FR at low loading in thin PP films ▪ Light stability function	▪ Effectiveness limited to thin section ▪ Not able to reach UL94, V2 or V0	[11]
Red phosphorus	▪ Effective at low levels ▪ Excellent mechanical properties	▪ Safety concern ▪ Discoloration of polymer	[14]
Intumescent (P, N)	▪ Halogen free ▪ Low smoking occurring ▪ Low toxicity	▪ Poor processabililty ▪ High price ▪ Volatility problems	[10]
Silicon compounds	▪ Environmentally friendly ▪ Anti-dripping properties ▪ Good processabililty	▪ Few features effective PP	[13]
Composite materials	▪ Synergistic effects with other flame retardants ▪ Reduce dripping ▪ Improved quality of char ▪ Reduced peak heat release rate	▪ Poor thermal stability ▪ Difficult to incorporate ▪ High cost	[50]

Adv. App. of Micro and Nano Clay II – Synthetic Polymer Composites   Materials Research Forum LLC
Materials Research Foundations 129 (2022) 79-107   https://doi.org/10.21741/9781644902035-4

12. Applications

The usage of polymer-clay composites for fire-resistant applications is becoming highly prevalent, especially now that it's been discovered that these composites can replace part of the conventional flame retardant because they can retain the fire safety ratings even at lower loadings. This leads to a better balance of traits for the nanocomposite in comparison to the micro-composite fire retardant product, as well as a lower cost for the fire retardant resin in some situations, provided that the organ clay is less expensive than the flame retardant it replaces [42]. It's worth noting that organ clay may replace traditional flame retardants on a weight-for-weight ratio of more than 1:1, so that 1 g of organ clay can replace more than 1 g of micro clay flame retardant, resulting in a lighter material. With a few exceptions, clay nanocomposite systems appear to be a virtually ubiquitous synergist for flame retardant compounds [43].

A flame retardant chemical has been made by intercalation process using common smectite clay montmorillonite and is termed flame retardant hide powder. To prepare flame-retardant leather, this hide powder is mixed with pigskin. Furthermore, leather's thermal stability is improved. Clay minerals are effective flame retardants in a variety of polymer materials, particularly leather, making them appropriate for the creation of upholstery material in the aerospace engineering industry [44].

Polymeric materials combined with finely distributed inorganic nanoparticles can be utilized to create materials that can resist the harsh environment of space and can be employed in space systems where fire retardancy and weight reduction are priorities. Because of the stringent aerospace requirements, fire retardancy and thermal stability of polymer nanocomposites have been investigated with great interest. Nanocomposite materials are a fantastic way to enhance physical and structural characteristics like the coefficient of thermal expansion. Large aperture telescopes and antennas used in aeronautical research benefit greatly from such characteristics [45].

Organic polymers coupled with finely dispersed nanoparticles can be used to make materials that can withstand the extreme conditions of space and can be utilized in space systems that promote fire retardancy and weight reduction. Fire retardancy, as well as thermal persistence of nanomaterials, has been discussed with significant attention due to the demanding aeronautical requirements [46].

The use of nanocomposite materials is the best approach to modify the physical and structural characteristics like the thermal expansion coefficient. These features are especially beneficial to large aperture telescopes and antennas used in the aerospace study [44].

Organically modified nano clay was incorporated into monomeric methyl methacrylate to create PMMA-based PNCs. Ultimately, the PNCs were produced and their flame retardant, thermal, and mechanical characteristics were examined. PNCs' flame retardant characteristics have improved, allowing them to be used in aerospace applications [47].

Organo-modified nanoclays have been used to create epoxy/nanoclay composites. The mechanical, thermal, and resistance characteristics of these PNCs were greatly improved, allowing them to be employed in a variety of applications in the aerospace, defense, and automobile sectors [48].

A new fire-resistant chemical containing phosphorus, nitrogen (melamine), and inorganic clay (montmorillonite, a natural clay) was tested with linear low-density polyethylene (LLDPE), a commonly used, low-cost polymer that has difficulties with high combustibility and burn to drip. Loadings of 25–35 percent MPPM in the polyethylene lowered the PHRR by 50–75 %, while loadings of 35 % MPPM achieved UL94-V0 without burning drips [49].

An emulsion of PU when reinforced with clay, can fulfill the demands of modern society. They are widely used in the textile, leather, paper, wood, and paper industry due to their non-toxicity, and non-flammability [50]. Organic polymers with finely dispersed inorganic nano-materials can produce materials that can withstand the harsh space environment and can be used in space systems where weight reduction and flame retardancy is required.

13. Summary

Flame retardants (FR) are substances that are applied to flammable materials to make them more fire-resistant. They're made to reduce the risk of fire if they come into touch with a heating element, such as a cigarette or a candle, or if there's an electrical failure. These flame retardants reduce the flammability of polymeric materials by forming the char layer on the surface of the polymer. This chapter provides detailed information about the flame retardancy of micro clay/nano clay polymeric materials. The advantages of the use of micro/ nano clay over conventional flame retardant have also been discussed in this chapter. Methods for the synthesis and characterization of polymer micro/ nano clay composites and their potential applications as a flame retardant in various fields of life have been studied in detail.

References

[1] A. Gonzalez, A. Dasari, B. Herrero, E. Plancher, J. Santaren, A. Esteban, S.H. Lim, Fire retardancy behavior of PLA based nanocomposites, Polymer Degradation and Stability. 97 (2012) 248-256. https://doi.org/10.1016/j.polymdegradstab.2011.12.021

[2] J. Alongi, F. Bosco, F. Carosio, A. Di Blasio, G. Malucelli, A new era for flame retardant materials?, Materials Today. 17 (2014) 152-153. https://doi.org/10.1016/j.mattod.2014.04.005

[3] K. Babu, G. Renden, R. Afriyie Mensah, N.K. Kim, L. Jiang, Q. Xu, A. Restas, R. Esmaeely Neisiany, M.S. Hedenqvist, M. Försth, A review on the flammability properties of carbon-based polymeric composites: State-of-the-art and future trends, Polymers. 12 (2020) 15-18. https://doi.org/10.3390/polym12071518

[4] W.K. Salih, Fire retardancy assessment of polypropylene composite filed with nano clay prepared from Iraqi bentonite, Journal of Physics: Conference Series. 1003 (2018). https://doi.org/10.1088/1742-6596/1003/1/012019

[5] H. Yang, B. Yu, X. Xu, S. Bourbigot, H. Wang, P. Song, Lignin-derived bio-based flame retardants toward high-performance sustainable polymeric materials, Green Chemistry. 22 (2020) 2129-2161. https://doi.org/10.1039/D0GC00449A

[6] J. Troitzsch, Flame Retardants., Kunststoffe - German Plastics. 74 (1984) 60-62.

[7] C. Nithiyapathi, K. Thirunavukkarasu, A. Daniel Das, D. Tamilvendan, Progression in Fire Retardant Properties of Polymer Composites-A Review, IOP Conference Series: Materials Science and Engineering. 1059 (2021). https://doi.org/10.1088/1757-899X/1059/1/012058

[8] S. Giraud, F. Rault, A. Cayla, F. Salaün, History and evolution of fire retardants for textiles. (2016) 15-17.

[9] A.K. Roy Choudhury, Flame- and fire-retardant finishes, in: Principles of Textile Finishing, 2017: pp. 195-244. https://doi.org/10.1016/B978-0-08-100646-7.00008-4

[10] H. Vahabi, F. Laoutid, M. Mehrpouya, M.R. Saeb, P. Dubois, Flame retardant polymer materials: An update and the future for 3D printing developments, Materials Science and Engineering: R: Reports. 144 (2021) 100-604. https://doi.org/10.1016/j.mser.2020.100604

[11] S. Bourbigot, S. Duquesne, Fire retardant polymers: Recent developments and opportunities, Journal of Materials Chemistry. 17 (2007) 2283-2300. https://doi.org/10.1039/b702511d

[12] M. Bar, R. Alagirusamy, A. Das, Flame retardant polymer composites, Fibers and Polymers. 16 (2015) 705-717. https://doi.org/10.1007/s12221-015-0705-6

[13] X.W. Cheng, R.C. Tang, J.P. Guan, S.Q. Zhou, An eco-friendly and effective flame retardant coating for cotton fabric based on phytic acid doped silica sol approach,

Progress in Organic Coatings. 141 (2020) 105-539.
https://doi.org/10.1016/j.porgcoat.2020.105539

[14] A.R. Horrocks, A. Sitpalan, B.K. Kandola, Design and characterisation of
bicomponent polyamide 6 fibres with specific locations of each flame retardant
component for enhanced flame retardancy, Polymer Testing. 79 (2019) 10-41.
https://doi.org/10.1016/j.polymertesting.2019.106041

[15] Q. Tai, R.K.K. Yuen, L. Song, Y. Hu, A novel polymeric flame retardant and
exfoliated clay nanocomposites: Preparation and properties, Chemical Engineering
Journal. 183 (2012) 542-549. https://doi.org/10.1016/j.cej.2011.12.095

[16] S. Araby, B. Philips, Q. Meng, J. Ma, T. Laoui, C.H. Wang, Recent advances in
carbon-based nanomaterials for flame retardant polymers and composites, Composites
Part B: Engineering. (2021) 108-675.
https://doi.org/10.1016/j.compositesb.2021.108675

[17] Y.-M. Li, S.-L. Hu, D.-Y. Wang, Polymer-based ceramifiable composites for flame
retardant applications: A review, Composites Communications. 21 (2020) 100-405.
https://doi.org/10.1016/j.coco.2020.100405

[18] A.Y. Snegirev, M.K. Handawy, V. V Stepanov, V.A. Talalov, Pyrolysis and
combustion of polymer mixtures: Exploring additivity of the heat release rate, Polymer
Degradation and Stability. 161 (2019) 245-259.
https://doi.org/10.1016/j.polymdegradstab.2019.01.037

[19] Z. Xu, W. Xing, Y. Hou, B. Zou, L. Han, W. Hu, Y. Hu, The combustion and
pyrolysis process of flame-retardant polystyrene/cobalt-based metal organic
frameworks (MOF) nanocomposite, Combustion and Flame. 226 (2021) 108-116.
https://doi.org/10.1016/j.combustflame.2020.11.013

[20] W. He, P. Song, B. Yu, Z. Fang, H. Wang, Flame retardant polymeric
nanocomposites through the combination of nanomaterials and conventional flame
retardants, Progress in Materials Science. 114 (2020) 100687.
https://doi.org/10.1016/j.pmatsci.2020.100687

[21] P.J. Allender, Flame retardant polymer aspects of design, Materials and Design. 8
(1987) 160-167. https://doi.org/10.1016/S0261-3069(87)90126-9

[22] A.A. Azeez, K.Y. Rhee, S.J. Park, D. Hui, Epoxy clay nanocomposites - processing,
properties and applications: A review, Composites Part B: Engineering. 45 (2013)
308-320. https://doi.org/10.1016/j.compositesb.2012.04.012

[23] R. Babu Valapa, S. Loganathan, G. Pugazhenthi, S. Thomas, T.O. Varghese, An Overview of Polymer-Clay Nanocomposites, Elsevier Inc., 2017. https://doi.org/10.1016/B978-0-323-46153-5.00002-1

[24] T.C. Mokhena, M.J. Mochane, J.S. Sefadi, S.V. Motloung, D.M. Andala, Thermal conductivity of graphite-based polymer composites, Impact of Thermal Conductivity on Energy Technologies. (2018) 181. https://doi.org/10.5772/intechopen.75676

[25] S. Ray, S.Y. Quek, A. Easteal, X.D. Chen, The potential use of polymer-clay nanocomposites in food packaging, International Journal of Food Engineering. 2 (2006). https://doi.org/10.2202/1556-3758.1149

[26] F. Ahmed, S. Kumar, N. Arshi, M.S. Anwar, L. Su-Yeon, G.-S. Kil, D.-W. Park, B.H. Koo, C.G. Lee, Preparation and characterizations of polyaniline (PANI)/ZnO nanocomposites film using solution casting method, Thin Solid Films. 519 (2011) 8375-8378. https://doi.org/10.1016/j.tsf.2011.03.090

[27] S. Ray, A.J. Easteal, Advances in polymer-filler composites: Macro to nano, Materials and Manufacturing Processes. 22 (2007) 741-749. https://doi.org/10.1080/10426910701385366

[28] J.W. Gilman, Flammability and thermal stability studies of polymer layered-silicate (clay) nanocomposites, Applied Clay Science. 15 (1999) 31-49. https://doi.org/10.1016/S0169-1317(99)00019-8

[29] S.P. da S. Ribeiro, L. dos S. Cescon, R.Q.C.R. Ribeiro, A. Landesmann, L.R. de M. Estevão, R.S.V. Nascimento, Effect of clay minerals structure on the polymer flame retardancy intumescent process, Applied Clay Science. 161 (2018) 301-309. https://doi.org/10.1016/j.clay.2018.04.037

[30] S. Leporatti, Polymer clay nano-composites, Polymers. 11 (2019) 58883. https://doi.org/10.3390/polym11091445

[31] C.M.L. Preston, G. Amarasinghe, J.L. Hopewell, R.A. Shanks, Z. Mathys, Evaluation of polar ethylene copolymers as fire retardant nanocomposite matrices, Polymer Degradation and Stability. 84 (2004) 533-544. https://doi.org/10.1016/j.polymdegradstab.2004.02.004

[32] L. Ahmed, B. Zhang, L.C. Hatanaka, M.S. Mannan, Application of polymer nanocomposites in the flame retardancy study, Journal of Loss Prevention in the Process Industries. 55 (2018) 381-391. https://doi.org/10.1016/j.jlp.2018.07.005

[33] F. Samyn, S. Bourbigot, C. Jama, S. Bellayer, Fire retardancy of polymer clay nanocomposites: Is there an influence of the nanomorphology?, Polymer Degradation

and Stability. 93 (2008) 2019-2024.
https://doi.org/10.1016/j.polymdegradstab.2008.02.013

[34] O. Bera, B. Pilic, J. Pavlicevic, M. Jovicic, B. Hollo, K.M. Szecsenyi, M. Spirkova, Preparation and thermal properties of polystyrene/silica nanocomposites, Thermochimica Acta. 515 (2011) 1-5. https://doi.org/10.1016/j.tca.2010.12.006

[35] N. Misra, V. Kumar, J. Bahadur, S. Bhattacharya, S. Mazumder, L. Varshney, Layered silicate-polymer nanocomposite coatings via radiation curing process for flame retardant applications, Progress in Organic Coatings. 77 (2014) 1443-1451. https://doi.org/10.1016/j.porgcoat.2014.04.027

[36] P. Wei, S. Bocchini, G. Camino, Nanocomposites combustion peculiarities. A case history: Polylactide-clays, European Polymer Journal. 49 (2013) 932-939. https://doi.org/10.1016/j.eurpolymj.2012.11.010

[37] D. Yi, R. Yang, C.A. Wilkie, Full scale nanocomposites: Clay in fire retardant and polymer, Polymer Degradation and Stability. 105 (2014) 31-41. https://doi.org/10.1016/j.polymdegradstab.2014.03.042

[38] D.O. Castro, Z. Karim, L. Medina, J.O. Häggström, F. Carosio, A. Svedberg, L. Wågberg, D. Söderberg, L.A. Berglund, The use of a pilot-scale continuous paper process for fire retardant cellulose-kaolinite nanocomposites, Composites Science and Technology. 162 (2018) 215-224. https://doi.org/10.1016/j.compscitech.2018.04.032

[39] J. Zhang, D.D. Jiang, C.A. Wilkie, Fire properties of styrenic polymer-clay nanocomposites based on an oligomerically-modified clay, Polymer Degradation and Stability. 91 (2006) 358-366. https://doi.org/10.1016/j.polymdegradstab.2005.04.040

[40] G.R. Udo wagenknecht, Bernard Kretzchmar, Investigation of Fire Retardant Properties of Polypropylene-Clay-Nanocomposites, 212 (2003) 207-212. https://doi.org/10.1002/masy.200390084

[41] J.D. Swann, Y. Ding, M.B. McKinnon, S.I. Stoliarov, Controlled atmosphere pyrolysis apparatus II (CAPA II): A new tool for analysis of pyrolysis of charring and intumescent polymers, Fire Safety Journal. 91 (2017) 130-139. https://doi.org/10.1016/j.firesaf.2017.03.038

[42] M. Rahman, F. Zahin, M.A.S.R. Saadi, A. Sharif, M.E. Hoque, Surface Modification of Advanced and Polymer Nanocomposites, in: N. Dasgupta, S. Ranjan, E. Lichtfouse (Eds.), Environmental Nanotechnology: Volume 1, Springer International Publishing, Cham, 2018: pp. 187-209. https://doi.org/10.1007/978-3-319-76090-2_6

[43] S. Sinha Ray, Recent Trends and Future Outlooks in the Field of Clay-Containing Polymer Nanocomposites, Macromolecular Chemistry and Physics. 215 (2014). https://doi.org/10.1002/macp.201400069

[44] S. Gul, A. Kausar, B. Muhammad, S. Jabeen, Research Progress on Properties and Applications of Polymer/Clay Nanocomposite, Polymer-Plastics Technology and Engineering. 55 (2016) 684-703. https://doi.org/10.1080/03602559.2015.1098699

[45] S. Chua, R. Fang, Z. Sun, M. Wu, Z. Gu, Y. Wang, J.N. Hart, N. Sharma, F. Li, D. Wang, Hybrid solid polymer electrolytes with two-dimensional inorganic nanofillers, Chemistry-A European Journal. 24 (2018) 18180-18203. https://doi.org/10.1002/chem.201804781

[46] A. Bhat, S. Budholiya, S.A. Raj, M.T.H. Sultan, D. Hui, A.U.M. Shah, S.N.A. Safri, Review on nanocomposites based on aerospace applications, Nanotechnology Reviews. 10 (2021) 237-253. https://doi.org/10.1515/ntrev-2021-0018

[47] J. Zhu, P. Start, K.A. Mauritz, C.A. Wilkie, Thermal stability and flame retardancy of poly(methyl methacrylate)-clay nanocomposites, Polymer Degradation and Stability. 77 (2002) 253-258. https://doi.org/10.1016/S0141-3910(02)00056-3

[48] H. Vahabi, M.R. Saeb, K. Formela, J.M.L. Cuesta, Flame retardant epoxy/halloysite nanotubes nanocomposite coatings: Exploring low-concentration threshold for flammability compared to expandable graphite as superior fire retardant, Progress in Organic Coatings. 119 (2018) 8-14. https://doi.org/10.1016/j.porgcoat.2018.02.005

[49] G. Makhlouf, M. Hassan, M. Nour, Y.K. Abdel-Monem, A. Abdelkhalik, Evaluation of fire performance of linear low-density polyethylene containing novel intumescent flame retardant, Journal of Thermal Analysis and Calorimetry. 130 (2017) 1031-1041. https://doi.org/10.1007/s10973-017-6418-x

[50] B.K. Kim, J.W. Seo, H.M. Jeong, Morphology and properties of waterborne polyurethane/clay nanocomposites, European Polymer Journal. 39 (2003) 85-91. https://doi.org/10.1016/S0014-3057(02)00173-8

Adv. App. of Micro and Nano Clay II – Synthetic Polymer Composites   Materials Research Forum LLC
Materials Research Foundations 129 (2022) 108-128   https://doi.org/10.21741/9781644902035-5

Chapter 5

Nano Clay Reinforcement of Thermoplastics for Engineering Applications

G. Santhosh[1*], G.P. Nayaka[2]

[1]Department of Mechanical Engineering, NMAM Institute of Technology, Nitte-574110, India

[2]Physical and Materials Chemistry Division, CSIR-National Chemistry Laboratory, Pune, Maharashtra-411008, India

*drsanthug@gmail.com

Abstract

Thermoplastics are widely used advanced materials in various fields as an essential product. The increasing interest in replacing thermosets due to their advanced properties such as high toughness, manufacturing time and processing possibilities. Thermoplastics exhibit poor mechanical properties in several engineering applications; hence the use of nano clay minerals enhances the mechanical property of thermoplastics to be used in many engineering applications. The reinforcing effect of clay minerals (nano-clay) in various thermoplastics such as polycarbonate (PC), polypropylene (PP), Acrylonitrile butadiene styrene (ABS) and high-density polyethylene (HDPE), which are popularly known as commodity plastics and are widely used in fabricating automotive parts and household articles. The frequently used clay minerals such as montmorillonite, bentonite, hectorite, halloysites and kaolinite as filler material in thermoplastics make them suitable in many engineering applications with enhanced mechanical properties. The integration of clay minerals into a polymer matrix allows both properties from clay minerals and polymer to be combined so as to exploit entirely new functionalities. In recent years, owing to a number of practical applications in the field of opto-electronics, a great deal of interest has been paid to study optical, dielectric and conduction behaviours of various polymeric materials. This chapter is designed to be a source for nano-clay reinforced thermoplastic nano-composite research including synthesis, characterization, structure/property relationship and applications considering optical and electrical conductivity which are discussed.

Keywords

Polymer, Thermoplastics, Clay Minerals, Nano-Composites

Contents

Nano Clay Reinforcement of Thermoplastics for Engineering Applications108

1.   Introduction

Polymer nano-composites (PNCs), overwhelmingly with nano-particles (NPs) have engrossed a lot of curiosity by the present scientists. As a general rule, the nanoparticles are in the range 1–100 nm. Nanotechnology offers curiosity and flexibility which isn't seen in many other field because of high surface region offered by the nanostructured materials [1]. Polymer NCs are of meticulous interest for optoelectronics, attributable to their excellent ability to exploit unique material behaviour, in conjunction with magnificent process-ability of the polymeric frameworks [1]. Clay incorporated PNCs attract researchers due to their low cost, achievable strategy, competent stiffening effect; besides, they show astoundingly unrivalled properties at exceptionally low filler dose when compared to conventional polymer composites. PNCs gaining ample attention of scholars as nano-materials offer extensive variety of attractive and enhanced properties in the field of opto-electronics. Particularly polycarbonates (PCs) are widely utilized in different applications [2].

In recent years increasing demands for the safe and reliable electrical gadgets prompts the investigation of novel materials. In this way, the remarkable increase in the improvement of novel protecting films has been accomplished, particularly in the field of nanocomposite. Clay incorporated polymer NCs are specifically used, because of their ease, plausible strategy, hardening impact; besides, they show incredibly prevalent properties at extremely low filler loadings.

2.   Introduction to clay nano-particles

Clay nano-particles are broadly examined clay minerals for PNCs. Clay minerals are obtained from common clay-mineral deposits, these clay minerals are having sheet-like structures which are also called as hydrous silicates [3].

Silicates are of mainly 4 categories with the crystalline arrangement as mentioned below.

- Kaolinite
- Montmorillonite
- Illite
- Chlorite

Montmorillonite (MMT) are the ones widely explored and utilized nano-clay to deliver excellent properties when used in polymeric matrix. This is because of its high aspect-ratio and its incomparable 'exfoliation' nature. MMT clay exhibit 2:1 arrangement with an octahedral alumina sheets (Al). The octahedral alumina sheets share oxygen atoms of tetrahedral silica sheets composed of SiO_4 groups connected to hexagonal frame of repeating units of Si_4O_{10}. The Al structure composed by closely packed oxygen atoms in two layers; between these layers octahedral aluminium particles are present [4] with halfway from the oxygen atoms. These layers called as tactoids of thickness of around 1 nm with the lateral scale of around 300 Å to a few microns [10]. The interlayer in MMT are responsible to balance the negative charges created by isomorph addition of crystals [5]. The Unmodified MMTs have hydrated Na^+ or K^+ particles, this hydration makes the MMT water soluble and eventually unable to coexist with all polymers that are water insoluble [6]. Hence, the clay minerals are modified to make them polymer compatible [7-9] with the inorganic cations exchange by alkylammonium, sulfonium, or phosphonium surfactants.

In general, clay polymer NCs can adopt two strategies based on infiltration into the interlayer: Figure 1. represent the morphology of clay and polymer system. In the event that the dispersion of polymeric chains paves the way to a limited extension of the clay sheets under 20-30 Å, the morphology is characterized as exfoliated; if the infiltration of the polymer prompts a delamination of silicate tactoids [11]. In majority of the studies, most of the properties change was seen in exfoliated NCs due to the high aspect ratio of the single tactoid. Nonetheless, the capacity of natural chains to intercalate inside the interlayer additionally relies upon the technique for preparation of the NCs since change of nano-clays is not sufficient to get a total exfoliation of the layers.

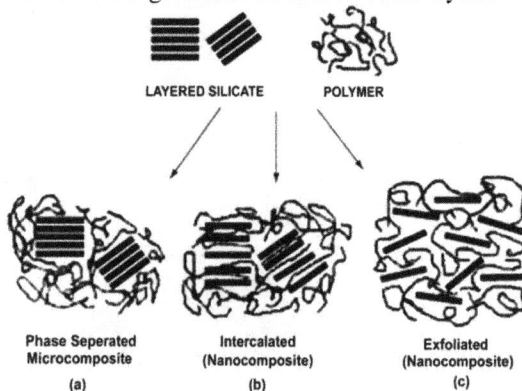

LAYERED SILICATE   POLYMER

Phase Seperated
Microcomposite
(a)

Intercalated
(Nanocomposite)
(b)

Exfoliated
(Nanocomposite)
(c)

Figure 1. Schematic representation of clay dispersion in polymer matrix [15]

Nano clay has become most important entities in the nano world for their exceptionally appreciated and established hold in tailoring the NCs properties by their unique intercalation aspects [12]. On such area where organic clay minerals are proving themselves to be efficient members of the NPs family is solar cells. These organic solar cells are prominent as good as low cost renewable energy source, however the use of these fillers are limited due to their sensitivity towards moisture which requires a proper encapsulation to improve their shielding properties [13, 14].

3.   Introduction to halloysite nanotubes

Halloysite nanotube (HNT) is a sort of aluminosilicate nanoclay, famous for their high aspect ratio and hallow configuration which can be found intermixed with Kaolinite and Montmorillonite. HNT with the molecular formula, $H_4Al_2O_9Si_2 \cdot 2H_2O$ have unique tubular structure make them suitable as nano-containers with the intention to store and adsorb with abundant –OH groups. The use of HNT can provide high mechanical and thermal strength [16-20]. The unique properties of HNTs make them most convenient in the field of composites to use them as reinforcing nanofillers [21]. But the interfacial decoration of HNTs is still a challenging task in the fabrication of PNCs. This task remain tough due to unavailability of promising functional groups on the outer surface of HNTs. However, HNTs can be used as a fire/flame retardant materials [22-24] compared to other NPs. HNTs are available at low cost [25] and ecologically disposal. Besides all the benefits, HNT is also recognized as the best dielectric filler which gained ample of attention in the field of micro-circuit components [26-31].

4.   Polymeric matrix

Polymers are the most conventional and widely accepted material due to their ability in fabricating different polymeric systems which increase the properties due to design flexibilities. Further, their low cost, low density makes them suitable for many commercial applications.

4.1   Introduction to polycarbonate

Polycarbonate (PC) belongs to π-conjugate species [33-35] and is widely adopted in the fields of transport, food-packaging and optics. The chemical structure of PC is presented in figure 2. PC is a highly transparent polymer which possesses unique properties that include optical, electrical and high glass transition temperature (T_g) with combined high impact and toughness [36-44]. Toyota researched [45] to know the importance of PC with nano-clay mineral, nylon and carbon nanotube (CNT) [46] and found astonishingly good properties. Initially nano-fillers were added to polymer matrix which focused only on mechanical properties. however, the recent research have found many opportunities to understand the behaviour of NCs in many fields with advanced properties [47].

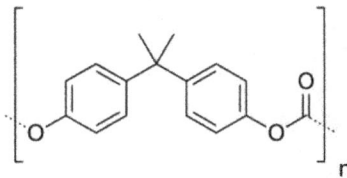

Figure 2. Schematic depictions of PC structure

4.2   Polycarbonate (PC)/Clay NCs (PCCNCs)

Polycarbonate is the most popularly utilized polymer in various fields, opto-electronics is one of the major areas where PC is used because of its dielectric and transparent nature. Though, importance is given to the fabrication of PC-functionalized clay NCs. Distinctive course of action and methodologies for the functionalized clay with surfactants have been associated with polymer-clay NCs. Regardless, the effect of functionalized clay and dissipating area of polymer-clay NCs stays unclear.

The execution of polymer-clay NCs depends upon a couple of parameters, for instance, the clay percentage and dispersing condition. Surface functionalized with silane blends has been considered in recent decades, and it is understood that silane alteration is assumed to improve filler dispersing in the matrix; this behaviour is responsible in achieving in the properties due to interactions between filler and matrix. The clay modifiers used as fire retardant materials with good mechanical properties which reduce the 'T_g' of polymer-clay NCs as a result of weaker organoclay-polymer interactions [48-56]. The inclusion of the nano-clay to extend the stiffness of pristine PC and awesome outcomes in tribological properties were also obtained [57-67].

5.   Optical analysis of PC/HNT NCs

The NCs reinforced with clay NPs is considered to have most valuable significance for various optical and opto-electronic applications [68]. Further, the NCs are utilized to protect inorganic solar cells, without affecting the parent nature of the host polymeric systems. The interest to fabricate the optical UV-shielding materials is one among the challenging area of research [69, 70]. UV-visible spectral study is more appropriate to understand the nature of polymer NCs on shielding applications.

5.1   UV-visible absorption studies

The spectra in Figure 3. depict the nature of absorption of NCs in the wavelength range 200–750 nm. Figure. 3. shows a clear and distinct peak from UV-A range, this peak correlated to the electronic transition highest occupied and lowest unoccupied molecular orbital's (HOMO-LUMO). Furthermore, the dispersed HNTs undergo exfoliation and being a reason to form tough peak at 340 nm which may be regarded due to strong interaction between HNT and carbonyl groups of PC [71, 72]. Eventually the NCs exhibits hypsochromic shift with the increase in conjugation between HNT and PC indicating the inter chain interaction and stacking. As can be seen from the Figure 3. the absorption edge of NCs shifts towards the lower wavelength caused by the energy difference in the valence band (VB) and conduction band (CB). Further the behaviour of NCs confirms the creation of conjugated groups in unsaturated bonds of PC [73-75]. Further the validation of NCs as UV-A shielding material is understood by the change in energy gap value to 3.75 to 4.13 eV.

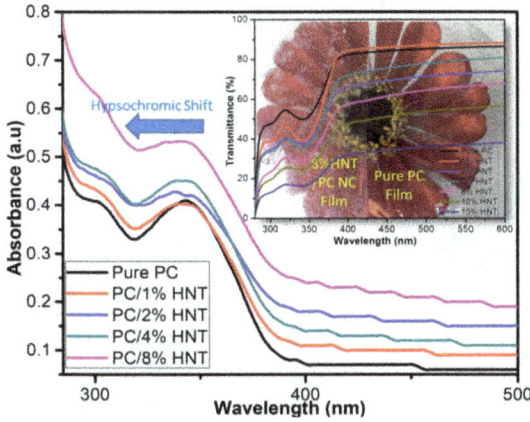

Figure 3. UV-Visible absorbance spectra of NCs with 0, 1, 2, 4 and 8 wt% of HNT. Inset: Transmittance spectra of NCs with 0, 1, 2, 4 and 8 wt% of HNT

5.2   Absorption co-efficient and optical band gap measurements

Absorption coefficient is an important parameter used to analyse the band-gap engineering of the NCs and effect of NPs in PC is evaluated by the following formulae [76];

$$\alpha = 2.303 \left(\frac{A}{d} \right) \tag{1}$$

where, d is the sample thickness [77].

The absorption coefficient result is presented in the Figure 4. The variation in the obtained results is tabulated in the Table 1. Pristine polymer exhibits a band-gap of 4.13 eV and for rest of the NCs the absorption edge is shifting towards lower values is ascribed for the complete dispersion of HNT in polymer, for the NCs having maximum HNT content exhibits band-gap energy of 3.75 eV [78].

The optical functionality of polymer-clay NCs is analysed by Tauc's direct and indirect band-gap methods [79].

$$(\alpha h\vartheta) = B(h\vartheta - E_g)^m \tag{2}$$

B: Tauc's constant, E_g: optical band gap, 'm' is an index, 'm' can be assigned 0.5 (allowed direct transition) and 2 (allowed indirect transitions)

The direct and indirect optical band gap (E_g) of NC films is obtained by extrapolating the linear portion to zero; results are tabulated in Table 1. The results obtained are presented in figure 5.

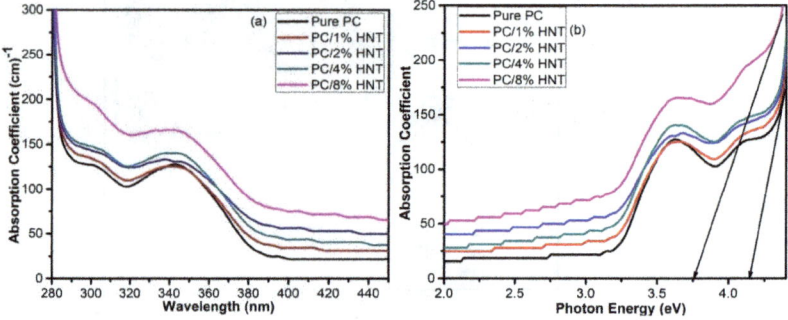

Figure 4. Absorption coefficients as function of (a) wavelength and (b) photon energy of NCs with 0, 1, 2, 4 and 8 wt% of HNT

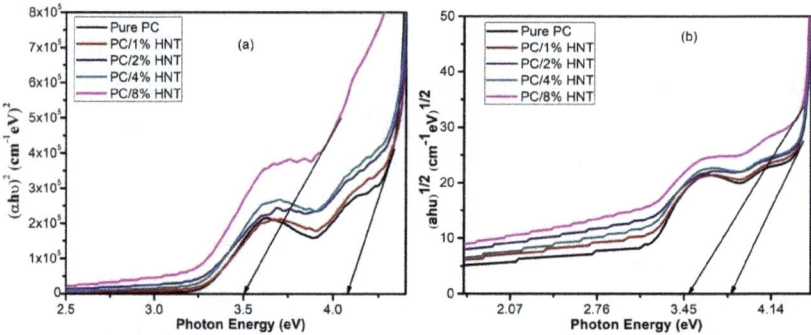

Figure 5. Tauc's plots of (a) indirect band gap and (b) direct band gap of NCs with 0, 1, 2, 4 and 8 wt% of HNT

It is observed from the results the E_g of pristine PC is 3.8 eV referring PC a good insulating material, however the values keep changing after HNT loading, the E_g of NCs decreases to 3.48 eV for highest HNT content in PC. This behaviour of decrease in E_g is due to the creation of localized occupied and localized unoccupied energy bands by the interaction of HNT in PC suggesting NCs a perfect material for PV application

The fabricated polymer-clay NC films were investigated by Urbach's energy (E_u), the E_u provides clear information about the optical electronic transitions of localized states adjacent to VB and extended state in the CB [80].

$$\alpha = \alpha_0\, e^{(E_g - E_u)} \tag{3}$$

α_0: constant, E_u: Urbach's energy.

The obtained E_u of NCs tabulated in the Table 1 and Figure 6. As we know the reinforcement of HNT brought many structural changes in PC, which is also responsible for the increase in E_u with increase in HNT loading, creating amorphous halos in the NC [81].

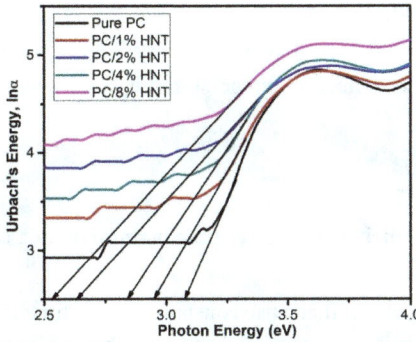

Figure 6. Urbach's energy as a function of photon energy of NCs with 0, 1, 2, 4 and 8 wt% of HNT

Table 1. Optical parameters of NCs with 0, 1, 2, 4 and 8 wt% of HNT

HNT content in PC (wt %)	Urbach's energy (E_u, eV)	Absorption edge (eV)	Band gap (eV)	
			Direct	Indirect
0	0.86	4.13	3.80	4.1
1	0.95	4.09	3.70	4.02
2	0.98	3.90	3.67	3.85
4	1.12	3.87	3.65	3.84
8	1.42	3.75	3.48	3.53

Adv. App. of Micro and Nano Clay II – Synthetic Polymer Composites   Materials Research Forum LLC
Materials Research Foundations 129 (2022) 108-128   https://doi.org/10.21741/9781644902035-5

6.   Electrical studies of PC/HNT NCs

The polymer-clay NCs were analysed for electric and dielectric behaviours. The HNTs used in PC offer many benefits with improved functionalities, further the HNTs provide increased design flexibility with reduced cost. Inorganic conductive nano-filler reinforced polymer NCs are considered promising for this particular application, which display an outstanding dielectric constant. The present investigation is an effort to fabricate polymer-clay NC films with good electric and dielectric results [82, 83].

6.1   Dielectric properties

The polymer-clay NC films were evaluated for dielectric properties, to understand the energy dissipation of polymer-clay NC films.

6.1.1 Dielectric constant

The dielectric constant (ε') of polymer-clay NC films evaluated using measured capacitance values due to dielectric charge stored in a capacitor and determined by the below equation;

$$\varepsilon' = \frac{Ct}{\varepsilon_0 A} \tag{4}$$

where, t: dielectric layer thickness, ε_o: free space permittivity, C: capacitance, A: electrode area.

Figure 7(a) shows the results of dielectric constant (ε') as a function of HNTs content. It is evident that ε' of the NCs is in increasing trend with the increase in filler loading, due to this behaviour charge storage capacity of the polymer system increases. Subsequently the NCs hold the energy related to the alignment of electric charges and electric field. Further, the enhancement in ε' is attributed to the development of small capacitor networks in the NCs. This behaviour of NCs is further validated by Maxwell−Wagner (MW) interface [84]. However, with applied frequencies the energy storage capacity of NCs is found to be decreasing due to misalignment of permanent and induced dipoles which is quite reverse at low frequencies leading to higher relaxation time with enhanced polarization. A comparable development is noticed in other polymeric composites, predicting the reality of NCs.

6.1.2 Dielectric loss

Figure 7(b) depicts the variation of dielectric loss (ε'') as a function of filler content. The ε'' of the NCs increases with the increase in filler content, HNT induce excessive polarization leading to the rise in ε''. Further, with the frequency change the ε'' of the NCs shows downtrend caused by the charge immobilization and field reversal. This behaviour of the NCs can be attributed to the reduced charge accumulation in the PC matrix.

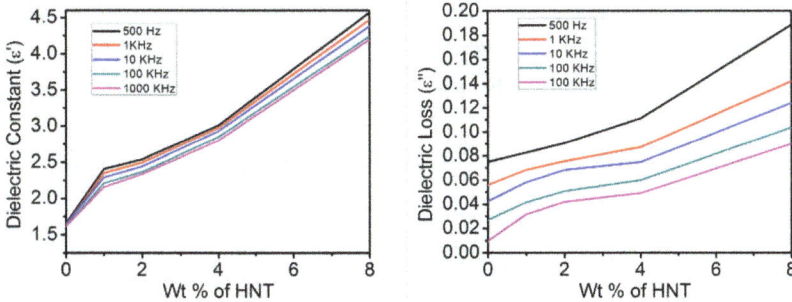

Figure 7. Plots of (a) dielectric constant and (b) dielectric loss as a function of HNT content in NCs at applied frequency

6.1.3 Loss tangent

Loss tangent (Tan δ) of PCCNCs are evaluated using dielectric constant (ε') and dielectric loss (ε'') using the below expression;

$$\varepsilon'' = \varepsilon' \, Tan \, (\delta) \tag{5}$$

where, δ: Phase angle

Figure 8(a) depicts the change in Tan δ with the variation in filler content. The Tan δ of NC show an increasing trend as the NPs reduces the inter-particle distance caused due to reduction of charge trapping sites.

6.1.4 AC conductivity

The electrical conductivity is an important parameter which measures the materials ability to conduct and to evaluate the charge transport nature of the NCs (heterogeneous systems). σ_{ac} the material property which depends on the frequency change, that varies in polymeric system due to the presence of charge carriers σ_{ac} of NCs is measured using the relation below [85-87];

$$\sigma_{ac} = d \frac{G_s}{A} \tag{6}$$

D: film thickness, A: area, G_s: conductance

The variation of σ_{ac} as a function of frequency is depicted in the Figure 8(b). it is noticed from the figure that the NCs exhibit an enhanced conductivity with HNTs loading resulting from the electronic interactions of NCs, further it induces a continues conducting channels.

Adv. App. of Micro and Nano Clay II – Synthetic Polymer Composites   Materials Research Forum LLC
Materials Research Foundations 129 (2022) 108-128   https://doi.org/10.21741/9781644902035-5

HNTs are rare clay minerals separates the localized states by filling the defect sites by lowering the potential barrier helps to promote charge carriers. The experiential results provide string evidence to the "ac universal law" which clearly indicate the charge migration via hopping mechanism. The results also indicate that at higher frequency electrons in the NCs get enough energy to travel from one state to another responsible for the increase in electrical conductivity in the applied field.

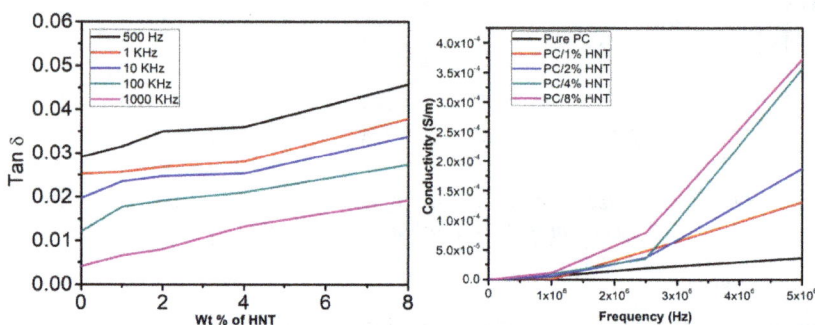

Figure 8. Plots of (a) loss tangent (Tan δ) as a function of HNT content in NCs and (b) AC conductivity as a function of applied frequency

Conclusion

This chapter briefs on research investigation of HNTs reinforced PNCs and their Opto-Electric behaviours. In this investigation, HNTs used as main reinforcing material with polycarbonate as matrix. The NC films were developed by solution casting method with 0, 1, 2, 4 and 8 wt%. The optical studies reveal the effect of NPs on photonic absorption behaviors exhibiting their use in UV-A shielding behavior. UV absorption spectra of PC films exhibit clear and broad absorption peak in the wavelength range of 320 – 350 nm. The optical band gap studies (direct and indirect) of PNCs reveal the photon- electron interaction indicating the reduced band gap. Optical band gap energy of NCs with 8 wt% HNT reduced to 3.70 eV. The electrical behavior of polymer films such as dielectric constant (ε'), dielectric loss (ε''), AC conductivity (σ_{ac}) are recorded for the NCs. This research investigation showed enhancement in overall electrical properties of NCs, the results obtained are very important in identifying the use of NCs in electrical applications. The NC films show improved electrical conductivity upon incorporation of HNT to PC, the NC with 8 wt% HNT shows σ_{ac} of 6.4×10^2 Scm^{-1}.

Materials Research Foundations 129 (2022) 108-128   https://doi.org/10.21741/9781644902035-5

References

[1] G. Santhosh, B.S. Madhukar, G.P. Nayaka, and Siddaramaiah, "Luminescent down-shifting aluminosilicate nanocomposites: an efficient UVA shielding material for photovoltaics," J. Mater. Sci. Mater. Electr., 29 (2018) 6720-6729. https://doi.org/10.1007/s10854-018-8658-3

[2] N.K. Subramani, K.N. Shilpa, S. Shivanna, and Siddaramaiah, "Highly flexible and visibly transparent poly (vinyl alcohol)/calcium zincate nanocomposite films for UVA shielding applications as assessed by novel ultraviolet photon induced fluorescence quenching," Macromolecules, 49 (2016) 2791-2801. https://doi.org/10.1021/acs.macromol.5b02282

[3] J. Zheng, R.W. Siegel, and C.G. Toney, "Polymer crystalline structure and morphology changes in nylon-6/ZnO nanocomposites," J. Polym. Sci. Part B Polym. Phys., 41 (2003) 1033-1050. https://doi.org/10.1002/polb.10452

[4] F. Uddin, "Clays, nanoclays, and montmorillonite minerals," Metall. Mater. Trans. A, 39 (2008) 2804-2814. https://doi.org/10.1007/s11661-008-9603-5

[5] S. Pavlidou and C.D. Papaspyrides, "A review on polymer-layered silicate nanocomposites," Prog. Polym. Sci., 33 (2008) 1119-1198. https://doi.org/10.1016/j.progpolymsci.2008.07.008

[6] L. Torre, M. Chieruzzi, and J.M. Kenny, "Compatibilization and development of layered silicate nanocomposites based of unsaturated polyester resin and customized intercalation agent," J. Appl. Polym. Sci., 115 (2010) 3659-3666. https://doi.org/10.1002/app.31461

[7] M. Alexandre and P. Dubois, "Polymer-layered silicate nanocomposites: preparation, properties and uses of a new class of materials," Mater. Sci. Eng. R Rep., 28 (2000) 1-63. https://doi.org/10.1016/S0927-796X(00)00012-7

[8] K.N. Shilpa, K.S. Nithin, S. Sachhidananda, and B.S. Madhukar, and Siddaramaiah, "Visibly transparent PVA/sodium doped dysprosia (Na2Dy2O4) nanocomposite films, with high refractive index: An optical study," J. Alloys Compd., 694 (2017) 884-891. https://doi.org/10.1016/j.jallcom.2016.10.004

[9] G. Stoclet, "Elaboration of poly (lactic acid)/halloysite nanocomposites by means of water assisted extrusion: structure, mechanical properties and fire performance," RSC Adv., 4 (2014) 57553-57563. https://doi.org/10.1039/C4RA06845A

[10] P. Kiliaris, C.D. Papaspyrides, Polymer/layered silicate (clay) nanocomposites: An overview of flame retardancy, Progress in Polymer Science, 35 (2010) 902-958. https://doi.org/10.1016/j.progpolymsci.2010.03.001

[11] W.O. Yah, A. Takahara, and Y.M. Lvov, "Selective modification of halloysite lumen with octadecylphosphonic acid: new inorganic tubular micelle," J. Am. Chem. Soc., 134 (2012) 1853-1859. https://doi.org/10.1021/ja210258y

[12] P. Yuan, D. Tan, and F.A. Bergaya, "Properties and applications of halloysite nanotubes: recent research advances and future prospects," Appl. Clay Sci., 112 (2015) 75-93. https://doi.org/10.1016/j.clay.2015.05.001

[13] A. Dey and S.K. De, "Large dielectric constant in zirconia polypyrrole hybrid nanocomposites," J. Nanosci. Nanotechnol., 7 (2007) 2010-2015. https://doi.org/10.1166/jnn.2007.759

[14] L.M. Clayton, M. Cinke, M. Meyyappan, and J. P. Harmon, "Dielectric properties of PMMA/soot nanocomposites," J. Nanosci. Nanotechnol., 7 (2007) 2494-2499. https://doi.org/10.1166/jnn.2007.428

[15] N. Raman, S. Sudharsan, K. Pothiraj, Synthesis and structural reactivity of inorganic-organic hybrid nanocomposites - A review, Journal of Saudi Chemical Society. 16 (2012) 339-352. https://doi.org/10.1016/j.jscs.2011.01.012

[16] H. Li, "Polypropylene fibers fabricated via a needleless melt-electrospinning device for marine oil-spill cleanup," J. Appl. Polym. Sci., 131 (2014). https://doi.org/10.1002/app.40080

[17] G. Griffini, M. Levi, and S. Turri, "Thin-film luminescent solar concentrators: A device study towards rational design," Renew. Energy, 78 (2015) 288-294. https://doi.org/10.1016/j.renene.2015.01.009

[18] M. Biswal, S. Mohanty, S.K. Nayak, and P.S. Kumar, "Effect of functionalized nanosilica on the mechanical, dynamic-mechanical and morphological performance of polycarbonate/nanosilica nanocomposites," Polym. Eng. Sci., 53 (2013) 1287-1296. https://doi.org/10.1002/pen.23388

[19] S. Pillai, K.R. Catchpole, T. Trupke, and M.A. Green, "Surface plasmon enhanced silicon solar cells," J. Appl. Phys., 101 (2007) 093105. https://doi.org/10.1063/1.2734885

[20] K. Znajdek, "Zinc oxide nanoparticles for improvement of thin film photovoltaic structures' efficiency through down shifting conversion," Opto-Electron. Rev., 25 (2017) 99-102. https://doi.org/10.1016/j.opelre.2017.05.005

[21] M. Eslamian, "Excitation by acoustic vibration as an effective tool for improving the characteristics of the solution-processed coatings and thin films," Prog. Org. Coat., 113 (2017) 60-73. https://doi.org/10.1016/j.porgcoat.2017.08.008

[22] G. Santhosh, G.P. Nayaka, J. Aranha, and Siddaramaiah, "Investigation on electrical and dielectric behaviour of halloysite nanotube incorporated polycarbonate nanocomposite films," Trans. Indian Inst. Met., 70 (2017) 549-555. https://doi.org/10.1007/s12666-016-1033-2

[23] Y.C. Cao, "Preparation of thermally stable well-dispersed water-soluble CdTe quantum dots in montmorillonite clay host media," J. Colloid Interface Sci., 368 (2012) 139-143. https://doi.org/10.1016/j.jcis.2011.11.044

[24] J. Gaume, C. Taviot-Gueho, S. Cros, A. Rivaton, S. Therias, and J.L. Gardette, "Optimization of PVA clay nanocomposite for ultra-barrier multilayer encapsulation of organic solar cells," Sol. Energy Mater. Sol. Cells, 99 (2012) 240-249. https://doi.org/10.1016/j.solmat.2011.12.005

[25] S. Cros, "Definition of encapsulation barrier requirements: A method applied to organic solar cells," Sol. Energy Mater. Sol. Cells, 95 (2011) S65-S69. https://doi.org/10.1016/j.solmat.2011.01.035

[26] E. Picard, A. Vermogen, J.F. Gérard, and E. Espuche, "Barrier properties of nylon 6-montmorillonite nanocomposite membranes prepared by melt blending: Influence of the clay content and dispersion state: Consequences on modeling," J. Membr. Sci., 292 (2007) 133-144. https://doi.org/10.1016/j.memsci.2007.01.030

[27] H. Kim, Y. Miura, and C.W. Macosko, "Graphene/polyurethane nanocomposites for improved gas barrier and electrical conductivity," Chem. Mater., 22 (2010) 3441-3450. https://doi.org/10.1021/cm100477v

[28] L. Cui, J.T. Yeh, K. Wang, F.C. Tsai, and Q. Fu, "Relation of free volume and barrier properties in the miscible blends of poly(vinyl alcohol) and nylon 6-clay nanocomposites film," J. Membr. Sci., 327 (2009) 226-233. https://doi.org/10.1016/j.memsci.2008.11.027

[29] H. Jing "Preparation and characterization of polycarbonate nanocomposites based on surface-modified halloysite nanotubes," Polym. J., 46 (2014) 307. https://doi.org/10.1038/pj.2013.100

[30] H. Kim and C.W. Macosko, "Processing-property relationships of polycarbonate/graphene composites," Polymer, 50 (2009) 3797-3809. https://doi.org/10.1016/j.polymer.2009.05.038

[31] C.Y. Lee and J.K. Kil, "Hydrophilic property by contact angle change of ion implanted polycarbonate," Rev. Sci. Instrum., 79 (2008) 02C508. https://doi.org/10.1063/1.2804906

[32] Khairina Azmi Zahidah, Saeid Kakooei, Mokhtar Che Ismail, Pandian Bothi Raja, Halloysite nanotubes as nanocontainer for smart coating application: A review, Progress in Organic Coatings, 111 (2017) 175-185. https://doi.org/10.1016/j.porgcoat.2017.05.018

[33] M. Okamoto, "Recent advances in polymer/layered silicate nanocomposites: an overview from science to technology," Mater. Sci. Technol., 22 (2006) 756-779. https://doi.org/10.1179/174328406X101319

[34] S. Iijima, "Helical microtubules of graphitic carbon," Nature, 354 (1991) 56. https://doi.org/10.1038/354056a0

[35] M. Yoonessi and J.R. Gaier, "Highly conductive multifunctional graphene polycarbonate nanocomposites," Acs Nano, 4 (2010) 7211-7220. https://doi.org/10.1021/nn1019626

[36] P.J. Yoon, D.L. Hunter, and D.R. Paul, "Polycarbonate nanocomposites. Part 1. Effect of organoclay structure on morphology and properties," Polymer, 44 (2003). 5323-5339. https://doi.org/10.1016/S0032-3861(03)00528-7

[37] P.J. Yoon, D.L. Hunter, and D.R. Paul, "Polycarbonate nanocomposites: Part 2. Degradation and color formation," Polymer, 44 (2003) 5341-5354. https://doi.org/10.1016/S0032-3861(03)00523-8

[38] P. Steurer, R. Wissert, R. Thomann, and R. Mülhaupt, "Functionalized graphenes and thermoplastic nanocomposites based upon expanded graphite oxide," Macromol. Rapid Commun., 30 (2009) 316-327. https://doi.org/10.1002/marc.200800754

[39] A.K. Geim, "Graphene: Status and prospects," Science, 324 (2009) 1530-1534. https://doi.org/10.1126/science.1158877

[40] S. Stankovich et al., "Graphene-based composite materials," Nature, 442 (2006) 282. https://doi.org/10.1038/nature04969

[41] J. Li and J.K. Kim, "Percolation threshold of conducting polymer composites containing 3D randomly distributed graphite nanoplatelets," Comp. Sci. Technol., 67 (2007) 2114-2120. https://doi.org/10.1016/j.compscitech.2006.11.010

[42] M. Maric and C.W. Macosko, "Improving polymer blend dispersion in mini-mixers," Polym. Eng. Sci., 41 (2001)118-130. https://doi.org/10.1002/pen.10714

[43] P. Spitael and C.W. Macosko, "Strain hardening in polypropylenes and its role in extrusion foaming," Polym. Eng. Sci., 44 (2004) 2090-2100. https://doi.org/10.1002/pen.20214

[44] M.S. Rama and S. Swaminathan, "Polycarbonate/clay nanocomposites via in situ melt polycondensation," Ind. Eng. Chem. Res., 49 (2010) 2217-2227. https://doi.org/10.1021/ie9015649

[45] Y. Yoo, K.Y. Choi, and J.H. Lee, "Polycarbonate/montmorillonite nanocomposites prepared by microwave-aided solid state polymerization," Macromol. Chem. Phys., 205 (2004) 1863-1868. https://doi.org/10.1002/macp.200400179

[46] J. Feng, J. Hao, J. Du, and R. Yang, "Effects of organoclay modifiers on the flammability, thermal and mechanical properties of polycarbonate nanocomposites filled with a phosphate and organoclays," Polym. Degrad. Stab., 97 (2012) 108-117. https://doi.org/10.1016/j.polymdegradstab.2011.09.019

[47] A. J. Hsieh, "Mechanical response and rheological properties of polycarbonate layered-silicate nanocomposites," Polym. Eng. Sci., 44 (2004) 825-837. https://doi.org/10.1002/pen.20074

[48] G. Ciric Marjanvic, "Progress in polyaniline composites with transition metal oxides," Fundam. Conjug. Polym. Blends Copolym. Compos. Synth. Prop. Appl., (2015) 119-162. https://doi.org/10.1002/9781119137160.ch2

[49] L. Peponi, D. Puglia, L. Torre, L. Valentini, and J. M. Kenny, "Processing of nanostructured polymers and advanced polymeric based nanocomposites," Mater. Sci. Eng. R Rep., 85 (2014) 1-46. https://doi.org/10.1016/j.mser.2014.08.002

[50] M. Kawasumi, N. Hasegawa, M. Kato, A. Usuki, and A. Okada, "Preparation and mechanical properties of polypropylene - clay hybrids," Macromolecules, 30 (1997) 6333-6338. https://doi.org/10.1021/ma961786h

[51] W.H. Press, B.P. Flannery, S.A. Teukilsky, and W.Y. Vettering, Numerical Recipes, Cambridge University Press, Cambridge, 1986.

[52] D. Henry, N. Eby, J. Goodge, and D. Mogk, "X-ray reflection in accordance with Bragg's Law," Sci. Educ. Research Cent., 2012.

[53] B.J. Ash, "Mechanical properties of Al2O3/polymethyl methacrylate nanocomposites," Polym. Compos., 23 (2002)1014-1025. https://doi.org/10.1002/pc.10497

[54] M. Joshi, A. Bhattacharyya, and S.W. Ali, "Characterization techniques for nanotechnology applications in textiles," 2008.

[55] Y.S. Nanayakkara, S. Perera, S. Bindiganavale, E. Wanigasekara, H. Moon, and D. W. Armstrong, "The effect of AC frequency on the electro wetting behavior of ionic liquids," Anal. Chem., 82 (2010) 3146-3154. https://doi.org/10.1021/ac9021852

[56] Y. Yuan and T.R. Lee, "Contact angle and wetting properties," in Surface science techniques, Springer, (2013) 3-34. https://doi.org/10.1007/978-3-642-34243-1_1

[57] D.K. Owens and R.C. Wendt, "Estimation of the surface free energy of polymers," J. Appl. Polym. Sci., 13 (1969) 1741-1747. https://doi.org/10.1002/app.1969.070130815

[58] U. Baishya and D. Sarkar, "Structural and optical properties of zinc sulphide-polyvinyl alcohol (ZnS-PVA) nanocomposite thin films: effect of Zn source concentration," Bull. Mater. Sci., 34 (2011) 1285-1288. https://doi.org/10.1007/s12034-011-0316-9

[59] P. Gill, T.T. Moghadam, and B. Ranjbar, "Differential scanning calorimetry techniques: applications in biology and nanoscience," J. Biomol. Tech. JBT, 21 (2010) 167.

[60] D.V. Rosato, N.R. Schott, and M.G. Rosato, Plastics Institute of America Plastics Engineering, Manufacturing & Data Handbook. Springer Science & Business Media, 2001. https://doi.org/10.1007/978-1-4615-1615-6

[61] V. Gupta, M. Sharma, and N. Thakur, "Optimization criteria for optimal placement of piezoelectric sensors and actuators on a smart structure: a technical review," J. Intell. Mater. Syst. Struct., 21 (2010) 1227-1243. https://doi.org/10.1177/1045389X10381659

[62] A.A. Zaky and R. Hawley, Dielectric solids. Routledge, Thoemms Press, 1970.

[63] R.P. Weber, K.S. Vecchio, and J.C.M. Suarez, "Dynamic behavior of gamma-irradiated polycarbonate," Mater. Rio J., 15 (2010) 218-224. https://doi.org/10.1590/S1517-70762010000200019

[64] A. Tayel, M.F. Zaki, A.B. El Basaty, and T.M. Hegazy, "Modifications induced by gamma irradiation to Makrofol polymer nuclear track detector," J. Adv. Res., 6 (2015) 219-224. https://doi.org/10.1016/j.jare.2014.01.005

[65] B.S. Yilbas, H. Ali, F.A. Sulaiman, and H.A. Qahtani, "Effect of mud drying temperature on surface characteristics of a polycarbonate PV protective cover," Sol. Energy, 143 (2017) 63-72. https://doi.org/10.1016/j.solener.2016.12.052

[66] G.B. Hadjichristov, V. Ivanov, and E. Faulques, "Reflectivity modification of polymethyl methacrylate by silicon ion implantation," Appl. Surf. Sci., 254 (2008) 4820-4827. https://doi.org/10.1016/j.apsusc.2008.01.115

[67] H.K. Chitte, N.V. Bhat, N.S. Karmakar, D.C. Kothari, and G.N. Shinde, "Synthesis and characterization of polymeric composites embedded with silver nanoparticles," World J. Nano Sci. Eng., 2 (2012) 19-24. https://doi.org/10.4236/wjnse.2012.21004

[68] E. Tavenner, P. Meredith, B. Wood, M. Curry, and R. Giedd, "Tailored conductivity in ion implanted polyetherether ketone," Synth. Met., 145 (2004) 183-190. https://doi.org/10.1016/j.synthmet.2004.05.005

[69] S.A. Nouh, "Physical changes associated with gamma doses of PM-555 solid-state nuclear track detector," Radiat. Meas., 38 (2004) 167-172. https://doi.org/10.1016/j.radmeas.2003.11.004

[70] M.A. Shams-Eldin, C. Wochnowski, M. Koerdt, S. Metev, A.A. Hamza, and W. Jüptner, "Characterization of the optical-functional properties of a waveguide written by an UV-laser into a planar polymer chip," Opt. Mater., 27 (2005) 1138-1148. https://doi.org/10.1016/j.optmat.2004.09.019

[71] M. Todica, T. Stefan, D. Trandafir, and S. Simon, "ESR and XRD investigation of effects induced by gamma radiation on PVA-TiO2 membranes," Open Phys., 11 (2013) 928-935. https://doi.org/10.2478/s11534-013-0276-3

[72] G. Rudko, A. Kovalchuk, V. Fediv, W.M. Chen and I.A. Buyanova, "Enhancement of polymer endurance to UV light by incorporation of semiconductor nanoparticles". Nanoscale Research Letters, 10 (2015) 81-88. https://doi.org/10.1186/s11671-015-0787-5

[73] A. Singhal, K.A. Dubey, Y.K. Bhardwaj, D. Jain, S. Choudhury and A.K. Tyagi, "UV-shielding transparent PMMA/In2O3 nanocomposite films based on In2O3 nanoparticles". RSC Advances, 3 (2013) 20913-20921. https://doi.org/10.1039/c3ra42244e

[74] S. Sugumaran, C.S. Bellan, D. Muthu, S. Raja, D. Bheeman and R. Rajamani, "Novel hybrid PVA-In ZnO transparent thin films and sandwich capacitor structure by dip coating method: preparation and characterizations," RSC Advances, 5 (2015) 10599-10610. https://doi.org/10.1039/C4RA14817G

[75] H. Althues, J. Henle and S. Kaskel, "Functional inorganic nanofillers for transparent polymers," Chemical Society Reviews, 36 (2007) 1454-1465. https://doi.org/10.1039/b608177k

[76] J.G. Liu and M. Ueda, "High refractive index polymers: Fundamental research and practical applications," Journal of Materials Chemistry, 19 (2009) 8907-8919. https://doi.org/10.1039/b909690f

[77] C. Lu, Y. Cheng, Y. Liu, F. Liu and B. Yang, "A facile route to ZnS-polymer nanocomposite optical materials with high nanophase content via g-ray irradiation initiated bulk polymerization," Advanced Materials, 18 (2006) 1188-1192. https://doi.org/10.1002/adma.200502404

[78] T.K. Leodidou, P. Margraf, W. Caseri, U.W. Suter and P. Walther, "Polymer sheets with a thin nanocomposite layer acting as a UV filter," Polymer Advanced Technology, 8 (1997) 505-512. https://doi.org/10.1002/(SICI)1099-1581(199708)8:8<505::AID-PAT678>3.0.CO;2-U

[79] N. Suzuki, Y. Tomita, K. Ohmori, M. Hidaka and K. Chikama, "Highly transparent ZrO2 nanoparticle-dispersed acrylate photopolymers for volume holographic recording," Optics Express, 14 (2006) 12712−12719. https://doi.org/10.1364/OE.14.012712

[80] J.G. Liu, Y. Nakamura, T. Ogura, Y. Shibasaki, S. Ando and M. Ueda, "Optically transparent sulfur - containing polyimide - TiO2 nanocomposite films with high refractive index negative pattern formation from poly (amic acid) - TiO2 nanocomposite film," Chemistry of Materials, 20 (2008) 273−281. https://doi.org/10.1021/cm071430s

[81] B.T. Liu, S.J. Tang, Y.Y. Yu and S.H. Lin, "High-refractive-index polymer/ inorganic hybrid films containing high TiO2 contents. Colloids and Surfaces," A: Physico-chemical and Engineering Aspects, 377 (2011) 138-143. https://doi.org/10.1016/j.colsurfa.2010.12.046

[82] H.W. Su, and W.C. Chen, "High refractive index polyimide-nanocrystalline-titania hybrid optical materials," Journal of Materials Chemistry, 18 (2008) 1139-1145. https://doi.org/10.1039/b717069f

[83] H. Zhang, Z. Cui, Y. Wang, K. Zhang, X. Ji, C. Lü, B. Yang and M. Gao, "From water-soluble CdTe nanocrystals to fluorescent nanocrystal-polymer transparent composites using polymerizable surfactants," Advanced Materials, 15 (2003) 777-780. https://doi.org/10.1002/adma.200304521

[84] A. Sionkowska, J. Skopińska and M. Wisniewski, "Photochemical stability of collagen/poly (vinyl alcohol) blends," Polymer Degradation and Stability, 83 (2004) 117-125. https://doi.org/10.1016/S0141-3910(03)00232-5

[85] Y. Chen, Z. Sun, Y. Yang and Q. Ke, "Heterogeneous photocatalytic oxidation of polyvinyl alcohol in water," Journal of Photochemistry and Photobiology A: Chemistry, 142 (2001) 85-89. https://doi.org/10.1016/S1010-6030(01)00477-4

[86] J. Davenas, A. Ltaief, V. Barlier, G. Boiteux, and A. Bouazizi, "Nanomaterials for photovoltaic conversion," Materials Sci. Engineering C, 28 (2008) 744-750. https://doi.org/10.1016/j.msec.2007.10.055

[87] H. Althues, R. Palkovits, A. Rumplecker, P. Simon, W. Sigle, M. Bredol, U. Kynast and S. Kaskel, "Synthesis and characterization of transparent luminescent ZnS:Mn/PMMA nanocomposites," Chemistry of Materials, 18 (2006) 1068-1072. https://doi.org/10.1021/cm0477422

Adv. App. of Micro and Nano Clay II – Synthetic Polymer Composites   Materials Research Forum LLC
Materials Research Foundations 129 (2022) 129-152   https://doi.org/10.21741/9781644902035-6

Chapter 6

Nano Clay-Polymer Composite for Water Treatment

Aliya Naz[1,2*], Abhiroop Chowdhury[3]

[1]Department of Environmental Science and Engineering, Indian Institute of Technology, Dhanbad, India

[2]Jindal School of Liberal Arts and Humanities, O.P. Jindal Global University, Sonipat, Haryana, India

[3]Jindal School of Environment and Sustainability, O.P. Jindal Global University, Sonipat, Haryana, India

* naazaliya6@gmail.com

Abstract

The usage of clay polymer composites has increased in the past few decades. Due to extensive availability, wide surface areas, and good adsorption efficiency clay polymers are gaining popularity amongst researchers to remove wide range of organic as well as inorganic pollutants from effluents. This book chapter sheds light on the characteristics, occurrence, types and synthesis of clay minerals used for preparing nano-clay polymer composites. It also highlights the types, applicability and efficiencies of particular nano clay polymers for the removal of dyes, paints, nutrients, potentially toxic elements and pharmaceutical contaminants from wastewater.

Keywords

Clay Polymer Composites, Wastewater, Advance Treatment, Potentially Toxic Elements, Adsorption

Contents

1.   Introduction

Wastewater generated from industries and municipalities are the major contributors for degrading the water quality of fresh water bodies. Fresh and uncontaminated drinking water is the foremost requirement for healthy human community. According to a recent WHO report, "about 2.1 billion people out of a total population of 7.7 billion lack access to clean, easily available water, and 4.5 billion people lack securely managed sanitation" [1,2].

Increasing water pollution all over the globe is the matter of grave concern for all the scientist and researchers. The most of available water treatment technologies are either complex or expensive. In developing nations, a domestic reverse osmosis system is too expensive for the socioeconomically deprived and low-income population to afford. As a result, the general public are compelled to consume contaminated water. Polluted groundwater as well as river water due to influx of untreated or partially treated industrial wastewater and the agricultural runoff have raised different health complications in human beings. Gastrointestinal, hepatic, renal impairments, increasing cases of cancer are the result of consumption of contaminated water with heavy metals, pesticides, and organic/inorganic chemicals/substances [3]. The rising demand for clean water and water scarcity throughout the globe has promulgated the usage of recycled/ treated water.

Most of the effective technologies to treat wastewater are based on the electrodialysis, electrolysis, ion-exchange, and reverse osmosis, which comprises an expensive, complex and sensitive technology. These expensive technologies are not very popular for large scale instalment in developing countries. The conventional physical and chemical based methods produced secondary contaminants which needs additional treatment, and made the process costly. The popularity of treatment methods are dependent on the sustainability, economic properties and complexity. In this view, adsorption is the most attractive process in developing countries due to its lower capital cost and its high efficiency to remove different types of contaminants.

The treatment of emerging pollutants in wastewater is also a challenge among researchers. The current scenario of the advancement of treatment processes for various micro-pollutants have cost-effective and less complex technological solutions. The search of cost effective and suitable water treatment progressed towards innovation of ecofriendly and budget friendly adsorbents. Natural and synthetic both adsorbents are found effective to remove water pollutants, but the bio-polymers are now been adopted at pilot scale and also receiving great attention. Bio-polymeric clay composites are biodegradable and best solution in front of synthetic and non-biodegradable polymers [4]. The development of synthesis and uses of clay polymers as an adsorbent is became popular in the last few decades. Clay/polymer composites offer tremendous improvement for various contaminants. With the advancement of scientific knowledge and technologies a substantial progress has been achieved in the synthesis of natural, synthetic and effective micro and nano-polymers. However, scientific investigation is focusing on understanding the mechanisms and the enhancement in the adsorption efficiency of clay micro as well as nano composite polymers.

In this chapter we are discussing about the nano-polymers, their synthesis and usage for treatment of wastewater, in order to remove organic and inorganic pollutants. Wastewater released from fertilizer, textile, paper, plastic, tannery, food, cosmetic, pharmaceutical industries contain different heavy metals, harmful dyes, organic pollutants and sometimes high concentration of nutrients such as phosphorus and ammonium. This chapter also deals with the progress of clay composites, their advantages, and limitations in order to treat wastewater released from various industries.

2.   Characteristics of clay materials

On the basis of chemical composition and morphological structure clay minerals have been categorized in various groups [5]. The most important physicochemical properties of clay composites are particle size, shape, their distribution, and surface area [6].

Clay polymers is formed by layered silicate minerals with layered structural units called tetrahedral or octahedral sheets [7-8]. Octahedral sheets are composed of Al or Mg in 6-fold coordination with oxygen. However, tetrahedral sheets are formed of Si-O$_2$linked with other tetrahedral sheets and hydroxyl [9]. The overall structure of clay polymers distinguished various properties and helps in adsorption of several organic/inorganic contaminants from wastewater [10]. On the basis of mineralogical composition, about thirty types of clay composites materials are introduced for different applications [6]. In recent two decades, montmorillonite (MT) among different clay minerals, has often been used in the removal of organic pigments and dyes due to its low cost, high surface area and high-cation exchange capacity [11]. Polymer/clay hydrogel composite have attracted scientist because of its relatively low budget, less installation cost and good adsorption capacity for both organic as well as inorganic pollutants [12].

3.   Types of clay minerals

Nano- clay are clay structures with a particle size in the scale of 10^{-9} m, composed of layered silicate matrix, which can be further sub-classified into montmorillonite, bentonite, kaolinite, hectorite, and halloysite. Nano-clay synthesized through variety of clay minerals, the most common clay mineral group is phyllositicates. Smectite class of clay is widely used as adsorbent in past two decades [13]. The major clay minerals are discussed as below:

3.1   Smectite

Smecites belong to phyllositicates group of clay mineral abundant in the temperate soil and sediments. This clay mineral gained popularity among researchers due to its physic-chemical characteristics-layered nature and huge specific surface, high cation exchange properties. These properties of smecites, provides good adsorption capacity for various inorganic as well as organic pollutants [14-15]. Smecites are basically 2 layered minerals which includes two tetrahedral sheets and an octahedral sheet through sharing of apical oxygens [13]. Based on the composition, smecites can be classified into 4 major subclass-Montmorilonite, Bentonite, nontronite, and saponite [13-14].

3.1.1 Montmorilonite

Montmorilonite (MT) is the most common and popular clay mineral used for the removal of contaminants from the wastewater. Structure of this mineral is the same as it is smecites group of clay, which consists an octahedral Al$_2$O$_3$ sheet packed in between two tetrahedral SiO$_2$ sheets [16]. MT based nano clay and modified MT clay is applied for the remediation of toxic elements, dyes, and pesticides [9]. The uses of MT based nano polymers is now attracting researchers to experiment its treatment efficiencies for the wastewater released from municipal and industries [6]. A large surface area and net negative surface charges of MT attracts cationic metals and metalloids (As, Cd, Cu, Pb, Ni, Cr etc.) and organic dyes for example- methylene blue (MB), crystal violet (CV), orange G (OG), and many others. Through electrostatic interaction [13].

3.1.2 Bentonite

The impure MT mineral is called Bentonite [13]. Bentonite is very popular among developing countries due to its availability, low cost and significantly good pollutant removal efficiencies [13]. Bentonite is being used for different purposes from last thousands of years, but their uses as an effective adsorbents starts after 19th century [17]. Bentonite is effectively adsorbed phenol from the water [18]. Recently the modified bentonites with alginate beads were used for the removal of methylin blue [19]. Wheras Zn^{2+} effectively removed from the contaminated water with alkaline Ca-bentonite [20]. In 1997, Ruiz and the reserach group, used a modified bentonite for the removal of Al from wastewater and found effective for the removal of cationinc metals from the auoues solution [21]. Copper is effectivelly removed from the water with Alginate-immobilized bentonite clay by [22]. The efficiency of Na-bentonite is studied for the remediation of various toxic elements-Cd, Cu and Pb, and it is reveled that the ionic structure of Na-Bentonite is very sensitive for the Cd whereas it was little impact on the Pb and Cu removal [23]. The ionic capacity was studied by analyzing zeta potential and sedimentation rate [23]. High nutrient concentration in eutrophic lakes affectes the aquatic life [24]. The removal of phosphate and ammonium is required in this case from the discharging wastewater into water bodies. According to Zamparas et al, bentonite-humic acid clay composites shown the adsorption rate for PO_4, 26.5 mg/g and for NH_4, 202.1 mg/g [25].

3.2   Kaolinite

Kaolinite [$Al_2Si_2O_5(OH)_4$] is 1:1 layered structure of clay minerals consists tetrahedral SiO_2 and Al_2O_3 [13]. The structure of this mineral do not consist lower valance cations, thus kaolinite do not have any net charge [13]. Kaolinite nano-clay have significantly smaller surface area in comparison to MT and other smectite. Because of these reasons, kaolinite is not very efficient for the remediation of dyes and toxic elements from the contaminated waters [26]. Although the adsorption capacity is low the modified kaolinite nanoclay have proven efficient for the remediation of toxic elements and organic pollutants [26-29].

3.3   Illite

Illite is 2:1 layered clay mineral, very similar to Smectite, in which silica tetrahedral sheet is substituted by Al_2O_3 and layers are tightly bonded together with high charge density. In illite, an interlayer K^+cation, is embedded in the holes forms a grid of oxygen atoms between adjoining layers [30-31]. The structure of illite composite is $K1-1.5Al_4[Si_{6.5}-7Al_1-1.5O_{20}](OH)_4$ [32]. Recently cesium an emerging pollutant has been adsorbed through illite nanoparticles.

3.4   Sepiolite

Sepiolite is fibrous clay, generally occur in the arid/semiarid zones of tropical areas [33]. Sepiolite is $H_2Mg_2O_9Si_3$ specific clay mineral, which chemical formulae is $Si_{12}Mg_8O_{30}(OH)_4(OH_2)4.8H_2O$ [34]. Sapolite silicate have large specific surface area and very less reduced ion exchange capacity [35].

Poly (vinylimidazole)/sepiolite nanoclay composites was synthesized by Kara et al. 2016, through *in-situ* polymerization for the remediation of Hg^{2+} in aqueous solution [36]. The mixed nano-clay polymer composites show a uniform, significantly dense, and smooth surface dispersion, which improves adsorption capacity in comparison to conventional sepiolite clay ([6,37].

4.   Nano-clay polymer composites

According to synthesis procedure of nano-clay polymers, it can be broadly divided into 3 major types. These are i) Conventional nano-clay composites ii) Intercalated nanoclay composites, and iii) Exfoliated nano-clay composites. The conventional composites are simple 2 layered clay minerals. Natural nanoclay composites is example of conventional clay composite. This nanoclay composites are easier to synthesize from the clay minerals. Oskui et al., worked on the natural nanoclay for the adsorption of trivalent chromium [Cr(III)] from the contaminated aqueous solution [37]. Natural nanoclay for this study were obtained from the Urmia, Iranian clay. The natural clay was sieved with 854 µm sieve in order to remove large particles [37].

The intercalated nanoclay prepared artificially through inserting polymer chain in between interlayers. Thus, the intercalated nanoclay composites consist of multiple layers of different polymers at an equal distance [38] (Fig.1). These nanoclay particles are more efficient for the adsorption of organic as well as inorganic cations in comparison to conventional polymer composites. Ghodke et al. compared conventional nano-clay and intercalcated polymers for the degradation of phenol from water [39]. TiO_2 and nanonoclay

Adv. App. of Micro and Nano Clay II – Synthetic Polymer Composites   Materials Research Forum LLC
Materials Research Foundations 129 (2022) 129-152   https://doi.org/10.21741/9781644902035-6

intercalated polymer exhibited 59% of phenol removal while the conventional nano-clay removed only 47% of phenol at the initial concentration of Phenol -500 mg/L [39].

The exfoliated nanoclay composites represents the insertion of extensive polymers between clay interlayers in delaminated and dispersion manner. Thus, process increases the affinity in-between polymer and nanoclay [40-41]. This class of nanoclay composites acquires high tensile strength, significantly high modules, heat dissertation, and easily biodegradable quality [42]. The exfoliated nanorods, nanotubes and nanosheets are commonly used for the remediation of organic dyes from water [41]. The nano-clays composites formed through the exfoliation of clay particles can be used for the synthesis of various advanced functional groups. The exfoliated nano-clay polymer composites with synthetic polymers, can be understand through the properties of synthetic materials via micro-regulation of the clay layers and other functional fragments [13].

A number of synthetic polymers are being used for the composition of nano-clay polymer composites. Few are: epoxies, poly vinyl chloride (PVC), polyurethanes, polyesters, polypropylene, and polystyrene. The most commonly used synthetic nano-clay polymer composites for the treatment of wastewater are Epoxy–clay nanocomposites, Polyurethane–clay nanocomposites, Polyimide–clay hybrid, Organoclay–polymer blends, Polystyrene–clay nanocomposite, Polysiloxane–clay nanocomposite, Poly(acrylic acid-co-acrylamide) nanoclay–polymer composite, Clay–biopolymer nanocomposites, Magnetic clay–polymer nanocomposites, Montmorillonite and hydroxyiron modified montmorillonite nanoclay, poly vinyl Alcohol (PVA)–nano-clay composits, Poly-acrylic acid hydrogel composite, Kappa-carrageenan/poly (vinyl alcohol) (PVA) nanocomposites hydrogels, Amino-functionalized attapulgite clay nanoparticle, N-vinyl-2-pyrrolidone/itaconic acid/organo clay nanocomposites hydrogels [42].

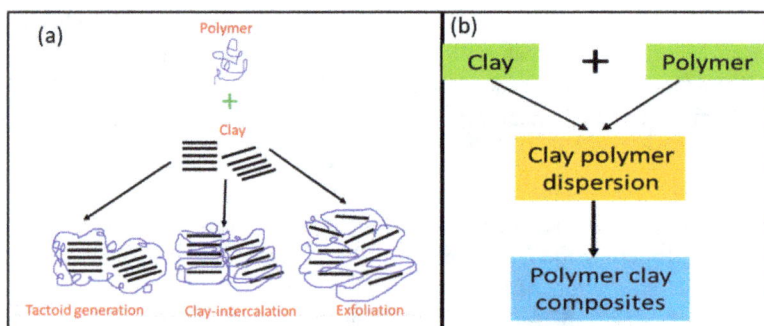

Figure 1. The synthesis of nanoclay polymers (a) Process of Tacoid generation, intercalations and exfoliation (b) SA common procedure of nano-clay synthesis

Adv. App. of Micro and Nano Clay II – Synthetic Polymer Composites   Materials Research Forum LLC
Materials Research Foundations 129 (2022) 129-152   https://doi.org/10.21741/9781644902035-6

5.   Wastewater treatment and clay polymers

Wastewater released in the form of industrial and municipal effluents contains various substances such as potentially toxic elements, aromatic hydrocarbons, artificial colors, chemicals, bacteria, weedicides, herbicides, insecticides, tannic acid, phosphate and ammonium [43-47]. These contaminated effluents cannot be treated through the conventional treatment processes. Pollutant removal from adsorption is most promising techniques used in the most of developed countries. Adsorption potential of nano-clay composite polymers have proven most efficient for a wide range of pollutants [6,48-49]. Figure 2 shows the effective removal of pollutants released from the major industries through nanoclay polymer composites (Figure 2). The clay composite polymers and their uses for the removal of different environmental pollution are listed in the Table 1.

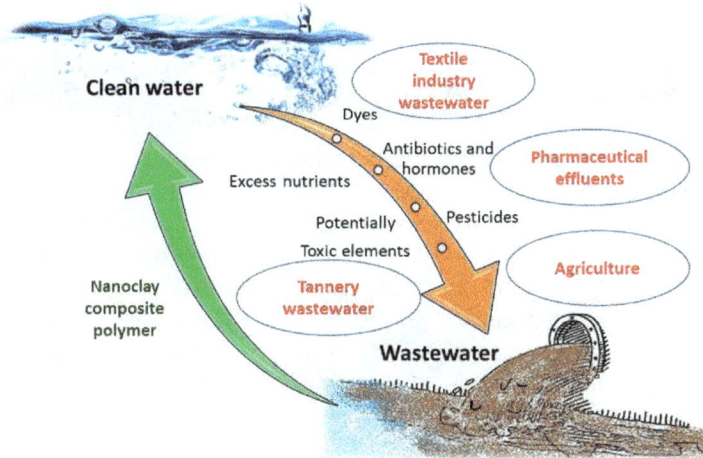

Figure 2: Graphical representation of wastewater tretment through nanoclay polymer composites for a wide range of wastewater

5.1   Inorganic pollutants

5.1.1   Toxic elements

Clay nano-polymers is promising adsorbent for remediation of potentially toxic elements from the water. A broad range of toxic elements for example- As, Cd, Pb, Cr have successfully removed from aqueous solution [13]. Clay-polymer metal complexes were prepared with bentonite clay and *Moringa oleifers* seed. This adsorbent used for preparing novel composite coagulant to remove cadmium, chromium and lead from contaminated water [50]. Modified nanoclay polimers are proven more effective in heavy metal removal from contaminated aquoes solution. Chitin-halloysite nanoclay hydrogel was synthesised and used for the removal of lead, copper and cadmium with adsorption rate of 8.2, 4.2 and 2.1 mg/g [51]. As per this study, the remediation of Pb was significant high in comparison of copper and cadmium. In contrast,nano-clay/TiO_2 composite reduced 90% of cadmium at 30 mgCd/l [52]. The synthesis and uses of biopolymer (chitosan)/nanoclay composites for the removal of Cu and Ni was investigated. The adsorption capacity for Cu and Ni was recorded 176 mg/g and 144 mg/g respectively [53]. Arsenic is one of the carcinogenic metalloid present in the As-affected groundwater [54]. Almost 230 million of populace in 108 nations are affected with the geogenic Arsenic contamination in the groundwater [55]. The removal of As from the drinking water through less complex, ecofriendly and cost effective method is a major challenge [56]. Arsenic has been significantly removed from the water through modified clay composites such as nanoclay-loaded electrospun fiber membrane,organically modified smectite adsorbents and hydroxyiron modified montmorillonite nanoclay [57-60]. Hexavalent chromium pollution through mining, effluents of ferrochrome, dye an tanneries are another most concerned issue amongst scientist [60]. Cr(VI) is carcinogenic, mutagenic and teratogenic, and its chronic exposure can cause cancer, tumor and even death [61-63]. The uses of Polysulfone mixed with montmorilloniteand modifed montmorillonitemembranes for extraction of hexavalent chromium from the aques solutionwas done by scientists [64-65]. According to their findings, the removal of Cr(VI) ions achieived with 99.3 to 95% rejection rates and flux of 102 and 271 l/m^2 per hr respectively [64-65]. Thus the toxic elements have been effectivelly removed from the aquores asolution with clay mineral matrix through various adsorption processes.

5.1.2 Nutrients

High phosphate and ammonium concentration in the wastewater and agricultural runoff cause eutrophication. The eutrophic water negatively impacts the aquatic flora and fauna [66-67]. The uses of nano-clay polymer composites are an ecofriendly water treatment technology for the remediation of high concentrations of nutrients from the municipal, industrial water as well as eutrophic lake. Yuan and Wu studied the efficiency of the Allophane nanoclay for the adsorption of phosphorus and suggseted 70 % removal at 14.2 mgP/l [68]. Likewise, ammonium also removed through adsorption by using nanoclay polymer composites [69-70].

Table 1. Uses of clay polymer composites for the removal of inorganic pollutants

S.no.	Pollutants	Clay polymers	Removal efficiency	Remarks	Ref.
1	Arsenic	Montmorillonite (MT) and hydroxyiron modified montmorillonite (HyFe-MT) nanoclay	The adsorption capacity were reported 0.0113 mg As/g for MT and 0.191 mg As/g for HyFe-MT	The adsorption capacity increased 5 times after iron modification in MT clay	[85]
2	Cadmium	nano-montmorillonite	The maximum adsorption capacity was estimated to be 17.61 mg/g at a Cd^{2+} concentration of 22.4 mg/L.	Adsorption experiments were carried out to examine the effects of the initial metal ion concentration (22.4–224 mg/L), pH (2.5–7.5), contact time (2–180 min) and temperature (15–40 °C).	[86]
3	Cadmium	novel polymer–clay composite	Maxiumum adsorption capacity was 1236 mg Cd/g	Kaolinite clay was treated with polyvinyl alcohol to produce a novel water-stable composite called polymer–clay composite adsorbent	[87]
4	Cobalt and Nickel	Sodium montmorillonite nanocomposites	89.17 removal of Co and 94.5% removal of Ni have been reported.	The Ni and Co removal efficiency were increased with mixing two nano composites Na-MT and polyacrylamide (PAM)	[88]
5	Copper (Cu II)	Cloisite nanoclay (Natural Na+-MT)	Highest dsorption capacity was recorded 4.23 mg Cu (II)/g	Natural Na+-MT (Na+-Cloisite nanoclay) was purchased	[89]
6	Trivalent Chromium	Iranian natural nanoclay	The adsorption efficiency was 99% at the initial Cr(III) concentration is less than or equal to 140 mg/L,	The raw clay from Urmia (Iran) used for synthesize nano-clay after screening with 854 μm sieve. This clay also used from long back for the clarification of grape syrup by locals.	[37]
7	Hexavalent Chromium	Cellulose-Clay Composite Biosorbent	The maximum adsorption capacity was recorded 22.2 mg/g	This is an eco-friendly adsorbent and sufficient to remove upto 100 mg/L of hexavalent chromium..	[90]
8	Lead and cadmium	PVA–nano-clay adsorbent (Kaolinite clay modified with polyvinyl alcohol)	The adsorption capacity was obtained 56.18 mg/g and 41.67 mg/g for Pb^{2+} and Cd^{2+} respectively	Increasing the adsorbent dosage over the range 0.1–1.0 g has decreased the equilibrium adsorption capacity, of the adsorbent.	[91]

9	Mercury (Hg II)	dithizone-montmorillonite (Dz-MT) composit	90% of aquoes Hg (II) (50 mg/L) was removed within 1 hr 30 min at 0.5 g/L of adsorbent dose	dithizone-montmorillonite (Dz-MT) composit was prepared through facile one-step method	[92]
10	Ammonium nitrogen	chitosan-bentonite film composite	The maximum adsorption was recorded 43.19% and 82.11% for chitosan and chitosan film composites respectively.	15 mg/l of ammonia nitrogen solution was adsorbed with 0.5g of chitosan and chitosan films.	[93]
11	Sulfate	Organo-nano-clay	The maximum adsorption capacity was 38.02 mg/g at 40 °C	The adsorbent was prepared through natural zeolite and cationic surfactant cetyltrimethylammonium bromide	[94]
12	Phosphorus	Allophane nanoclay	14.2 to 4.2 mg/L of P concentration can be removed through it, with a removal efficiency of 70%.	Allophane is an economical and it is easily available from road side deposits. These materials are safe to use. Allophane nanoclay is recoverable after use. Phosphorus can be eliminated from wastewater and the adsorbed P can be recycled to farmland.	[68]

5.2 Organic pollutants

5.2.1 Dyes and paints

Dyes and paints contain wide range of pollutants which can affect the exposed population through drinking water and food. The wastewater released from textile, pulp and papers, printing press, cosmetics, pharmaceutical and leather manufacturing industries contains huge amounts of dyes [71]. Untreated or partially treated wastewater containing dyes discharged into aquatic bodies can adversely affect the living organism exposed to these [72]. The acute and chronic impacts of dyes are skin allergy, metamorphosis, tumor and cancer [73]. The adsorption through nano-composites clay is being used by various researchers. Most common dye, methylene blue is been adsorbed through thin film membrane prepared with mixed matrix nanoclay/chitosan on poly-vinylidene difluoride microfiltration base, Kappa-carrageenan biopolymer-based nanocomposite hydroge, PVA/CMC/halloysite nanoclay bio composites, nZVI nanoclay, iron impregnated nanoclay, and nano-clay composite $PLA/C_{20}A$ [74-78]. Orange G dye has been successfully removed from the aqueus solution through nanoclay composites [79]. Nano zero valent iron mixed clay composites also proven effcient for the removal of orange G [80].

5.2.2 Pharmaceutical effluents

Pharmaceutical waste contains antibiotics, hormones. Most of the pharmaceutical medicine's residue cannot be removed from the water through conventional water treatment unites [81]. These residues increase the resistance in the pathogenic bacteria, which leads a big treat on human health [82]. A study in Europe quantified the antibiotic contamination in the treated water, and found at least one antibiotics out of 53 studied antibiotics in the treated effluents of seven European countries. The major antibiotics reported were ciprofloxacin, ofloxacin, enrofloxacin, orbifloxacin, azithromycin, clarithromycin, sulfapyridine, sulfamethoxazole, trimethoprim, nalidixic acid, pipemidic acid, oxolinic acid, cefalexin, clindamycin, metronidazole, ampicillin, and tetracycline [83]. The nanomaterials are proven very efficient in the removal of antibiotics and hormones form the contaminated water [84].

Table 2. Uses of clay polymer composites for the removal of organic pollutants

S.no.	Pollutants	Clay polymers	Removal efficiency	Remarks	Ref.
1	Bromophenol blue dye	Kaolinite -poly (acrylamide co-acrylic acid)	50 mg/l of dye concentration reduced about 90% with adsorbent dose of 7.5 g/l and temperature 25 ^{0}C	Kaolinite crosslinker added into acrylic acid and washed multiple times with deionized water, subsequently dried, powdered and screened.	[95]
2	Brilliant Green	Poly-acrylic acid hydrogel composite	-	This adsorbent was prepared by prepared by incorporation of kaoline clay.	[96]
3	Crystal violet	Kappa-carrageenan/poly (vinyl alcohol) (PVA) nanocomposites hydrogels	dye removal was 151mg per g of adsorbent per miute	This adsorbent was prepared through Na-MT nanoclay. The mixture PVA and MT was interlinked with freezing–thawing method and with K$^+$. The nanocomposites showed a relatively improved adsorption capacity in comparison to clay-free hydrogel.	[97]

4	Methylene blue (MB)	chitin–clay microsphere composites	99.99% of MB removal efficiency was achieved with adsorption rate of 10mg/g in 10 min.	Clay nanosheets were inserted into a nanofiber weaved chitin microsphere matrix.	[98]
5	Methylene blue	Amino-functionalized attapulgite clay nanoparticle	Removed MB with efficiency of 215.73/120 mg/g per minute	Economic and ecofreindly materilas were used to prepare this adsorbent nanoparticle. The regeneration capacity of this nanoclay adsorbent was good and it can be reused atleast five times without lossing its adsorption capacity.	[99]
6	Methylene blue	Zeolite-reduced graphene oxide	Removal effieceincy was 53.3 mg per g of adsorbent per minute	This nano-clay polymer first prepared from halloysite nanotubes by mixing with graphene oxide through a hydrothermal reaction.	[100]
7	Malachite green		Removal effieceincy was 48.6 mg per g of adsorbent per minute		
8	Methylene blue and methyl orange	Montmorillonite (MT) and Layered double hydroxides	Removal efficiency of MB and methyl orange was about 74% and 88%, respectively.	This nanoclay composites were prepared through uniformly layerd clay polymers nanocomposites.	[101]
9	Safranin-T	N-vinyl-2-pyrrolidone/itaconic acid/organo clay nanocomposites hydrogels	adsorption capacity was recorded 550/50 mg/g per minute	Used clay for synthesis-Organo clay Technique of synthesis-by free radical polymerization.	[102]
10	Atrazine, an herbicide	Allophanic clays and nano-clay	93–96% of Atrazine was removed	This adsorbent can be used for the removal of other similar composition of pesticides and weedicides	[103]
11	Congo Red	Na-MT nanoclay filed hydrogel of poly acrylic acid	110/40 mg/g per minute	The composite hydrogel showed good adsorption efficiency for dyes at both low concentration from 2.5-50 mg/L, as well as at high initial concentration of 100-600 mg/L.	[104, 105]

Conclusion

This chapter shed light into the utility of nano-clay composites in removal of organic and inorganic pollutants. Nano clay particles can be derived from several classes of clay such as Smectite, Montmorilonite, Bentonite, Kaolinite, Illite, Sepiolite etc. The nano- clay composites are observed to be efficient in removing dyes- Bromophenol blue dye, Brilliant

Green, Crystal violet, Methylene blue (MB), Malachite green, Methylene blue and methyl orange, Safranin-T. It is also found to be efficient in removing other organic pollutants from wastewater such as Phosphorus, Atrazine (herbicide). Heavy metal pollutants such as Trivalent chromium, Hexavalent Chromium, Cobalt and Nickel can also be successfully removed from the effluent.

The wide presence of clay minerals, its cost-effective polymerization process makes this material the next most important candidate for future water treatment process development.

References

[1] R.A. Mandour, Human health impacts of drinking water (surface and ground) pollution Dakahlyia Governorate, Egypt, Appl. Water Sci. 2 (2012) 157-163. https://doi.org/10.1007/s13201-012-0041-6

[2] A. Gupta, E. Ruebush, Aquasight: Automatic water impurity detection utilizing convolutional neural networks. arXiv preprint arXiv:1907 (2019) 07573.

[3] S. Wasi, S.Tabrez, M. Ahmad, Toxicological effects of major environmental pollutants: an overview, Environ. Monit. Assess.185 (2013) 2585-2593. https://doi.org/10.1007/s10661-012-2732-8

[4] A.B.W.S. Casariego, B.W.S. Souza, M.A. Cerqueira, J.A. Teixeira, L.Cruz, R. Díaz, A.A. Vicente, Chitosan/clay films' properties as affected by biopolymer and clay micro/nanoparticles' concentrations, Food Hydrocoll. 23 (2009) 1895-1902. https://doi.org/10.1016/j.foodhyd.2009.02.007

[5] M.S. Nazir, M.H.M. Kassim, L. Mohapatra, M.A. Gilani, M.R. Raza, K. Majeed, Characteristic properties of nanoclays and characterization of nanoparticulates and nanocomposites. In Nanoclay reinforced polymer composites, Springer, Singapore, 2016, pp. 35-55. https://doi.org/10.1007/978-981-10-1953-1_2

[6] F. Guo, S.Aryana, Y.Han, Y. Jiao, A review of the synthesis and applications of polymer-nanoclay composites, Appl. Sci.8 (2018) 1696. https://doi.org/10.3390/app8091696

[7] G. Rytwo, Clay minerals as an ancient nanotechnology: historical uses of clay organic interactions, and future possible perspectives, Macla 9, (2008). 15-17.

[8] S.M. Lee, D. Tiwari, Organo and inorgano-organo-modified clays in the remediation of aqueous solutions:An overview, Appl. Clay Sci., 59 (2012) 84-102. https://doi.org/10.1016/j.clay.2012.02.006

[9] M.K. Uddin, A review on the adsorption of heavy metals by clay minerals, with special focus on the past decade, Chem. Eng. J. 308 (2017) 438-462. https://doi.org/10.1016/j.cej.2016.09.029

[10] I. Savic, S.I.S.I. Stojiljkovic, I. Savic, D. Gajic, Industrial application of clays and clay minerals, Clays and Clay Minerals: Geological Origin, Mechanical Properties and Industrial Applications; Wesley, LR, Ed, (2014) 379-402.

[11] Y. Shi, Z. Xue, X. Wang, L.Wang, A.Wang, Removal of methylene blue from aqueous solution by sorption on lignocellulose-g-poly (acrylic acid)/montmorillonite three-dimensional cross-linked polymeric network hydrogels. Polym. Bull.70 (2013)1163-1179. https://doi.org/10.1007/s00289-012-0898-4

[12] L. Wang, J.P. Zhang, A. Q. Wang Removal of methylene blue from aqueous solution using chitosan-g-poly(acrylic acid)/montmorillonite super adsorbent nanocomposite. Colloids Surf. A. 322 (2008) 47-53. https://doi.org/10.1016/j.colsurfa.2008.02.019

[13] T. Zhang, W. Wang, Y. Zhao, H. Bai, T. Wen, S. Kang, , ... S. Komarneni, Removal of heavy metals and dyes by clay-based adsorbents: From natural clays to 1D and 2D nano-composites, Chem. Eng. J. (2020) 127574. https://doi.org/10.1016/j.cej.2020.127574

[14] J.T. Kloprogge, S. Komarneni, J.E. Amonette,. Synthesis of smectite clay minerals: a critical review, Clays Clay Miner. 47 (1999) 529-554. https://doi.org/10.1346/CCMN.1999.0470501

[15] R.A. Schoonheydt, Smectite-type clay minerals as nanomaterials. Clays Clay Miner. 50 (2002) 411-420. https://doi.org/10.1346/000986002320514136

[16] M. Valášková, Clays, clay minerals and cordierite ceramics-A review, (2015).

[17] DD. Eisenhour, R.K. Brown, Bentonite and its impact on modern life. Elements, 5 (2009) 83-88. https://doi.org/10.2113/gselements.5.2.83

[18] F. A. Banat, B. Al-Bashir, S. Al-Asheh, O. Hayajneh, Adsorption of phenol by bentonite, Environ. Pollut. 107 (2000) 391-398. https://doi.org/10.1016/S0269-7491(99)00173-6

[19] L.M. Pandey, Enhanced adsorption capacity of designed bentonite and alginate beads for the effective removal of methylene blue, Appl. Clay Sci. 169 (2019) 102-111. https://doi.org/10.1016/j.clay.2018.12.019

[20] H. Zhang, Z. Tong, T. Wei, Y. Tang,. Removal characteristics of Zn (II) from aqueous solution by alkaline Ca-bentonite, Desalination. 276 (2011)103-108. https://doi.org/10.1016/j.desal.2011.03.026

[21] R. Ruiz, C. Blanco, C. Pesquera, F. Gonzalez, I. Benito, J. L. Lopez, Zeolitization of a bentonite and its application to the removal of ammonium ion from waste water. Appl. Clay Sci. 12 (1997) 73-83. https://doi.org/10.1016/S0169-1317(96)00038-5

[22] W.S. Tan, A.S.Y. Ting, Alginate-immobilized bentonite clay: adsorption efficacy and reusability for Cu (II) removal from aqueous solution, Bioresour.Technol. 160 (2014) 115-118. https://doi.org/10.1016/j.biortech.2013.12.056

[23] D.A. Glatstein, F.M. Francisca, Influence of pH and ionic strength on Cd, Cu and Pb removal from water by adsorption in Na-bentonite, Appl. Clay Sci. 118 (2015) 61-67. https://doi.org/10.1016/j.clay.2015.09.003

[24] V.H. Smith, D.W. Schindler, Eutrophication science: where do we go from here? Trends Ecol. Evol. 24 (2009) 201-207. https://doi.org/10.1016/j.tree.2008.11.009

[25] M. Zamparas, M. Drosos, Y. Georgiou, Y. Deligiannakis, I. Zacharias, A novel bentonite-humic acid composite material Bephos™ for removal of phosphate and ammonium from eutrophic waters, Chem. Eng. J. 225 (2013). 43-51. https://doi.org/10.1016/j.cej.2013.03.064

[26] K.G. Bhattacharyya, S. S. Gupta, Adsorption of a few heavy metals on natural and modified kaolinite and montmorillonite: a review, Adv.Colloid Interface Sci. 140 (2008) 114-131. https://doi.org/10.1016/j.cis.2007.12.008

[27] Ö. Yavuz, Y. Altunkaynak, F. Güzel, Removal of copper, nickel, cobalt and manganese from aqueous solution by kaolinite, Water Res. 37 (2003) 948-952. https://doi.org/10.1016/S0043-1354(02)00409-8

[28] U.F. Alkaram, A. A. Mukhlis, A. H. Al-Dujaili, The removal of phenol from aqueous solutions by adsorption using surfactant-modified bentonite and kaolinite, J. Hazard. Mater. 169 (2009) 324-332. https://doi.org/10.1016/j.jhazmat.2009.03.153

[29] Q. Zhang, Z. Yan, J. Ouyang, Y. Zhang, H. Yang, D. Chen, Chemically modified kaolinite nanolayers for the removal of organic pollutants. Appl. Clay Sci. 157 (2018) 283-290. https://doi.org/10.1016/j.clay.2018.03.009

[30] N. Kumari, C. Mohan, Basics of clay minerals and their characteristic properties. (2021) https://doi.org/10.5772/intechopen.97672

[31] L. Chen, Y. Zhao, H.Bai, Z. Ai, P. Chen, Y. Hu,........ S. Komarneni, Role of Montmorillonite, Kaolinite, or Illite in Pyrite Flotation: Differences in Clay Behavior

Based on Their Structures, Langmuir. 36 (2020) 10860-10867. https://doi.org/10.1021/acs.langmuir.0c02073

[32] R. Zhen, Y.S. Jiang, F.F. Li, B. Xue, A study on the intercalation and exfoliation of illite, Res.Chem.Intermed. 43 (2017) 679-692. https://doi.org/10.1007/s11164-016-2645-1

[33] A.U. Gehring, P. Keller, B. Frey, J. Luster, The occurrence of spherical morphology as evidence for changing conditions during the genesis of a sepiolite deposit, Clay Miner. 30 (1995) 83-86. https://doi.org/10.1180/claymin.1995.030.1.10

[34] J. Abdo, H. AL-Sharji, E. Hassan, Effects of nano-sepiolite on rheological properties and filtration loss of water-based drilling fluids, Surf Interface Anal., 48 (2016) 522-526. https://doi.org/10.1002/sia.5997

[35] A. Esteban-Cubillo, R. Pina-Zapardiel, J. S., Moya, M. F., Barba, C. Pecharromán, The role of magnesium on the stability of crystalline sepiolite structure, J. Eur. Ceram. Soc.28 (2008)1763-1768. https://doi.org/10.1016/j.jeurceramsoc.2007.11.022

[36] A. Kara, N. Tekin, A. Alan, A. Şafaklı, Physicochemical parameters of Hg (II) ions adsorption from aqueous solution by sepiolite/poly (vinylimidazole), J. Environ. Chem. Eng. 4 (2016) 1642-1652. https://doi.org/10.1016/j.jece.2016.02.028

[37] F.N. Oskui, H. Aghdasinia, M.G. Sorkhabi, Adsorption of Cr (III) using an Iranian natural nanoclay: applicable to tannery wastewater: equilibrium, kinetic, and thermodynamic, Environ. Earth Sci.78 (2019) 106. https://doi.org/10.1007/s12665-019-8104-8

[38] R. Mukhopadhyay, N. De, Nano clay polymer composite: synthesis, characterization, properties and application in rainfed agriculture. Glob. J. Biosci. Biotechnol. 3 (2014) 133-138.

[39] S. Ghodke, S. Sonawane, R. Gaikawad, K.C. Mohite, TIO2/Nanoclay nanocomposite for phenol degradation in sonophotocatalytic reactor. Can J Chem Eng. 90 (2012) 1153-1159. https://doi.org/10.1002/cjce.20630

[40] J. Weiss, P. Takhistov, D.J. McClements, Functional materials in food nanotechnology, J. Food Sci. 71 (2006). R107. https://doi.org/10.1111/j.1750-3841.2006.00195.x

[41] A.Wong, S.F., Wijnands, T. Kuboki, C.B. Park, Mechanisms of nanoclay-enhanced plastic foaming processes: Effects of nanoclay intercalation and exfoliation, J. Nanoparticle Res. 15 (2013) 1-15. https://doi.org/10.1007/s11051-013-1815-y

[42] R. Mukhopadhyay, D. Bhaduri, B. Sarkar, R. Rusmin, D. Hou, R. Khanam, , ... & Y. S. Ok, Clay-polymer nanocomposites: Progress and challenges for use in sustainable water treatment. J. Hazard, Mater. 383 (2020) 121125. https://doi.org/10.1016/j.jhazmat.2019.121125

[43] T. Robinson, G. McMullan, R. Marchant, P. Nigam, Remediation of dyes in textile effluent: a critical review on current treatment technologies with a proposed alternative, Bioresour. Technol.77 (2001) 247-255. https://doi.org/10.1016/S0960-8524(00)00080-8

[44] S. Khan, A. Malik, Environmental and health effects of textile industry wastewater. In Environmental deterioration and human health, Springer, Dordrecht (2014). pp. 55-71. https://doi.org/10.1007/978-94-007-7890-0_4

[45] M. Jaishankar, T. Tseten, N. Anbalagan, B. B. Mathew, K. N. Beeregowda, Toxicity, mechanism and health effects of some heavy metals, Interdiscip. Toxicol. 7 (2014) 60. https://doi.org/10.2478/intox-2014-0009

[46] L.C. Castillo-Carvajal, J.L. Sanz-Martín, B.E. Barragán-Huerta, Biodegradation of organic pollutants in saline wastewater by halophilic microorganisms: a review, Environ. Sci. Pollut. 21 (2014) 9578-9588. https://doi.org/10.1007/s11356-014-3036-z

[47] A. Amari, F.M. Alzahrani, K. Mohammedsaleh Katubi, N. S. Alsaiari, M.A. Tahoon, F.B. Rebah, Clay-polymer nanocomposites: Preparations and utilization for pollutants removal, Materials. 14 (2021) 1365. https://doi.org/10.3390/ma14061365

[48] M. Mohapi, J. S. Sefadi, M. J. Mochane, S. I. Magagula, K. Lebelo, Effect of LDHs and Other Clays on Polymer Composite in Adsorptive Removal of Contaminants: A Review, 10 (2020) 957. https://doi.org/10.3390/cryst10110957

[49] R. Gahlot, K. Taki, M. Kumar, Efficacy of nanoclays as the potential adsorbent for dyes and metal removal from the wastewater: a review, Environ. Nanotechnol. Monit. Manag. 14 (2020) 100339. https://doi.org/10.1016/j.enmm.2020.100339

[50] K. Ravikumar, J. Udayakumar, Preparation and characterisation of green clay-polymer nanocomposite for heavy metals removal, Chem Ecol. 36 (2020) 270-291. https://doi.org/10.1080/02757540.2020.1723559

[51] K.D. Nguyen, T.T.C. Trang, T. Kobayashi, Chitin-halloysite nanoclay hydrogel composite adsorbent to aqueous heavy metal ions, J. Appl. Polym. Sci. 136 (2019) 47207. https://doi.org/10.1002/app.47207

[52] H. Sharififard, M. Ghorbanpour, S. Hosseinirad, Cadmium removal from wastewater using nano-clay/TiO2 composite: kinetics, equilibrium and thermodynamic study, Adv. Environ. Techn. 4 (2018) 203-209.

[53] U. Malayoglu, Removal of heavy metals by biopolymer (chitosan)/nanoclay composites, Sep. Sci. Techno. l53 (2018) 2741-2749. https://doi.org/10.1080/01496395.2018.1471506

[54] J. Podgorski, M. Berg, Global threat of arsenic in groundwater, Science 368 (2020) 45-850. https://doi.org/10.1126/science.aba1510

[55] E. Shaji, M. Santosh, K.V. Sarath, P. Prakash, V. Deepchand, B.V. Divya, Arsenic contamination of groundwater: A global synopsis with focus on the Indian Peninsula, Geosci. Front.12 (2020) 101079. https://doi.org/10.1016/j.gsf.2020.08.015

[56] L. Weerasundara, Y.S. Ok, J. Bundschuh, Selective removal of arsenic in water: A critical review, Environ. Pollut. 268 (2021) 115668. https://doi.org/10.1016/j.envpol.2020.115668

[57] D.A. Almasri, T. Rhadfi, M.A. Atieh, G. McKay, S. Ahzi, High performance hydroxyiron modified montmorillonite nanoclay adsorbent for arsenite removal, Chem. Eng. Sci. 335 (2018) 1-12. https://doi.org/10.1016/j.cej.2017.10.031

[58] R. Mukhopadhyay, K.M. Manjaiah, S.C. Datta, B. Sarkar, Comparison of properties and aquatic arsenic removal potentials of organically modified smectite adsorbents, J. Hazard. Mater. 377 (2019) 124-131. https://doi.org/10.1016/j.jhazmat.2019.05.053

[59] E.M.B. Dela Peña, K. Araño, M.L. Dela Cruz, P.A. de Yro, L.J.L. Diaz, The design of a bench-scale adsorbent column based on nanoclay-loaded electrospun fiber membrane for the removal of arsenic in wastewater, Water. Environ. J . (2021) https://doi.org/10.1111/wej.12683

[60] J.J. Coetzee, N. Bansal, E.M. Chirwa, Chromium in environment, its toxic effect from chromite-mining and ferrochrome industries, and its possible bioremediation, Expos. Health. 12 (2020) 51-62. https://doi.org/10.1007/s12403-018-0284-z

[61] A. Naz, B. K. Mishra, S. K. Gupta, Human health risk assessment of chromium in drinking water: a case study of Sukinda chromite mine, Odisha, India, Expos. Health. 8 (2016) 253-264. https://doi.org/10.1007/s12403-016-0199-5

[62] A. Naz, A. Chowdhury, B. K. Mishra, S. K. Gupta, Metal pollution in water environment and the associated human health risk from drinking water: A case study of Sukinda chromite mine, India, Hum. Ecol. Risk. Assess. 22 (2016) 1433-1455. https://doi.org/10.1080/10807039.2016.1185355

[63] A. Chowdhury, A. Naz, S.K. Maiti, Bioaccumulation of potentially toxic elements in three mangrove species and human health risk due to their ethnobotanical uses, Environ. Sci. Pollut. Res. 28 (2021) 33042-33059. https://doi.org/10.1007/s11356-021-12566-w

[64] L. Jacob, S. Joseph, L.A. Varghese, Polysulfone/MMT mixed matrix membranes for hexavalent chromium removal from wastewater, Arab. J. Sci. Engg. 45 (2020) 7611-7620. https://doi.org/10.1007/s13369-020-04711-3

[65] N. Abdullah, N. Yusof, W.J., Lau, J. Jaafar, A.F. Ismail, Recent trends of heavy metal removal from water/wastewater by membrane technologies. J. Ind. Eng. Chem. 76 (2019) 17-38. https://doi.org/10.1016/j.jiec.2019.03.029

[66] J. Cloern, T. Krantz, J.E. Duffy Eutrophication. Encyclopedia of Earth. Washington DC, USA: Environmental Information Coalition, National Council for Science and the Environment. Retrieved from http://www. eoearth. org/article/Eutrophication. (2007)

[67] J. Luo, R. Pu, H. Duan, R. Ma, Z. Mao, Y. Zeng,..... Q. Xiao, Evaluating the influences of harvesting activity and eutrophication on loss of aquatic vegetations in Taihu Lake, China, Int. J. Appl. Earth Obs. Geoinf. 87 (2020) 102038. https://doi.org/10.1016/j.jag.2019.102038

[68] G. Yuan, L. Wu, , Allophane nanoclay for the removal of phosphorus in water and wastewater. Science and Technology of Advanced Materials, 8 (2007), 60. https://doi.org/10.1016/j.stam.2006.09.002

[69] P.V. Haseena, K. S,.Padmavathy, P.R. Krishnan, G. Madhu, Adsorption of ammonium nitrogen from aqueous systems using chitosan-bentonite film composite, Procedia Technol. 24 (2016) 733-740. https://doi.org/10.1016/j.protcy.2016.05.203

[70] F. Mazloomi, M. Jalali, Adsorption of ammonium from simulated wastewater by montmorillonite nanoclay and natural vermiculite: experimental study and simulation, Environ. Monit. Assess.189 (2017) 1-19. https://doi.org/10.1007/s10661-017-6080-6

[71] Z. Carmen, S. Daniela, Textile organic dyes-characteristics, polluting effects and separation/elimination procedures from industrial effluents-a critical overview Rijeka: IntechOpen (2012) pp. 55-86.. https://doi.org/10.5772/32373

[72] M. Ismail, K. Akhtar, M. I. Khan, T. Kamal, M. A Khan,....... S.B. Khan, Pollution, toxicity and carcinogenicity of organic dyes and their catalytic bio-remediation. Curr. Pharm. Des. 25 (2019) 3645-3663. https://doi.org/10.2174/1381612825666191021142026

[73] A. Rana, K. Qanungo,. Orange G dye removal from aqueous-solution using various adsorbents: A mini review (2021) Materials Today: Proceedings. https://doi.org/10.1016/j.matpr.2021.04.230

[74] P. Daraei, S.S. Madaeni, E. Salehi, N. Ghaemi, H.S. Ghari, M.A. Khadivi, E. Rostami, Novel thin film composite membrane fabricated by mixed matrix nanoclay/chitosan on PVDF microfiltration support: Preparation, characterization and performance in dye removal, J. Membr. Sci. 436 (2013) 97-108. https://doi.org/10.1016/j.memsci.2013.02.031

[75] G.R. Mahdavinia, A. Baghban, S. Zorofi, A. Massoudi, Kappa-carrageenan biopolymer-based nanocomposite hydrogel and adsorption of methylene blue cationic dye from water, J. Mater. Environ. Sci. 5 (2014) 330-337.

[76] S. Radoor, J. Karayil, J. Parameswaranpillai, S. Siengchin, Adsorption of methylene blue dye from aqueous solution by a novel PVA/CMC/halloysite nanoclay bio composite: Characterization, kinetics, isotherm and antibacterial properties, J. Environ. Health sci. 18 (2020) 1311-1327. https://doi.org/10.1007/s40201-020-00549-x

[77] M.M. Tarekegn, R.M., Balakrishnan, A.M. Hiruy, A.H. Dekebo, Removal of methylene blue dye using nano zerovalent iron, nanoclay and iron impregnated nanoclay-a comparative study, RSC Advances. 11 (2021) 30109-30131. https://doi.org/10.1039/D1RA03918K

[78] M. Andrade-Guel, C. Cabello-Alvarado, R.L. Romero-Huitzil, O.S. Rodríguez-Fernández, C. A. Ávila-Orta, G. Cadenas-Pliego..... J. Cepeda-Garza, Nanocomposite PLA/C20A Nanoclay by Ultrasound-Assisted Melt Extrusion for Adsorption of Uremic Toxins and Methylene Blue Dye, Nanomaterials, 11 (2021) 2477. https://doi.org/10.3390/nano11102477

[79] M.A. Salam, S.A. Kosa, A.A. Al-Beladi, Application of nanoclay for the adsorptive removal of Orange G dye from aqueous solution, J. Mol. Liq. 241 (2017) 469-477. https://doi.org/10.1016/j.molliq.2017.06.055

[80] A. A. Al-Beladia, S.A. Kosaa, R.A. Wahabb, M.A. Salama, Removal of Orange G dye from water using halloysite nanoclay-supported ZnO nanoparticles, Desalination Water Treat. 196 (2020) 287-298. https://doi.org/10.5004/dwt.2020.25923

[81] C.W. Pai, D. Leong, C.Y. Chen, G.S. Wang, Occurrences of pharmaceuticals and personal care products in the drinking water of Taiwan and their removal in conventional water treatment processes, Chemosphere. 256 (2020) 127002. https://doi.org/10.1016/j.chemosphere.2020.127002

[82] S.A. Kraemer, A. Ramachandran, G.G. Perron, Antibiotic pollution in the environment: from microbial ecology to public policy, Microorganisms, 7 (2019) 180. https://doi.org/10.3390/microorganisms7060180

[83] S. Rodriguez-Mozaz, I. Vaz-Moreira, S.V., Della Giustina, M. Llorca, D.Barceló, , S. Schubert, ... C.M. Manaia, Antibiotic residues in final effluents of European wastewater treatment plants and their impact on the aquatic environment, Environ. Int. 140 (2020) 105733. https://doi.org/10.1016/j.envint.2020.105733

[84] M. Cerro-Lopez, M.A. Méndez-Rojas, Application of nanomaterials for treatment of wastewater containing pharmaceuticals. In Ecopharmacovigilance. Springer, Cham. 2017 pp. 201-219. https://doi.org/10.1007/698_2017_143

[85] D. A. Almasri, T. Rhadfi, M. A. Atieh, G. McKay, S. Ahzi, High performance hydroxyiron modified montmorillonite nanoclay adsorbent for arsenite removal. Chem Eng J. 335 (2018) 1-12. https://doi.org/10.1016/j.cej.2017.10.031

[86] W. Liu, C. Zhao, S.Wang, L. Niu, Y. Wang, S. Liang, Z. Cui, Adsorption of cadmium ions from aqueous solutions using nano-montmorillonite: kinetics, isotherm and mechanism evaluations. Research on Chemical Intermediates, 44 (2018) 1441-1458. https://doi.org/10.1007/s11164-017-3178-y

[87] E. I. Unuabonah, B. I. Olu-Owolabi, E. I. Fasuyi, K. O. Adebowale, Modeling of fixed-bed column studies for the adsorption of cadmium onto novel polymer-clay composite adsorbent. J. Hazard. Mater. 179 (2010) 415-423. https://doi.org/10.1016/j.jhazmat.2010.03.020

[88] K. Moreno-Sader, A. García-Padilla, A. Realpe, M. Acevedo-Morantes, J. B.Soares, Removal of heavy metal water pollutants (Co2+ and Ni2+) using polyacrylamide/sodium montmorillonite (PAM/Na-MMT) nanocomposites, ACS Omega. 4 (2019) 10834-10844. https://doi.org/10.1021/acsomega.9b00981

[89] M. Soleimani, Z. H. Siahpoosh, Determination of Cu (II) in water and food samples by Na+-cloisite nanoclay as a new adsorbent: Equilibrium, kinetic and thermodynamic studies. J. Taiwan Inst. Chem. Eng. 59 (2016) 413-423. https://doi.org/10.1016/j.jtice.2015.09.009

[90] A.S.K. Kumar, S. Kalidhasan, V. Rajesh, N. Rajesh, Application of cellulose-clay composite biosorbent toward the effective adsorption and removal of chromium from industrial wastewater, Ind. Eng. Chem. Res. 51 (2012) 58-69. https://doi.org/10.1021/ie201349h

[91] E. I. Unuabonah, B. I. Olu-Owolabi, K. O. Adebowale, L. Z. Yang, Removal of lead and cadmium ions from aqueous solution by polyvinyl alcohol-modified kaolinite clay: a novel nano-clay adsorbent. Adsorpt. Sci. Technol. 26 (2008) 383-405. https://doi.org/10.1260/0263-6174.26.6.383

[92] S. Elhami, S. Shafizadeh, Removal of Mercury (II) using modified Nanoclay. Mater. Today: Proc., 3 (2016) 2623-2627. https://doi.org/10.1016/j.matpr.2016.06.005

[93] P.V. Haseena, K.S., Padmavathy, P.R., Krishnan, G. Madhu, Adsorption of ammonium nitrogen from aqueous systems using chitosan-bentonite film composite. Procedia Technol. 24 (2016) 733-740. https://doi.org/10.1016/j.protcy.2016.05.203

[94] W. Chen, H.C. Liu, Adsorption of sulfate in aqueous solutions by organo-nano-clay: Adsorption equilibrium and kinetic studies. Journal of Central South University, 21 (2014) 1974-1981. https://doi.org/10.1007/s11771-014-2145-7

[95] A.A. El-Zahhar, N.S. Awwad, E.E. El-Katori, Removal of bromophenol blue dye from industrial waste water by synthesizing polymer-clay composite, J. Mol. Liq. 199 (2014) 454-461. https://doi.org/10.1016/j.molliq.2014.07.034

[96] S.R., Shirsath, A.P. Patil, R. Patil, J.B. Naik, P.R. Gogate, S.H. Sonawane, Removal of Brilliant Green from wastewater using conventional and ultrasonically prepared poly (acrylic acid) hydrogel loaded with kaolin clay: a comparative study, Ultrason Sonochem 20 (2013) 914-923. https://doi.org/10.1016/j.ultsonch.2012.11.010

[97] H. Hosseinzadeh, S. Zoroufi, G.R. Mahdavinia, Study on adsorption of cationic dye on novel kappa-carrageenan/poly (vinyl alcohol)/montmorillonite nanocomposite hydrogels, Polym. Bull. 72 (2015)1339-1363. https://doi.org/10.1007/s00289-015-1340-5

[98] R. Xu, J. Mao, N. Peng, X. Luo, C. Chang, Chitin/clay microspheres with hierarchical architecture for highly efficient removal of organic dyes, Carbohydrate polymers, 188 (2018)143-150. https://doi.org/10.1016/j.carbpol.2018.01.073

[99] Q. Zhou, Q.Gao, W. Luo, C. Yan, Z. Ji, P. Duan, One-step synthesis of amino-functionalized attapulgite clay nanoparticles adsorbent by hydrothermal carbonization of chitosan for removal of methylene blue from wastewater, Colloids Surf, A Physicochem Eng Asp. 470 (2015) 248-257. https://doi.org/10.1016/j.colsurfa.2015.01.092

[100] J. Zhu, Y. Wang, J. Liu, Y. Zhang, Facile one-pot synthesis of novel spherical Zeoliter GO composites for cationic dyes adsorption. Ind. Eng. Chem. Res. 53 (2014) 13711-13717. https://doi.org/10.1021/ie502030w

[101] K. Zhou, Q. Zhang, B.Wang, J. Liu, P. Wen, Z. Gui, Y. Hu, The integrated utilization of typical clays in removal of organic dyes and polymer nanocomposites, J. Clean. Prod. 81 (2014) 281-289. https://doi.org/10.1016/j.jclepro.2014.06.038

[102] G. Çöle, M.K. Gök, G. Güçlü, Removal of basic dye from aqueous solutions using a novel nanocomposite hydrogel: N-vinyl 2-pyrrolidone/itaconic acid/organo clay, Water Air Soil Pollut. 224 (2013) 1-16. https://doi.org/10.1007/s11270-013-1760-5

[103] M. Cea, P. Cartes, G. Palma, M.L. Mora, Atrazine efficiency in an andisol as affected by clays and nanoclays in ethylcellulose controlled release formulations, Revista de la ciencia del suelo y nutrición vegetal, 10 (2010) 62-77. https://doi.org/10.4067/S0718-27912010000100007

[104] R. Bhattacharyya, S. K. Ray Removal of congo red and methyl violet from water using nano clay filled composite hydrogels of poly acrylic acid and polyethylene glycol, Chem Eng J. 260 (2015) 269-283. https://doi.org/10.1016/j.cej.2014.08.030

[105] A. Naz, A. Chowdhury Pollutant extraction from water and soil using Montmorillonite clay-polymer composite: A rapid review, Mater. Today: Proc. (2021) https://doi.org/10.1016/j.matpr.2021.10.366

Adv. App. of Micro and Nano Clay II – Synthetic Polymer Composites   Materials Research Forum LLC
Materials Research Foundations 129 (2022) 153-182   https://doi.org/10.21741/9781644902035-7

Chapter 7

Nontronite-Starch based Nano-Composites and Applications

A. Bajpai*, M. Markam, V. Raj

Department of Chemistry, Government Science College, Jabalpur 482001, India

Email (abs_112@rediffmail.com*, mona.markam@gmail.com, vaniraj654@gmail.com)

Abstract

Polymer nano-composites consist of two or more constituents with at least one being of nanoscale dimension. Being naturally abundant, affordable, non-toxic and biocompatible, clay-based minerals and biopolymers are advantageous to afford eco-friendly nanocomposites, especially useful for biological applications. Starch, a common polysaccharide, finds traditional use in the food industry, and has recently become relevant in several advanced technologies. Nontronite, an iron rich smectite clay, still remains underexplored in the context of nanocomposite preparation. This book chapter attempts to provide a brief overview of syntheses and applications of nanocomposites based on nontronite, starch and polysaccharide-clay.

Keywords

Starch, Nontronite, Biopolymers, Smectite, Polymer Nano-Composites, Bionanocomposites

Contents

1.   Introduction

The industrial revolution has been a major contributor for the discovery of fossil fuel-derived synthetic polymers that have popularly come to be known as plastics. However, their non-biodegradable nature is now one of the serious concerns. This has motivated advances in industrial chemistry research that aims at synthesis of efficient, multifunctional materials for technologically and commercially relevant applications [1, 2]. Bioplastics, being derived from biobased renewable resources and biodegradable, are a promising alternative to plastics. However, bioplastics improvements must be made to achieve properties/performance comparable to traditional plastics. One of the strategies is through their modification into composites. Polymer composites comprise two or more constituents and typically involve a polymer matrix and a filler to enhance mechanical strength thereby improving the resultant properties when compared to the individual constituents. It is found that addition of nanoparticles (NPs) can greatly improve composite properties. These composites are termed as nano-composites as at least one of the components has a particle size in nano dimension. Recently, polymer nano-composites have gained impetus because of their innovative and exclusive properties, e.g., gas barrier performance, high dimensional stability, optical clarity, reduced gas permeability, heat deflection temperature, and flame retardancy [3-16].

Several clays have been explored for their ability to form high-performing polymer nano-composites. This is because clay-based minerals are advantageous as supporting materials due to their natural abundance, low cost, non-toxic, biocompatible and eco-friendly characteristics. Phyllosilicates are a common class of layered silicates and have been studied for the preparation of polymer nano-composites [17]. Layered silicates and biodegradable polymers afford environmentally friendly nanocomposites [19], and exhibit significantly enhanced properties, such as high moduli, low gas permeation and flammability, enhanced strength, thermal resistance and biodegradability, as compared to the individual constituents, [4, 19-22].

Polysaccharides are structural biopolymers and energy store houses for microorganisms, animals, and plants, and are naturally abundant and inexpensive [23]. Compared to other biopolymers, they are found to possess greater thermal stability, namely, polynucleotides and proteins [24-26]. Polysaccharides can have linear chain or branched structures, varied polydispersity, and diverse molecular weights. Presence of one or more types of functional groups such as carboxyl, hydroxyl and amine offers possibility for easy chemical modification. They have high levels of chirality and may or may not be soluble in water. Further, they are biodegradable, biocompatible, environmentally friendly and nonimmunogenic [27]. Such a great variety of molecular structures and discrete properties offers a wide scope to afford highly-functional nanocomposites. Indeed, a number of reviews and book chapters detail the progress in the field of polysaccharide-based nanocomposites. Some recent reviews focus on the use of modified polysaccharides in applications such as adsorbents for remediation of heavy metals and dyes [28], wastewater remediation [29], soil conditioners and fertilizer carriers [30], biomedical applications [31-39], edible and biodegradable films/coatings for food packaging [40-42]. One of the most common polysaccharides, starch, has conventional use in the food industry. Advanced technologies have explored its applications in agriculture, health, medicine, paper, textile, fine chemicals, petroleum and construction engineering sectors [43, 44].

The objective of this book chapter is to review recent literature describing the studies and applications of polysaccharide-clay based nano-composites and nontronite-starch based nanocomposites.

2.   Polymer nano-composites (PNCs)

PNCs are gaining scientific and industrial impetus due to several advantages when compared to conventional materials. PNCs have higher strength-to-weight ratio, products are customizable, corrosion/erosion resistant and manufacturing processes are flexible with lower cost. PNCs are more resistant to thermal degradation as compared to biopolymers [45]. They exhibit improved strength, flame retardation, and mechanical, barrier, electrical and optical properties [46-50]. PNCs have been developed as superabsorbent nano-composites for removal of dyes [51-53] and heavy metal ions [54, 55], for enhanced oil recovery [56]. PNC films find application as smart and active food packaging [57-60], slow-release formulations of pesticides [61], insulation material [62], active packaging [57, 58] and glucose biosensors [63].

Several methods have been reported for the preparation of PNCs. They can be prepared by the addition of NPs. However, uniform dispersal of NPs cannot be attained because of aggregation on mixing in the polymer matrix. Despite theoretical strategies available for prevention of aggregation of NPs, often trial and error methods are employed in practice. A very recent review describes the theories to expedite the fabrication of PNCs [64]. Additive manufacturing (AM) methods permit introduction of different degrees of flexibility in PNCs that is an advantageous feature for numerous industries such as aerospace, construction, electronics, biomedical, mechanical, telecommunication, and defence. However, there are limited studies in the synthesis and fabrication of polymer composites through AM technologies [65].

2.1   Clays

Clay minerals are ubiquitous, non-toxic and inexpensive. Their unique surface properties have been advantageous in fields of construction, engineering and medicine for a long time. They are being employed as 'green' materials for environmental remediation in the 21st century due to huge surface area, acidic surface, interlayer cations, and high cation exchange capacity (CEC). Acid-activated clay minerals adsorb dyes, heavy metals, and phenolic compounds. Organoclays adsorb hydrophobic contaminants from wastewater. Chlorinated organic compounds can be removed by redox processes using structural ferric ions in nontronite. Pillared clays can remove nonbiodegradable refractory contaminants from wastewater through the catalysis of aqueous oxidative processes [66].

Natural clays are typically created from volcanic ash through *in situ* modification or from volcanic rocks through hydrothermal modification [67]. Fig. 1 and 2 show the classification of clays in general and smectite/montmorillonite clays in particular, respectively.

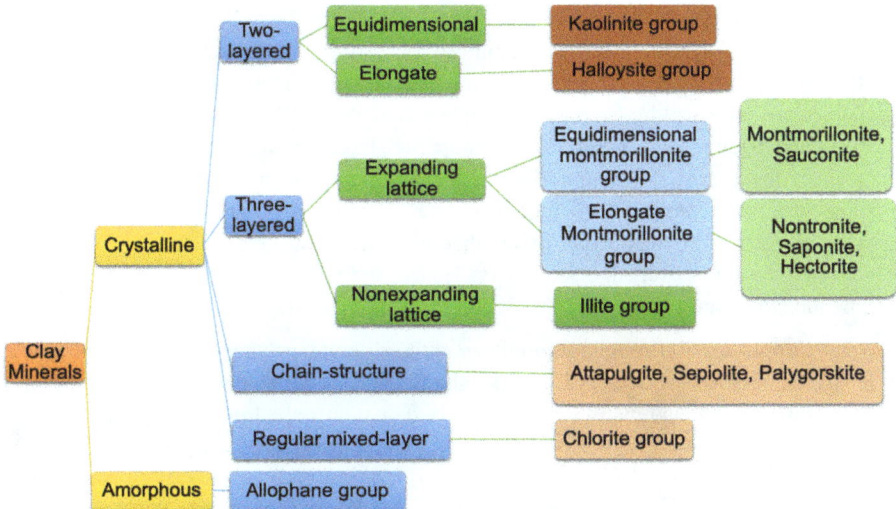

Fig. 1. General classification of clays

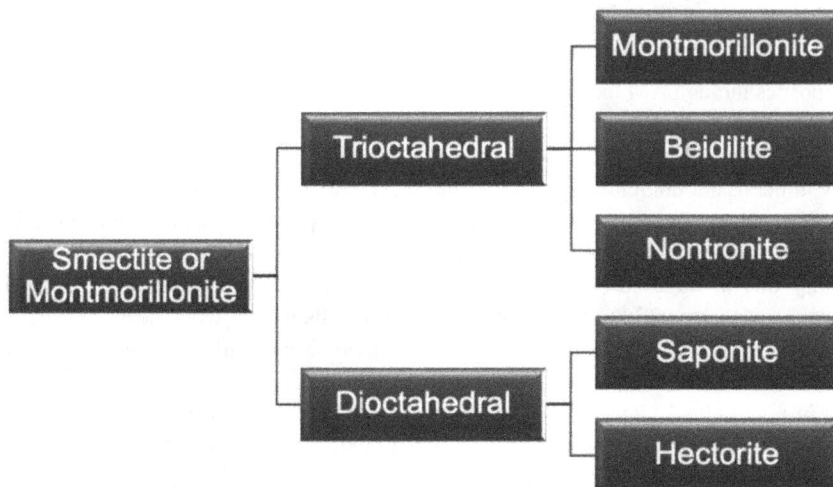

Fig. 2. Classification of smectite clays

Smectites have small particle size, good swelling and adsorption behaviour, and high CEC. They are 2:1-layer silicates - two tetrahedral sheets embed one octahedral sheet. Di- and trioctahedral smectites both are naturally occurring. Dioctahedral smectites are in abundance in nature, e.g., montmorillonite, beidellite and nontronite. Permanent negative charges are produced by isomorphous replacement of Al^{3+} by Fe^{2+} or Mg^{2+} in octahedral sheets and in tetrahedral sheets Al^{3+} substitute Si^{4+}. Exchange of cations in interlayer spaces balances the charge. In montmorillonite, replacements occur in octahedral sheets. Al^{3+} substitutes tetrahedral Si^{4+} in beidellite and Fe^{3+} substitutes it in nontronite. Magnitude of layer charge and its octahedral or tetrahedral distribution greatly affect CEC, swelling behaviour and ion exchange selectivity. Hence, the adsorption of biopolymers to form functional bionano-composites (BNCs) is affected by the clay type [68].

Nano-composites of bentonite, kaolinite, montmorillonite, palygorskite, saponite, sepiolite [69], cloisite 30B (nanoclay) [70], halloysite [71] are widely reported but to the extent of our knowledge only a few results were reported on nontronite [68, 72].

2.1.1 Nontronite

Nontronite belongs to the smectite group, having a monoclinic crystal system. Colour ranges from green, olive-green, yellow-green, yellow, orange, or brown. Lustre can be resinous, waxy, or dull. This green waxy mineral was originally named by Johann Jakob Bernhardi and Rudolph Brandes in 1822 as "chloropal" derived from "chlor" and "opal"

Adv. App. of Micro and Nano Clay II – Synthetic Polymer Composites   Materials Research Forum LLC
Materials Research Foundations 129 (2022) 153-182   https://doi.org/10.21741/9781644902035-7

to suggest resemblance with common opal. In 1827, Pierre Berthier named it after a locality believed to be in Nontron, Dordogne, France. However, the actual locality is in Le Manderau, Saint-Pardoux-la-Rivière, near Nontron. Molecular formula of nontronite is $Na_{0.3}Fe_2((Si,Al)_4O_{10})(OH)_2 \cdot nH_2O$ [73] and Ti, Mg, Ca are present as common impurities. Nontronite is usually found along with minerals such as opal, quartz, mica, hornblende, olivine, kaolinite and pyroxenes. It is a weathering product of some ultramafic igneous rocks such as basalts and kimberlites. It also occurs in hydrothermally altered mineral deposits, inadequately drained volcanic ash soils, contact metamorphosed limestones and some mid ocean ridge basalts.

Table 1.   Locations of occurrence of nontronite

S. No.	Country	Place
01	Australia	Burra copper mine, Tamworth copper mine, Attunga, New South Wales
02	France	Saint-Pardoux, Froland, Norway
03	Germany	Saxony, Geilsdorf and Wolkenstein
04	Madagascar	Faratsiho
05	Russia	Okhansk
06	Sweden	Vittensten
07	USA	California, Crestmore, North Carolina, Riverside, Spruce Pine District, Washington, Woody Kern.

Nontronites are common in terrestrial and marine environments formed from iron-rich silicates by weathering or hydrothermal alteration. They are dioctahedral smectites, where two trivalent cations (Al^{3+} or Fe^{3+}) surround each octahedral OH or O ion. Fe (ca. 30%) is present as Fe_2O_3 and Al (ca <12%) exists as Al_2O_3. The microstructure of nontronite has not been fully understood because the experimental amount of tetrahedral Fe^{3+} determined from various techniques of analysis and computations exceeds theoretical values. Nontronites are an important source of Fe. Industrial minerals nanomaghemite and nanowüstite, generally employed in biomedicines and semiconductors, are derived from natural nontronite. High-resolution SEM and TEM coupled with EDX and ESR spectroscopies show that mineral microregions containing nanomaghemite and nanowüstite consist of mixed valencies of Fe that resemble in characteristic electronic and magnetic properties of Fe^{2+}-Fe^{3+} minerals [74].

Adv. App. of Micro and Nano Clay II – Synthetic Polymer Composites   Materials Research Forum LLC
Materials Research Foundations 129 (2022) 153-182   https://doi.org/10.21741/9781644902035-7

2.2   Clay nanocomposites

Clay minerals are valuable in development of novel functional materials because their interlayer functional groups are a key factor. They offer numerous possibilities for applications as catalysts, photocatalysts, drug delivery systems, adsorbents, and nano-composites [75].

Structure of polymer clay nano-composite are of following types [76, 77], Fig. 3:

1. Phase-separated polymer clay nano-composite: the organic macromolecular chains cannot penetrate between the unmodified inorganic clay silicate layers, hence, the aggregates or stacks of clay are distributed in polymer matrix. The characteristics are not better than conventional micro-composites.

2. Intercalated polymer clay nano-composite: the organic polymer chains penetrate into the clay interlayers, alternate organic macromolecules and inorganic layers become arranged into ordered multiple layer structures with spacing of a few nanometers.

3. Exfoliated or delaminated polymer clay nano-composite: the clay layers are well dispersed and randomly oriented in a polymer matrix with more than 8.8 nm average d-spacing depending on the quantity of nanoclay loaded.

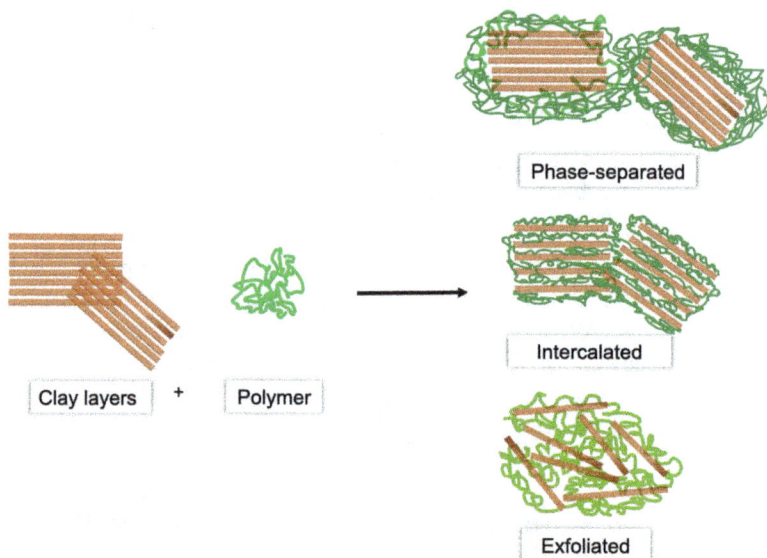

Fig. 3. Schematic representation of types of clay-polymer nanocomposite

Since the early 2000s, great advances in hybridization and modification have been observed in the development of nano-composites and hierarchical assemblies. Several strategies to introduce organic, inorganic and inorganic-organic functional entities among the clay lamellar layers have been reported [75]. Adsorption of biopolymers on the clay surfaces forms hybrid materials having properties of both components. A variety of nano-composites have been prepared from clay minerals, especially smectites, by strategic adsorption of functional biomolecules. The earliest users of nano-composites were automotive and aviation industries due to the need to reduce vehicle weight and better wear resistant parts. Thus, enabling lesser fuel consumption and exhaust gases emission, this indirectly reduces CO_2 emissions and environmental pollution [78, 79].

2.3   Polysaccharide based nano-composites

Polysaccharide-clay nano-composites are composed of inorganic clay dispersed in nanoscale, termed as nanoclay, as fillers in polysaccharide matrix. Fig. 4 depicts various applications of nanoclay-polysaccharide nanocomposites.

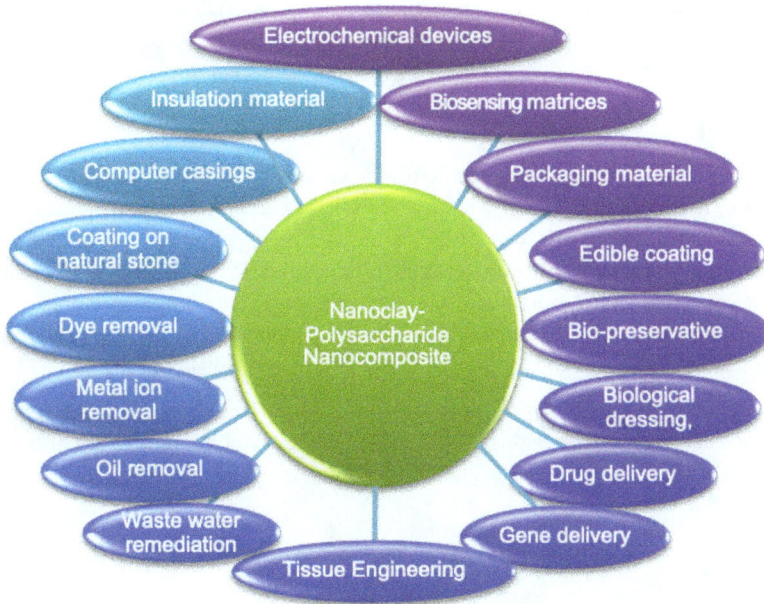

Fig. 4. Applications of nanoclay-polysaccharide nanocomposites

Recently, biopolymer-based nanocomposites, often termed as bionano-composites (BNCs), have gained importance owing to outstanding mechanical and thermal properties and possibility of biodegradation. BNCs consist of biopolymers and an inorganic component and at least one component must be of nanometer scale. Many polysaccharides can be used to prepare nanocomposites. The other components may be inorganic components, metal oxides, metal, carbon atoms, and functional materials. Though polysaccharides often are insoluble in common solvents that limits processing, some dissolution media like ionic solvents may dissolve polysaccharides [80]. Polysaccharide-nanoclay based nano-composites bestow diverse applications, like drug release in cancer therapy [81], biodegradable antimicrobial active food packaging films [82, 83], drug delivery [84] and insulating building materials [62]. They are also applied to remove harmful microbes, inorganic and organic pollutants from wastewater [85]. BNC films show decreased water uptake and water sorption ability as compared to the neat biopolymer component [48].

As compared to conventional polysaccharide films, nanobased polysaccharide packaging possesses better functional, mechanical and thermal properties besides superior gas and water barriers. However, it is imperative to consider migration safety for food applications [40].

2.3.1 Starch

Interest in starch is unsurprising as it meets sustainability requirements, i.e., natural abundance, low cost, non-toxicity, biodegradability, necessary for development of novel nano-composites. Further, it can afford cohesive films, a property critical for many practical applications. Additionally, starch is amenable to modifications in numerous different ways due to the presence of hydroxyl groups of anhydroglucose units enabling development of task-specific starch-based advanced functional nano-composite materials [86-88].

Plants produce and store carbohydrates in the form of starch as granules comprising amorphous and crystalline regions [89]. Fig. 5 shows some common sources of starch.

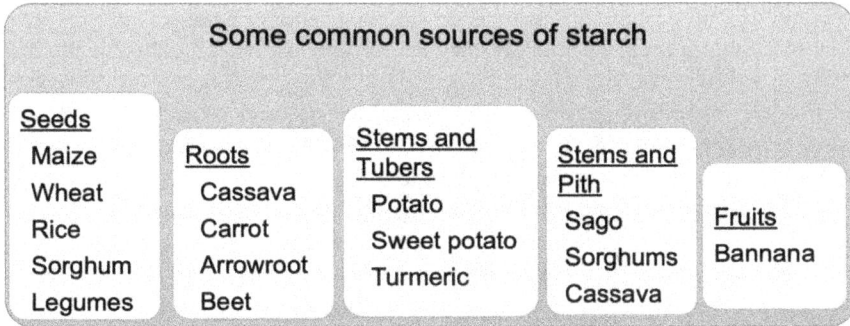

Fig. 5. Some common sources of starch.

Starch is a polysaccharide also known as amylum composed of glucose monomeric units and α-D-glucan chains of two types: linear amylose (α-1, 4-glucopyranosyl units) and amylopectin (branching at α-1,6 position) [87, 90], Fig. 6.

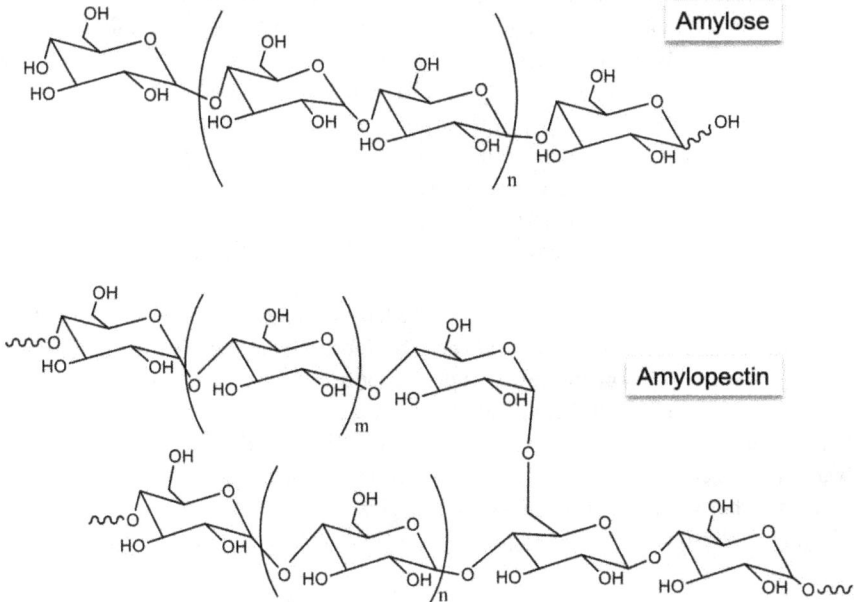

Fig. 6. Starch: chemical structure.

Adv. App. of Micro and Nano Clay II – Synthetic Polymer Composites   Materials Research Forum LLC
Materials Research Foundations 129 (2022) 153-182   https://doi.org/10.21741/9781644902035-7

Properties of starch depend on its source as each plant species produces macromolecular glucose chains of specific length and/or amylopectin:amylose ratio. Protein and fat content of storage organs is also found to vary significantly. Given this functional diversity based on starch source, researchers have a vast library of diverse starch forms to choose from for a specific application [91]. Starch films have been found to exhibit good barrier properties and better mechanical strength as compared to other biopolymer films, hence, starch is deemed a good raw material for the food packaging industry. Mechanical properties of the films have been shown to be influenced by the amylose:amylopectin ratio [92-103]. Indeed, starch has been used for the development of nano-composites for applications such as thermal insulation [104], gene delivery [105] and biomedical applications [106]. Starch-based nano-composites have also been extensively studied for packaging and pollution remediation [107].

2.3.2 Starch-based nano-composites food packaging

Continuing population growth is causing conflicts over water, energy, food, infrastructure and many other concerns. It has not only necessitated increased food production but also preservation of food for longer durations. In this context, food packaging ensures food safety and nanotechnology has made significant contributions to implement improvements in food packaging [108]. Food packaging films should be transparent, durable and offer a barrier against moisture and gases [89]. NPs have been synthesised from starch by green methods [109, 110]. Starch has good film forming ability, however, films based on entirely starch exhibit poor mechanical strength and barrier properties [111]. High surface hydrophilicity of starch is another disadvantage. Starch has a high melting temperature due to inter- and intramolecular hydrogen bonds and degrades before melting. Hence, melt processing operations, e.g., injection molding, extrusion led to degradation before melting. Consequently, plasticizer or chemical modification is required to mitigate this challenge [112-118]. Cellulose nanocrystals and clay NPs have been used to form starch ester films. These nano-composite films are shown to possess improved mechanical strength [89, 119]. Barrier properties depend on the degree of dispersion of the NPs within the polymer matrix. Solubility coefficient and diffusion measurements are used to determine the permeability of nano-composite films. Reduction in solubility coefficient indicates substantial decrease of water vapor permeability of the film. Introduction of nanoclay into starch films is shown to decrease the water vapor transport by diffusion [120]. Some recent reports on food packaging applications of nano-composites based on starch are collated in Table 2.

Table 2. Starch based nano-composites for food packaging applications

S. No.	Method of preparation	Materials used	Application/Properties	Ref.
1.	Baking	Cassava starch and organically modified nanoclay foams	Packaging for foods with low water content	[121]
2.	Casting	Crosslinked wheat starch reinforced by Na-MMT and TiO	Hybrid nanocomposite for environmentally friendly packaging material	[122]
3.	Casting	Cassava starch and nanoclay	Packaging films	[120]
4.	Biomimetic method	Corn starch, MMT and dopamine	Nacre-mimetic nanocomposite for food packaging and biological dressing	[123]
5.	*In situ* gelatinization	Thermoplastic starch, poly(lactic acid) and clay	Highly exfoliated nano-composites for green packaging	[124]
6.	Mechanical stirring	Starch, MMT and cellulose nanofibers	Biodegradable and green film for food packaging and preservation	[91]
7.	Melt blending	Thermoplastic cassava starch, halloysite clay nanotubes, poly(butylene adipate-co-terephthalate)	Packaging, agricultural and biomedical applications	[123]
8.	Melt intercalation	Thermoplastic pea starch, ethylene-vinyl acetate, bentonite	Food packaging, agriculture mulch film	[126]
9.	Ag NPs synthesized by reduction with *Mentha* leaves extract	Carboxymethyl cellulose, guar gum, Ag NPs	Antimicrobial nanocomposite films as edible coating for strawberries	[127]
10	Ag NPs synthesized by reduction with pomegranate seed extract	Starch and Ag NPs	Nanocomposite for active packaging	[110]

11.	Solution casting	Cationic starch, MMT and cellulose nanocrystals	BNCs for packaging of low water content foods	[128]
12.	Solution casting	Starch, PVA, halloysite nanotube	BNC for packaging with good water and gas barrier properties	[129]
13.	Solution mixing	Non-granular thermoplastic maize starch and MMT	Biocomposite films with superior mechanical and structural properties	[130]
14.	Solvent casting	Rice starch, chitosan, Ag and ZnO NPs	Antibacterial and antifungal nanocomposite films for packaging of peach fruits	[131]
15.	Casting	Potato starch, sorbitol/glycerol halloysite nanoclay	BNCs with high potential for food packaging	[132]

2.3.3 Starch-based nano-composite for pollution remediation

Contamination of the aquatic environment with dyes, heavy metals, medicines and plastics is a side effect of rapid global technological and economic development [133]. Industries that significantly contribute toward heavy metal pollution include mining, steel, metal coatings/platings, battery and electronics, fertilizers, leather tanning, and pesticides [134]. Heavy metals accumulate in living organisms due to non-biodegradable nature causing several disorders and diseases [135].

Starch-based nano-composites derived from potato and corn have been shown to efficiently remove heavy metal ions [136]. Efficiency of starch-based adsorbents is shown to be enhanced by incorporation of crown ethers [137], SnO_2 [138], and monolithic rectorite [139]. Natural clay minerals improve the capacity of starch matrix for heavy metal ion removal from aqueous solutions [140]. This is because biopolymers self-assemble through H-bonding and clay becomes exfoliated facilitating the exchange of metal ions in clay galleries [141, 142]. Modification of starch with sodium montmorillonite (Na-MMT) has numerous advantages: (i) surface area of MMT enhances the accessible sites in Na-MMT/starch, (ii) adsorption efficiency is increased by mobility of Na^+ ions [143], (iii) hydroxyl groups of starch augment the surface functionality [144]. However, the potential of starch/Na-MMT has not been exploited for heavy metal ion removal or wastewater treatment and is limited to development of packaging films. Na-MMT nano-composites are based on starch derived from cassava, corn potato, and bean starch. MMT nano-composites derived from bean starch are found to exhibit efficient heavy metal ion remediation [133].

Adv. App. of Micro and Nano Clay II – Synthetic Polymer Composites   Materials Research Forum LLC
Materials Research Foundations 129 (2022) 153-182   https://doi.org/10.21741/9781644902035-7

2.4   Nontronite based nano-composites

Nano-composites formed by intercalation of organic compounds in smectites are of interest because of their variety of applications. Adsorption of biopolymers on the clay surface affords new hybrid materials with combined properties of both components [145]. Various clay-based BNCs can be prepared by deliberate adsorption of biomolecules by clay [68]. Polypyrrole (PPy) was formed *in situ* to prepare functional polymer-clay nano-composites from MMT, Fe rich hectorite, smectite, nontronite and artificial saponites with tetrahedral substituted Fe^{3+}. Location and amount of Fe controls the content of monomer and spontaneous polymerization or over-oxidation of pyrrole. Very high Fe content in the clay can spontaneously induce fast monomer polymerization before the intercalation. PPy-clay nano-composites show mixed ionic–electronic conductivity. It is suggested that PPy or other conducting polymers and Fe-rich clays having active Fe^{2+}/Fe^{3+} centres could be used to develop electrode materials for cathodes of lithium-ion batteries and sensors [72].

Shewanella putrefaciens CN32 reduced fulvic acid and humic acid sorbed on nontronite [146]. Incubation of Fe-rich smectite (nontronite NAu-2) with insoluble and soluble lignin under anaerobic condition in presence and absence of an Fe^{3+}-reducing bacterium *Shewanella putrefaciens* CN32 is shown that structural Fe^{3+} in nontronite abiotically reduced lignin. Nontronite became nearly insoluble and did not transform upon reduction. Thus, the authors inferred that lignin significantly affects coupled Fe and C biogeochemical cycles by promoting microbe-mineral interactions in sediments and soils [147].

Nontronite produces hydroxyl radicals effectively and sustainably for degradation of 1,4-dioxane. It could be reused through biological or chemical regeneration of Fe^{2+}. Hence, it may be a promising remediation agent for organic contaminants in aqueous environment [148]. Nontronite, kaolinite and sepiolite have been investigated for carbon nanotube nano-composite synthesis. For kaolinite and sepiolite, catalytically active iron particles were required whereas for nontronite Fe present in its matrix was effective [149].

Shewanella putrefaciens CN32 biologically reduced structural Fe^{3+} in nontronite (NAu-2) intercalated chitosan. This nanocomposite was applicable for remediation of Cr^{6+} from contaminated groundwater and soil by combination of redox reactions and sorption [150].

Conclusion

At present, plastics have dominated the material world since they offer a wide range of properties required for a diverse range of applications from household to advanced technologies. However, concern for the environment has compelled scientists to explore alternative materials that are based on renewable resources and undergo biodegradation after active use. In this context, biopolymers are a promising class of materials, however, they suffer from several drawbacks. Consequently, hybrid materials with suitable properties have evolved as a better alternative. In particular, polymer composites coupled with incorporation of additives offers a high degree of synthetic pliability due to virtually unlimited combinatorial possibilities. Nano-composites are hybrid materials comprising two or more components where at least one is of nanoscale dimensions. High surface area of nanoparticles facilitates interaction with organic polymers, which in turn leads to the superior nano-composite properties. Desired performance of a given nano-composite requires the characteristics of particles and polymer to match along with a homogeneous dispersion of nanoparticles throughout the polymer matrix. Thus, it is necessary to preclude particle aggregation, which depends on the fundamental characteristics of the filler and polymer matrix. Generally, nano-composite design is done by trial-and-error method as it involves complex phenomena. Starch is an abundant natural biopolymer, which can be used to prepare various nano-composites with clay minerals. Starch nano-composites find applications in diverse fields, like pollution control, biomedical applications, food packaging and coatings. Fe-rich clay mineral nontronite is suitable for formation of nano-composites because presence of iron may provide special characteristics to nano-composites. Appropriate biopolymers along with Fe^{3+}-reducing bacteria may produce nano-composites with conducting or catalytic properties.

References

[1] H.H. Murray, Traditional and new applications for kaolin, smectite, and palygorskite: a general overview, Appl. Clay Sci. 17 (2000) 207-221. https://doi.org/10.1016/S0169-1317(00)00016-8

[2] A. Al-Futaisi, A. Jamrah, A. Al-Rawas, S. Al-Hanai, Adsorption capacity and mineralogical and physico-chemical characteristics of Shuwaymiyah palygorskite (Oman), Environ. Geol. 51 (2007) 1317-1327. https://doi.org/10.1007/s00254-006-0430-y

[3] Y. Kojima, A. Usuki, M. Kawasumi, A. Okada, Y. Fukushima, T. Kurauchi, O. Kamigaito, Mechanical properties of nylon 6-clay hybrid, J. Mater. Res. 8 (1993) 1185-1189. https://doi.org/10.1557/JMR.1993.1185

[4] P.C. Lebaron, Z. Wang, T.J. Pinnavaia, Polymer-layered silicate nanocomposites: an overview, Appl. Clay Sci. 15 (1999) 11-29. https://doi.org/10.1016/S0169-1317(99)00017-4

[5] E. Manias, A. Touny, L. Wu, K. Strawhecker, B. Lu, T.C. Chung, Polypropylene/montmorillonite nanocomposites. Review of the synthetic routes and

materials properties, Chem. Mater. 13 (2001) 3516-3523.
https://doi.org/10.1021/cm0110627

[6] T. Lan, P.D. Kaviratna, T.J.Pinnavaia, Mechanism of clay tactoid exfoliation in epoxy-clay nanocomposites, Chem. Mater. 7 (1995) 2144-2150.
https://doi.org/10.1021/cm00059a023

[7] Y.I. Tien, K.H. Wei, Hydrogen bonding and mechanical properties in segmented montmorillonite/polyurethane nanocomposites of different hard segment ratios, Polymer. 42 (2001) 3213-3221. https://doi.org/10.1016/S0032-3861(00)00729-1

[8] J.C. Huang, Z.K. Zhu, J. Yin, X.F. Qian, Y.Y. Sun, Poly(etherimide)/montmorillonite nanocomposites prepared by melt intercalation: morphology, solvent resistance properties and thermal properties, Polymer. 42 (2001) 873-877.
https://doi.org/10.1016/S0032-3861(00)00411-0

[9] T. Agag, T. Takeichi, Polybenzoxazine-montmorillonite hybrid nanocomposites: synthesis and characterization, Polymer. 41 (2000) 7083-7090.
https://doi.org/10.1016/S0032-3861(00)00064-1

[10] G. Galgali, C. Ramesh, A. Lele, Rheological study on the kinetics of hybrid formation in polypropylene nanocomposites, Macromolecules. 34 (2001) 852-858.
https://doi.org/10.1021/ma000565f

[11] X. Fu, S. Qutubuddin, Polymer-clay nanocomposites: exfoliation of organophilic montmorillonite nanolayers in polystyrene, Polymer. 42 (2001) 807-813.
https://doi.org/10.1016/S0032-3861(00)00385-2

[12] M. Okamoto, S. Morita, Y.H. Kim, T. Kotaka, H. Tateyama, Synthesis and structure of smectic clay/poly(methyl methacrylate) and clay/polystyrene nanocomposites via in situ intercalative polymerization, Polymer. 41 (2000) 3887-3890.
https://doi.org/10.1016/S0032-3861(99)00655-2

[13] M. Okamoto, S. Morita, T. Kotaka, Dispersed structure and ionic conductivity of smectic clay/polymer nanocomposites, Polymer. 42 (2001) 2685-2688.
https://doi.org/10.1016/S0032-3861(00)00642-X

[14] P.B. Messersmith, E.P. Giannelis, Synthesis and barrier properties of poly(ε-caprolactone) layered silicate nanocomposites, J. Polym. Sci. Part A Polym. Chem. 33 (1995) 1047-1057. https://doi.org/10.1002/pola.1995.080330707

[15] M. Kawasumi, N. Hasegawa, M. Kato, A. Usuki, A. Okada, Preparation and mechanical properties of polypropylene-clay hybrids, Macromolecules. 30 (1997) 6333-6338. https://doi.org/10.1021/ma961786h

[16] N. Hasegawa, M. Kawasumi, M. Kato, A. Usuki, A. Okada, Preparation and Mechanical Properties of Polypropylene - Clay Hybrids Using a Maleic anhydride-modified polypropylene oligomer. J. Appl. Polym. Sci. 67, 8792. https://doi.org/10.1002/(SICI)1097-4628(19980103)67:1<87::AID-APP10>3.0.CO;2-2

[17] R. E. Grim, Crystal Structures of Clay Minerals and their X-Ray Identification, G.W. Brindley, G. Brown (Eds.), Mineralogical Society, London, 1980, Volume 18, Issue 1, pp 84-85.

[18] S. Sinha Ray, Environmentally friendly nanofillers as reinforcements for composites, Environmentally Friendly Polymer Nanocomposites, Woodhead Publishing Ltd., 2013, pp 41-73. https://doi.org/10.1533/9780857097828.1.41

[19] M. Biswas, S.S. Ray, Recent progress in synthesis and evaluation of polymer-montmorillonite nanocomposites, Adv. Polym. Sci. 155 (2001) 167-221. https://doi.org/10.1007/3-540-44473-4_3

[20] S. Sinha Ray, M. Okamoto, Polymer/layered silicate nanocomposites: a review from preparation to processing, Prog. Polym. Sci. 28 (2003) 1539-1641. https://doi.org/10.1016/j.progpolymsci.2003.08.002

[21] S.S. Ray, M. Bousmina, Biodegradable polymers and their layered silicate nanocomposites: In greening the 21st century materials world, Prog. Mater. Sci. 50 (2005) 962-1079. https://doi.org/10.1016/j.pmatsci.2005.05.002

[22] R.A. Vaia, G. Price, P.N. Ruth, H.T. Nguyen, J. Lichtenhan, Polymer/layered silicate nanocomposites as high performance ablative materials, Appl. Clay Sci. 15 (1999) 67-92. https://doi.org/10.1016/S0169-1317(99)00013-7

[23] J. F. Kennedy, R. M. Alanís, Polysaccharides: structural diversity and functional versatility, second ed., Marcel Dekker, New York, Carbohydrate Polymers, 62(3), 2005, 301-301. https://doi.org/10.1016/j.carbpol.2005.05.027

[24] I.I. Muhamad, N.A.M. Lazim, S. Selvakumaran, Natural polysaccharide-based composites for drug delivery and biomedical applications, M S Hasnain, A. K. Nayak (Eds.), Natural Polysaccharides in Drug Delivery and Biomedical Applications, Elsevier Inc., 2019, pp 419-440. https://doi.org/10.1016/B978-0-12-817055-7.00018-2

[25] Z. Shriver, S. Raguram, R. Sasisekharan, Glycomics: A pathway to a class of new and improved therapeutics, Nat. Rev. Drug Discov. 3 (2004) 863-873. https://doi.org/10.1038/nrd1521

[26] R. A. Gross, B. Kalra, Biodegradable polymers for the environment, Science. 297 (2002) 803-7. https://doi.org/10.1126/science.297.5582.803

[27] J. Venkatesan, S. Anil, S.K. Singh, S.K. Kim, Preparations and Applications of Alginate Nanoparticles, J. Venkatesan, S Anil and S-K Kim (Eds.), Elsevier Inc., 2017, pp 251-268. https://doi.org/10.1016/B978-0-12-809816-5.00013-X

[28] S. Yadav, A. Yadav, N. Bagotia, A.K. Sharma, S. Kumar, Adsorptive potential of modified plant-based adsorbents for sequestration of dyes and heavy metals from wastewater - A review, J. Water Process Eng. 42 (2021) 102148. https://doi.org/10.1016/j.jwpe.2021.102148

[29] X. Qi, X. Tong, W. Pan, Q. Zeng, S. You, J. Shen, Recent advances in polysaccharide-based adsorbents for wastewater treatment, J. Clean. Prod. 315 (2021) 128221. https://doi.org/10.1016/j.jclepro.2021.128221

[30] M.M. Ghobashy, The application of natural polymer-based hydrogels for agriculture, Hydrogels Based Nat. Polym. (2020) 329-356.. https://doi.org/10.1016/B978-0-12-816421-1.00013-6

[31] S.K. Dubey, Dheeraj, S. Hejmady, A. Alexander, S. Tiwari, G. Singhvi, Graft-modified polysaccharides in biomedical applications, A Nayak, M S Hasnain, T Aminabhavi (Eds.), Tailor-Made Polysaccharides Biomed. Appl., Elsevier Inc., 2020, pp 69-100. https://doi.org/10.1016/B978-0-12-821344-5.00004-7

[32] P. Prasher, M. Sharma, M. Mehta, S. Satija, A.A. Aljabali, M.M. Tambuwala, K. Anand, N. Sharma, H. Dureja, N.K. Jha, G. Gupta, M. Gulati, S.K. Singh, D.K. Chellappan, K.R. Paudel, P.M. Hansbro, K. Dua, Current-status and applications of polysaccharides in drug delivery systems, Colloid Interface Sci. Commun. 42 (2021) 100418. https://doi.org/10.1016/j.colcom.2021.100418

[33] S. Ahmad, M. Ahmad, K. Manzoor, R. Purwar, S. Ikram, A review on latest innovations in natural gums based hydrogels: Preparations & applications, Int. J. Biol. Macromol. 136 (2019) 870-890. https://doi.org/10.1016/j.ijbiomac.2019.06.113

[34] R. Malviya, P.K. Sharma, S.K. Dubey, Modification of polysaccharides: Pharmaceutical and tissue engineering applications with commercial utility (patents), Mater. Sci. Eng. C. 68 (2016) 929-938. https://doi.org/10.1016/j.msec.2016.06.093

[35] B. Maji, Introduction to natural polysaccharides, Funct. Polysaccharides Biomed. Appl. (2019) 1-31. https://doi.org/10.1016/B978-0-08-102555-0.00001-7

[36] I.I. Muhamad, N.A.M. Lazim, S. Selvakumaran, Natural polysaccharide-based composites for drug delivery and biomedical applications, Nat. Polysaccharides Drug

Deliv. Biomed. Appl. (2019) 419-440. https://doi.org/10.1016/B978-0-12-817055-7.00018-2

[37] A.M. Morales-Burgos, E. Carvajal-Millan, N. Sotelo-Cruz, A.C. Campa-Mada, A. Rascón-Chu, Y. Lopez-Franco, J. Lizardi-Mendoza, Polysaccharides in Alternative Methods for Insulin Delivery, Biopolym. Grafting Synth. Prop. (2018) 175-197. https://doi.org/10.1016/B978-0-323-48104-5.00004-4

[38] A.A. Aly, M.K. El-Bisi, Grafting of Polysaccharides: Recent Advances, Biopolym. Grafting Synth. Prop. (2018) 469-519. https://doi.org/10.1016/B978-0-323-48104-5.00011-1

[39] V. Kumar, S. Nagar, P. Sharma, Opportunity of plant oligosaccharides and polysaccharides in drug development, Carbohydrates Drug Discov. Dev. (2020) 587-639. https://doi.org/10.1016/B978-0-12-816675-8.00015-4

[40] A. Orsuwan, R. Sothornvit, Polysaccharide Nanobased Packaging Materials for Food Application, Elsevier Inc., 2018. https://doi.org/10.1016/B978-0-12-811516-9.00007-5

[41] H.E. Tahir, Z. Xiaobo, G.K. Mahunu, M. Arslan, M. Abdalhai, L. Zhihua, Recent developments in gum edible coating applications for fruits and vegetables preservation: A review, Carbohydr. Polym. 224 (2019) 115141. https://doi.org/10.1016/j.carbpol.2019.115141

[42] G. Sason, A. Nussinovitch, Hydrocolloids for edible films, coatings, and food packaging, Handb. Hydrocoll. (2021) 195-235. https://doi.org/10.1016/B978-0-12-820104-6.00023-1

[43] J. Shuren, Production and use of modified starch and starch derivatives in China, in: R.H. Howeler, SL. Tan (Eds.), Proc. 6th Reg. Work., Ho Chi Minh City, Vietnam, 2000: pp. 553.-563.

[44] K. Ali, P. K. Roy, C. M. Hossain, D. Dutta, R. Vichare, M. R. Biswal, Starch-based nanomaterials in drug delivery applications, in: H.Bera, C. M. Hossain, S. Saha (Eds.), Biopolymer-Based Nanomaterials in Drug Delivery and Biomedical Applications, Academic Press,(2021), pp 31-56, ISBN 9780128208748. https://doi.org/10.1016/B978-0-12-820874-8.00023-3

[45] F.Z. Arrakhiz, K. Benmoussa, R. Bouhfid, A. Qaiss, Pine cone fiber/clay hybrid composite: Mechanical and thermal properties, Mater. Des. 50 (2013) 376-381. https://doi.org/10.1016/j.matdes.2013.03.033

[46] F. Wypych, F. Bergaya, R.A. Schoonheydt, From polymers to clay polymer nanocomposites, Dev. Clay Sci. 9 (2018) 331-359. https://doi.org/10.1016/B978-0-08-102432-4.00010-X

[47] M. Adamu, M.R. Rahman, S. Hamdan, Formulation optimization and characterization of bamboo/polyvinyl alcohol/clay nanocomposite by response surface methodology, Compos. Part B Eng. 176 (2019) 107297. https://doi.org/10.1016/j.compositesb.2019.107297

[48] B. Dogaru, B. Simionescu, M. Popescu, International Journal of Biological Macromolecules Synthesis and characterization of κ -carrageenan bio-nanocomposite films reinforced with bentonite nanoclay, Int. J. Biol. Macromol. 154 (2020) 9-17. https://doi.org/10.1016/j.ijbiomac.2020.03.088

[49] S. Mamaghani Shishavan, T. Azdast, S. Rash Ahmadi, Investigation of the effect of nanoclay and processing parameters on the tensile strength and hardness of injection molded Acrylonitrile Butadiene Styrene-organoclay nanocomposites, Mater. Des. 58 (2014) 527-534. https://doi.org/10.1016/j.matdes.2014.02.014

[50] S. S. Ray, Environmentally friendly nanofillers as reinforcements for composites. Environmentally Friendly, Polym. Nanocompos. (2013) 41-73. https://doi.org/10.1533/9780857097828.1.41

[51] M. V. Nagarpita, P. Roy, S.B. Shruthi, R.R.N. Sailaja, Synthesis and swelling characteristics of chitosan and CMC grafted sodium acrylate-co-acrylamide using modified nanoclay and examining its efficacy for removal of dyes, Int. J. Biol. Macromol. 102 (2017) 1226-1240. https://doi.org/10.1016/j.ijbiomac.2017.04.099

[52] M.A. Adebayo, J.I. Adebomi, T.O. Abe, F.I. Areo, Removal of aqueous Congo red and malachite green using ackee apple seed-bentonite composite, Colloids Interface Sci. Commun. 38 (2020) 100311. https://doi.org/10.1016/j.colcom.2020.100311

[53] Q. Wang, Y. Wang, L. Chen, A green composite hydrogel based on cellulose and clay as efficient absorbent of colored organic effluent, Carbohydr. Polym. 210 (2019) 314-321. https://doi.org/10.1016/j.carbpol.2019.01.080

[54] Y. Bulut, G. Akçay, D. Elma, I.E. Serhatli, Synthesis of clay-based superabsorbent composite and its sorption capability, J. Hazard. Mater. 171 (2009) 717-723. https://doi.org/10.1016/j.jhazmat.2009.06.067

[55] S. Zhu, S. Wang, X. Yang, S. Tufail, C. Chen, X. Wang, J. Shang, Green sustainable and highly efficient hematite nanoparticles modified biochar-clay granular composite

for Cr(VI) removal and related mechanism, J. Clean. Prod. 276 (2020) 123009. https://doi.org/10.1016/j.jclepro.2020.123009

[56] X. Hu, Y. Ke, Y. Zhao, S. Lu, Q. Deng, C. Yu, F. Peng, Synthesis, characterization and solution properties of β-cyclodextrin-functionalized polyacrylamide/montmorillonite nanocomposites, Colloids Surfaces A Physicochem. Eng. Asp. 560 (2019) 336-343. https://doi.org/10.1016/j.colsurfa.2018.10.035

[57] V.G.L. Souza, J.R.A. Pires, É.T. Vieira, I.M. Coelhoso, M.P. Duarte, A.L. Fernando, Activity of chitosan-montmorillonite bionanocomposites incorporated with rosemary essential oil: From in vitro assays to application in fresh poultry meat, Food Hydrocoll. 89 (2019) 241-252. https://doi.org/10.1016/j.foodhyd.2018.10.049

[58] M. Koosha, S. Hamedi, Intelligent Chitosan/PVA nanocomposite films containing black carrot anthocyanin and bentonite nanoclays with improved mechanical, thermal and antibacterial properties, Prog. Org. Coatings. 127 (2019) 338-347. https://doi.org/10.1016/j.porgcoat.2018.11.028

[59] K. El Bourakadi, M.E.M. Mekhzoum, A. el kacem Qaiss, R. Bouhfid, Processing and Biomedical Applications of Polymer/Organo-modified Clay Bionanocomposites, in: Sarat Kumar Swain and Mohammad Jawaid (Eds.), A volume in Micro and Nano Technologies, Elsevier Inc., 2019, pp 405-428. https://doi.org/10.1016/B978-0-12-816771-7.00021-1

[60] K. Majeed, M. Jawaid, A. Hassan, A. Abu Bakar, H.P.S. Abdul Khalil, A.A. Salema, I. Inuwa, Potential materials for food packaging from nanoclay/natural fibres filled hybrid composites, Mater. Des. 46 (2013) 391-410. https://doi.org/10.1016/j.matdes.2012.10.044

[61] S. Nir, Y. El-Nahhal, T. Undabeytia, G. Rytwo, T. Polubesova, Y. Mishael, O. Rabinovitz, B. Rubin, Clays, Clay Minerals, and Pesticides, Dev. Clay Sci. 5 (2013) 645-662. https://doi.org/10.1016/B978-0-08-098259-5.00022-6

[62] Y. Brouard, N. Belayachi, D. Hoxha, N. Ranganathan, S. Méo, Mechanical and hygrothermal behavior of clay - Sunflower (Helianthus annuus) and rape straw (Brassica napus) plaster bio-composites for building insulation, Constr. Build. Mater. 161 (2018) 196-207. https://doi.org/10.1016/j.conbuildmat.2017.11.140

[63] R.M. Apetrei, P. Camurlu, The effect of montmorillonite functionalization on the performance of glucose biosensors based on composite montmorillonite/PAN nanofibers, Electrochim. Acta. 353 (2020) 136484. https://doi.org/10.1016/j.electacta.2020.136484

[64] I. Colijn, K. Schroën, Thermoplastic bio-nanocomposites: From measurement of fundamental properties to practical application, Adv. Colloid Interface Sci. 292 (2021) 102419. https://doi.org/10.1016/j.cis.2021.102419

[65] A. Al Rashid, S.A. Khan, S. G. Al-Ghamdi, M. Koç, Additive manufacturing of polymer nanocomposites: Needs and challenges in materials, processes, and applications, J. Mater. Res. Technol. 14 (2021) 910-941. https://doi.org/10.1016/j.jmrt.2021.07.016

[66] U.C. Ugochukwu, Characteristics of clay minerals relevant to bioremediation of environmental contaminated systems, in: A.L. Mariano Mercurio, Binoy Sarkar (Eds.), Modif. Clay Zeolite Nanocomposite Mater., Elsevier, 2019: pp. 219-242. https://doi.org/10.1016/B978-0-12-814617-0.00006-2

[67] A. Wang, W. Wang, Introduction. Nanomaterials from Clay Minerals. in: A. Wang, W. Wang (Eds.). Micro and Nano Technologies, Elsevier, (2019), pp 1-20. https://doi.org/10.1016/B978-0-12-814533-3.00001-6

[68] E. Koutsopoulou, I. Koutselas, G.E. Christidis, A. Papagiannopoulos, I. Marantos, Effect of layer charge and charge distribution on the formation of chitosan - smectite bionanocomposites, Appl. Clay Sci. 190 (2020) 105583. https://doi.org/10.1016/j.clay.2020.105583

[69] A. Gil, L. Santamaría, S.A. Korili, M.A. Vicente, L.V. Barbosa, S.D. de Souza, L. Marçal, E.H. de Faria, K.J. Ciuffi, A review of organic-inorganic hybrid clay based adsorbents for contaminants removal: Synthesis, perspectives and applications, J. Environ. Chem. Eng. 9 (2021) 105808. https://doi.org/10.1016/j.jece.2021.105808

[70] B. Coppola, N. Cappetti, L. Di Maio, P. Scarfato, L. Incarnato, Layered silicate reinforced polylactic acid filaments for 3D printing of polymer nanocomposites, RTSI 2017 - IEEE 3rd Int. Forum Res. Technol. Soc. Ind. Conf. Proc. (2017) 3-6. https://doi.org/10.1109/RTSI.2017.8065892

[71] S. Saadat, G. Pandey, M. Tharmavaram, V. Braganza, D. Rawtani, Nano-interfacial decoration of Halloysite Nanotubes for the development of antimicrobial nanocomposites, Adv. Colloid Interface Sci. 275 (2020) 102063. https://doi.org/10.1016/j.cis.2019.102063

[72] S. Letaïef, P. Aranda, E. Ruiz-Hitzky, Influence of iron in the formation of conductive polypyrrole-clay nanocomposites, Appl. Clay Sci. 28 (2005) 183-198. https://doi.org/10.1016/j.clay.2004.02.008

[73] https://www.mindat.org/min-2924.html (accessed 13 September 2021)

[74] J. Cervini-Silva, E. Palacios, V. Gómez-Vidales, Nontronite as natural source and growth template for (nano)maghemite [γ-Fe2O3] and (nano)Wüstite [Fe1−xO], Appl. Clay Sci. 156 (2018) 178-186. https://doi.org/10.1016/j.clay.2018.02.009

[75] A. Gil , L. Santamaría , S.A. Korili , M.A. Vicente , L.V. Barbosa , S.D. de Souza , L. Marçal, E.H. de Faria, K.J. Ciuffi, A review of organic-inorganic hybrid clay based adsorbents for contaminants removal: Synthesis, perspectives and applications, J. Environ. Chem. Eng. 9 (2021) 105808. https://doi.org/10.1016/j.jece.2021.105808

[76] F. Bergaya, C. Detellier, J.-F. Lambert, G. Lagaly, Introduction to Polymer - Clay Nanocomposites, F. Bergaya, G. Lagaly (Eds.), Handbook of Clay Science, (2013) 655-677. https://doi.org/10.1016/B978-0-08-098258-8.00020-1

[77] R. Mukhopadhyay, N. De, Nano clay polymer composite: synthesis, characterization, properties and application in Rainfed Agriculture, Global J. Bio-Sci. Biotechnol. 3 (2014) 133-138.

[78] A.K. Naskar, J.K. Keum, R.G. Boeman, Polymer matrix nanocomposites for automotive structural components, Nat. Nanotechnol. 11 (2016) 1026-1030. https://doi.org/10.1038/nnano.2016.262

[79] A. Bhat, S. Budholiya, S.A. Raj, M.T.H. Sultan, D. Hui, A.U.M. Shah, S.N.A. Safri, Review on nanocomposites based on aerospace applications, Nanotechnol. Rev. 10 (2021) 237-253. https://doi.org/10.1515/ntrev-2021-0018

[80] Y.K. Kadokawa J, Shimohigoshi R, Yamashita K, Synthesis of chitin and chitosan stereoisomers by thermostable a-glucan phosphorylase catalyzed enzymatic polymerization of a-D-glucosamine 1-phosphate., Org Biomol Chem. 13 (2015) 4336-43. https://doi.org/10.1039/C5OB00167F

[81] R.R. Palem, K.M. Rao, G. Shimoga, R.G. Saratale, S.K. Shinde, G.S. Ghodake, S.H. Lee, Physicochemical characterization, drug release, and biocompatibility evaluation of carboxymethyl cellulose-based hydrogels reinforced with sepiolite nanoclay, Int. J. Biol. Macromol. 178 (2021) 464-476. https://doi.org/10.1016/j.ijbiomac.2021.02.195

[82] S. Memiş, F. Tornuk, F. Bozkurt, M.Z. Durak, Production and characterization of a new biodegradable fenugreek seed gum based active nanocomposite film reinforced with nanoclays, Int. J. Biol. Macromol. 103 (2017) 669-675. https://doi.org/10.1016/j.ijbiomac.2017.05.090

[83] R. Sharma, S.M. Jafari, S. Sharma, Antimicrobial bio-nanocomposites and their potential applications in food packaging, Food Control. 112 (2020) 107086. https://doi.org/10.1016/j.foodcont.2020.107086

[84] D. Qureshi, K.P. Behera, D. Mohanty, S.K. Mahapatra, S. Verma, P. Sukyai, I. Banerjee, S.K. Pal, B. Mohanty, D. Kim, K. Pal, Synthesis of novel poly (vinyl alcohol)/tamarind gum/bentonite-based composite films for drug delivery applications, Colloids Surfaces A Physicochem. Eng. Asp. 613 (2021) 126043. https://doi.org/10.1016/j.colsurfa.2020.126043

[85] R. Mukhopadhyay, D. Bhaduri, B. Sarkar, R. Rusmin, D. Hou, R. Khanam, S. Sarkar, J. Kumar Biswas, M. Vithanage, A. Bhatnagar, Y.S. Ok, Clay-polymer nanocomposites: Progress and challenges for use in sustainable water treatment, J. Hazard. Mater. 383 (2020) 121125. https://doi.org/10.1016/j.jhazmat.2019.121125

[86] L. Corre, D. Bras, J.A. Dufresne, Starch nanoparticles: a review. Biomacromolecules 11 (5) (2010) 1139-1153. https://doi.org/10.1021/bm901428y

[87] E. Ogunsona, E. Ojogbo, T. Mekonnen, Advanced material applications of starch and its derivatives, Eur. Polym. J. 108 (2018) 570-581. https://doi.org/10.1016/j.eurpolymj.2018.09.039

[88] S. Agarwal, Major factors affecting the characteristics of starch based biopolymer films, Eur. Polym. J. 160 (2021) 110788. https://doi.org/10.1016/j.eurpolymj.2021.110788

[89] E. Ojogbo, J. Jardin, T.H. Mekonnen, Robust and sustainable starch ester nanocomposite films for packaging applications, Ind. Crops Prod. 160 (2021) 113-153. https://doi.org/10.1016/j.indcrop.2020.113153

[90] N. Soykeabkaew, C. Thanomsilp, O. Suwantong, A review: Starch-based composite foams, Compos. Part A Appl. Sci. Manuf. 78 (2015) 246-263. https://doi.org/10.1016/j.compositesa.2015.08.014

[91] Á. L. Santana, M. Angela A. Meireles, New Starches are the Trend for Industry Applications: A Review, Food Public Heal. 4 (2014) 229-241. https://doi.org/10.5923/j.fph.20140405.04

[92] B. Ghanbarzadeh, H. Almasi, A.A. Entezami, Physical properties of edible modified starch/carboxymethyl cellulose films, Innov. Food Sci. Emerg. Technol. 11 (2010) 697-702. https://doi.org/10.1016/j.ifset.2010.06.001

[93] A.M. Nafchi, R. Nassiri, S. Sheibani, F. Ariffin, A.A. Karim, Preparation and characterization of bionanocomposite films filled with nanorod-rich zinc oxide, Carbohydr. Polym. 96 (2013) 233-239. https://doi.org/10.1016/j.carbpol.2013.03.055

[94] M.N. Abdorreza, L.H. Cheng, A.A. Karim, Food Hydrocoll. 25 (2011) 56-60. https://doi.org/10.1016/j.foodhyd.2010.05.005

[95] A. M. Nafchi, L.H. Cheng, A.A. Karim, Effects of plasticizers on thermal properties and heat sealability of sago starch films, Food Hydrocoll. 25 (2011) 56-60. https://doi.org/10.1016/j.foodhyd.2010.05.005

[96] M. Darder, P. Aranda, E. Ruiz-Hitzky, Bionanocomposites: A new concept of ecological, bioinspired, and functional hybrid materials, Adv. Mater. 19 (2007) 1309-1319. https://doi.org/10.1002/adma.200602328

[97] L.C. Geoffrey A Ozin, André Arsenault, Nanochemistry: A Chemical Approach to Nanomaterials, Second ed., Royal Society of Chemistry, Cambridge, 2009.

[98] A.M Nafchi, M. Moradpour, M. Saeidi, A.K. Alias, Thermoplastic starches: Properties, challenges, and prospects, Starch/Staerke. 65 (2013) 61-72. https://doi.org/10.1002/star.201200201

[99] R. Alebooyeh, A. Mohammadinafchi, M. Jokar, The Effects of ZnOnanorodson the Characteristics of Sago Starch Biodegradable Films, J. Chem. Heal. Risk. 2 (2012) 13-16.

[100] K. Shameli, M. Bin Ahmad, W.M.Z.W. Yunus, A. Rustaiyan, N.A. Ibrahim, M. Zargar, Y. Abdollahi, Green synthesis of silver/montmorillonite/chitosan bionanocomposites using the UV irradiation method and evaluation of antibacterial activity, Int. J. Nanomedicine. 5 (2010) 875-887. https://doi.org/10.2147/IJN.S13632

[101] R. Yamaoki, T. Tsujino, S. Kimura, Y. Mino, M. Ohta, Detection of organic free radicals in irradiated Foeniculi fructus by electron spin resonance spectroscopy, J. Nat. Med. 63 (2009) 28-31. https://doi.org/10.1007/s11418-008-0284-6

[102] W.J. Jia, C.B. Liu, L. Yang, J.L. Yang, L.Y. Fan, M.J. Huang, H. Zhang, G.T. Chao, Z.Y. Qian, B. Kan, A.L. Huang, K. Lei, C.Y. Gong, J. Zhao, J.M. Zhang, H.X. Deng, M.J. Tu, Y.Q. Wei, Synthesis, characterization, and thermal properties of biodegradable polyetheresteramide-based polyurethane, Mater. Lett. 60 (2006) 3686-3692. https://doi.org/10.1016/j.matlet.2006.03.089

[103] Y. Chen, S. Zhou, H. Yang, G. Gu, L. Wu, Preparation and characterization of nanocomposite polyurethane, J. Colloid Interface Sci. 279 (2004) 370-378. https://doi.org/10.1016/j.jcis.2004.06.074

[104] R. Babalola, A.O. Ayeni, P.S. Joshua, A.A. Ayoola, U.O. Isaac, U. Aniediong, V.E. Efeovbokhan, J.A. Omoleye, Synthesis of thermal insulator using chicken feather fibre in starch-clay nanocomposites, Heliyon. 6 (2020) 05384. https://doi.org/10.1016/j.heliyon.2020.e05384

[105] A. Tariq, S.A. Bhawani, M. Nisar, M.R. Asaruddin, K.M. Alotaibi, Starch-based nanocomposites for gene delivery, in: Polysaccharide-Based Nanocomposites Gene Deliv. Tissue Eng., Elsevier, 2021: pp. 263-277. https://doi.org/10.1016/B978-0-12-821230-1.00007-4

[106] T. Bai, B. Zhu, H. Liu, Y. Wang, G. Song, C. Liu, C. Shen, Biodegradable poly(lactic acid) nanocomposites reinforced and toughened by carbon nanotubes/clay hybrids, Int. J. Biol. Macromol. 151 (2020) 628-634. https://doi.org/10.1016/j.ijbiomac.2020.02.209

[107] M. Iman, T.K. Maji, Effect of crosslinker and nanoclay on starch and jute fabric based green nanocomposites, Carbohydr. Polym. 89 (2012) 290-297. https://doi.org/10.1016/j.carbpol.2012.03.012

[108] M.S. Mohseni, M.A. Khalilzadeh, M. Mohseni, F.Z. Hargalani, M.I. Getso, V. Raissi, O. Raiesi, Green synthesis of Ag nanoparticles from pomegranate seeds extract and synthesis of Ag-Starch nanocomposite and characterization of mechanical properties of the films, Biocatal. Agric. Biotechnol. 25 (2020) 101569. https://doi.org/10.1016/j.bcab.2020.101569

[109] M.A. El-Sheikh, A novel photosynthesis of carboxymethyl starch-stabilized silver nanoparticles, Sci. World J. (2014) 1-11. https://doi.org/10.1155/2014/514563

[110] B. Kumar, K. Smita, L. Cumbal, A. Debut, R.N. Pathak, Sonochemical synthesis of silver nanoparticles using starch: A comparison, Bioinorg. Chem. Appl. (2014) 1-8. https://doi.org/10.1155/2014/784268

[111] L. do Val Siqueira, C.I.L.F. Arias, B.C. Maniglia, C.C. Tadini, Starch-based biodegradable plastics: methods of production, challenges and future perspectives, Curr. Opin. Food Sci. 38 (2021) 122-130. https://doi.org/10.1016/j.cofs.2020.10.020

[112] C. Mutungi, F. Rost, C. Onyango, D. Jaros, H. Rohm, Crystallinity, thermal and morphological characteristics of resistant starch type iii produced by hydrothermal treatment of debranched Cassava starch, Starch/Staerke. 61 (2009) 634-645. https://doi.org/10.1002/star.200900167

[113] D.F. Apopei, M. V. Dinu, E.S. Drăgan, Graft copolymerization of acrylonitrile onto potatoes starch by ceric ion, Dig. J. Nanomater. Biostructures. 7 (2012) 707-716.

[114] H. Namazi, F. Fathi, A. Dadkhah, Hydrophobically modified starch using long-chain fatty acids for preparation of nanosized starch particles, Sci. Iran. 18 (2011) 439-445. https://doi.org/10.1016/j.scient.2011.05.006

[115] E. Ojogbo, R. Blanchard, T. Mekonnen, Hydrophobic and Melt Processable Starch-Laurate Esters: Synthesis, Structure-Property Correlations, J. Polym. Sci. Part A Polym. Chem. 56 (2018) 2611-2622. https://doi.org/10.1002/pola.29237

[116] E. Ojogbo, E.O. Ogunsona, T.H. Mekonnen, Chemical and physical modifications of starch for renewable polymeric materials, Mater. Today Sustain. 7-8 (2020) 100028. https://doi.org/10.1016/j.mtsust.2019.100028

[117] E.O. Ogunsona, T.H. Mekonnen, Multilayer assemblies of cellulose nanocrystal - polyvinyl alcohol films featuring excellent physical integrity and multi-functional properties, J. Colloid Interface Sci. 580 (2020) 56-67. https://doi.org/10.1016/j.jcis.2020.07.012

[118] M. Shayan, H. Azizi, I. Ghasemi, M. Karrabi, Effect of modified starch and nanoclay particles on biodegradability and mechanical properties of cross-linked poly lactic acid, Carbohydr. Polym. 124 (2015) 237-244. https://doi.org/10.1016/j.carbpol.2015.02.001

[119] J. Li, M. Zhou, G. Cheng, F. Cheng, Y. Lin, P.X. Zhu, Fabrication and characterization of starch-based nanocomposites reinforced with montmorillonite and cellulose nanofibers, Carbohydr. Polym. 210 (2019) 429-436. https://doi.org/10.1016/j.carbpol.2019.01.051

[120] C.M.O. Müller, J.B. Laurindo, F. Yamashita, Effect of nanoclay incorporation method on mechanical and water vapor barrier properties of starch-based films, Ind. Crops Prod. 33 (2011) 605-610. https://doi.org/10.1016/j.indcrop.2010.12.021

[121] D.K.M. Matsuda, A.E.S. Verceheze, G.M. Carvalho, F. Yamashita, S. Mali, Baked foams of cassava starch and organically modified nanoclays, Ind. Crops Prod. 44 (2013) 705-711. https://doi.org/10.1016/j.indcrop.2012.08.032

[122] A.R. Yousefi, B. Savadkoohi, Y. Zahedi, M. Hatami, K. Ako, Fabrication and characterization of hybrid sodium montmorillonite/TiO2 reinforced cross-linked wheat starch-based nanocomposites, Int. J. Biol. Macromol. 131 (2019) 253-263. https://doi.org/10.1016/j.ijbiomac.2019.03.083

[123] J. Li, M. Zhou, F. Cheng, Y. Lin, P. Zhu, Bioinspired approach to enhance mechanical properties of starch based nacre-mimetic nanocomposite, Carbohydr. Polym. 221 (2019) 113-119. https://doi.org/10.1016/j.carbpol.2019.05.090

[124] B. Ayana, S. Suin, B.B. Khatua, Highly exfoliated eco-friendly thermoplastic starch (TPS)/poly (lactic acid)(PLA)/clay nanocomposites using unmodified nanoclay, Carbohydr. Polym. 110 (2014) 430-439. https://doi.org/10.1016/j.carbpol.2014.04.024

[125] K.M. Dang, R. Yoksan, E. Pollet, L. Avérous, Morphology and properties of thermoplastic starch blended with biodegradable polyester and filled with halloysite nanoclay, Carbohydr. Polym. 242 (2020) 116392. https://doi.org/10.1016/j.carbpol.2020.116392

[126] V. Sessini, M.P. Arrieta, J.M. Raquez, P. Dubois, J.M. Kenny, L. Peponi, Thermal and composting degradation of EVA/Thermoplastic starch blends and their nanocomposites, Polym. Degrad. Stab. 159 (2019) 184-198. https://doi.org/10.1016/j.polymdegradstab.2018.11.025

[127] V. Kanikireddy, K. Varaprasad, M.S. Rani, P. Venkataswamy, B.J. Mohan Reddy, M. Vithal, Biosynthesis of CMC-Guar gum-Ag nanocomposites for inactivation of food pathogenic microbes and its effect on the shelf life of strawberries, Carbohydr. Polym. 236 (2020) 116053. https://doi.org/10.1016/j.carbpol.2020.116053

[128] K. Vaezi, G. Asadpour, S.H. Sharifi, Bio nanocomposites based on cationic starch reinforced with montmorillonite and cellulose nanocrystals: Fundamental properties and biodegradability study, Int. J. Biol. Macromol. 146 (2020) 374-386. https://doi.org/10.1016/j.ijbiomac.2020.01.007

[129] Z. Waheed, Y. Dong, N. Han, S. Liu, Water and gas barrier properties of polyvinyl alcohol (PVA)/ starch (ST)/ glycerol (GL)/ halloysite nanotube (HNT) bionanocomposite films : Experimental characterisation and modelling approach, Compos. Part B. 174 (2019) 107033. https://doi.org/10.1016/j.compositesb.2019.107033

[130] G. Mansour, M. Zoumaki, A. Marinopoulou, D. Tzetzis, M. Prevezanos, S.N. Raphaelides, Characterization and properties of non-granular thermoplastic starch-Clay biocomposite films, Carbohydr. Polym. 245 (2020) 116629. https://doi.org/10.1016/j.carbpol.2020.116629

[131] M. Kaur, A. Kalia, A. Thakur, Effect of biodegradable chitosan-rice-starch nanocomposite films on post-harvest quality of stored peach fruit, Starch/Staerke. 69 (2017) 1600208. https://doi.org/10.1002/star.201600208

[132] F. Sadegh-Hassani, A. Mohammadi Nafchi, Preparation and characterization of bionanocomposite films based on potato starch/halloysite nanoclay, Int. J. Biol. Macromol. 67 (2014) 458-462. https://doi.org/10.1016/j.ijbiomac.2014.04.009

[133] Á. García-Padilla, K.A. Moreno-Sader, Á. Realpe, M. Acevedo-Morantes, J.B.P. Soares, Evaluation of adsorption capacities of nanocomposites prepared from bean starch and montmorillonite, Sustain. Chem. Pharm. 17 (2020). https://doi.org/10.1016/j.scp.2020.100292

[134] K. Moreno-Sader, A. García-Padilla, A. Realpe, M. Acevedo-Morantes, J.B.P. Soares, Removal of Heavy Metal Water Pollutants (Co2+ and Ni2+) Using Polyacrylamide/Sodium Montmorillonite (PAM/Na-MMT) Nanocomposites, ACS Omega. 4 (2019) 10834-10844. https://doi.org/10.1021/acsomega.9b00981

[135] Y.S. Al-Degs, M.I. El-Barghouthi, A.A. Issa, M.A. Khraisheh, G.M. Walker, Sorption of Zn(II), Pb(II), and Co(II) using natural sorbents: Equilibrium and kinetic studies, Wat. Res. 40 (2006) 2645-2658. https://doi.org/10.1016/j.watres.2006.05.018

[136] T.H. Tran, H. Okabe, Y. Hidaka, K. Hara, Removal of metal ions from aqueous solutions using carboxymethyl cellulose/sodium styrene sulfonate gels prepared by radiation grafting, Carbohydr. Polym. 157 (2017) 335-343. https://doi.org/10.1016/j.carbpol.2016.09.049

[137] B.M. Ibrahim, N.A. Fakhre, Crown ether modification of starch for adsorption of heavy metals from synthetic wastewater, Int. J. Biol. Macromol. 123 (2019) 70-80. https://doi.org/10.1016/j.ijbiomac.2018.11.058

[138] M. Naushad, T. Ahamad, G. Sharma, A.H. Al-Muhtaseb, A.B. Albadarin, M.M. Alam, Z.A. ALOthman, S.M. Alshehri, A.A. Ghfar, Synthesis and characterization of a new starch/SnO2 nanocomposite for efficient adsorption of toxic Hg2+ metal ion, Chem. Eng. J. 300 (2016) 306-316. https://doi.org/10.1016/j.cej.2016.04.084

[139] F. Wang, P.R. Chang, P. Zheng, X. Ma, Monolithic porous rectorite/starch composites: fabrication, modification and adsorption, Appl. Surf. Sci. 349 (2015) 251-258. https://doi.org/10.1016/j.apsusc.2015.05.013

[140] M. Şölener, S. Tunali, A.S. Özcan, A. Özcan, T. Gedikbey, Adsorption characteristics of lead(II) ions onto the clay/poly(methoxyethyl)acrylamide (PMEA) composite from aqueous solutions, Desalination. 223 (2008) 308-322. https://doi.org/10.1016/j.desal.2007.01.221

[141] A.M. Atta, H.A. Al-Lohedan, Z.A. ALOthman, A.A. Abdel-Khalek, A.M. Tawfeek, Characterization of reactive amphiphilic montmorillonite nanogels and its application for removal of toxic cationic dye and heavy metals water pollutants, J. Ind. Eng. Chem. 31 (2015) 374-384. https://doi.org/10.1016/j.jiec.2015.07.012

[142] A.U. Rajapaksha, K.S. Dilrukshi Premarathna, V. Gunarathne, A. Ahmed, M. Vithanage, Sorptive removal of pharmaceutical and personal care products from water and wastewater, Pharm. Pers. Care Prod. Waste Manag. Treat. Technol. Emerg. Contam. Micro Pollut. (2019) 213-238. https://doi.org/10.1016/B978-0-12-816189-0.00009-3

[143] Y. Abdellaoui, M.T. Olguín, M. Abatal, B. Ali, S.E. Díaz Méndez, A.A. Santiago, Comparison of the divalent heavy metals (Pb, Cu and Cd) adsorption behavior by montmorillonite-KSF and their calcium- and sodium-forms, Superlattices Microstruct. 127 (2019) 165-175. https://doi.org/10.1016/j.spmi.2017.11.061

[144] M.E. Mahmoud, G.M. Nabil, M.M. Zaki, M.M. Saleh, Starch functionalization of iron oxide by-product from steel industry as a sustainable low cost nanocomposite for removal of divalent toxic metal ions from water, Int. J. Biol. Macromol. 137 (2019) 455-468. https://doi.org/10.1016/j.ijbiomac.2019.06.170

[145] E.I. Unuabonah, A. Taubert, Clay-polymer nanocomposites (CPNs): Adsorbents of the future for water treatment, Appl. Clay. Sci. 99 (2014) 83-92. https://doi.org/10.1016/j.clay.2014.06.016

[146] H. Zuo, R. Kukkadapu, Z. Zhu, S. Ni, L. Huang, Q. Zeng, C. Liu, H. Dong, Role of clay-associated humic substances in catalyzing bioreduction of structural Fe(III) in nontronite by Shewanella putrefaciens CN32, Sci. Tot. Env. 741 (2020) 140213. https://doi.org/10.1016/j.scitotenv.2020.140213

[147] Y. Sheng, H. Dong, R.K. Kukkadapu, S. Ni, Q. Zeng, J. Hu, E. Coffin, S. Zhao, A.J. Sommer, R.M. McCarrick, G.A. Lorigan, Lignin-enhanced reduction of structural Fe(III) in nontronite: Dual roles of lignin as electron shuttle and donor, Geochim. Cosmochim. Acta. 307 (2021) 1-21. https://doi.org/10.1016/j.gca.2021.05.037

[148] Q. Zeng, H. Dong, X. Wang, T. Yu, W. Cui, Degradation of 1, 4-dioxane by hydroxyl radicals produced from clay minerals, J. Hazard. Mater. 331 (2017) 88-98. https://doi.org/10.1016/j.jhazmat.2017.01.040

[149] K. Pastorková, K. Jesenák, M. Kadlečíková, J. Breza, M. Kolmačka, M. Čaplovičová, F. Lazišťan, M. Michalka, The growth of multi-walled carbon nanotubes on natural clay minerals (kaolinite, nontronite and sepiolite), Appl. Surf. Sci. 258 (2012) 2661-2666. https://doi.org/10.1016/j.apsusc.2011.10.114

[150] R. Singh, H. Dong, Q. Zeng, L. Zhang, K. Rengasamy, Hexavalent chromium removal by chitosan modified-bioreduced nontronite, Geochim. Cosmochim. Acta. 210 (2017) 25-41. https://doi.org/10.1016/j.gca.2017.04.030

Chapter 8

Polyethylene Terephthalate-Clay Nanocomposites

Pinku Chandra Nath[1], Nishithendu Bikash Nandi[2], and Biplab Roy[3]*

[1]Department of Bio Engineering, National Institute of Technology Agartala, Jirania, India-799046

[2]Department of Chemistry, National Institute of Technology Agartala, Jirania, Tripura, India-799046

[3]Department of Chemical Engineering, National Institute of Technology Agartala, Jirania, India-799046

* biplab.diatm24@gmail.com

Abstract

This chapter provides a comprehensive overview of clay-based PET nanocomposites. The different methods applied to prepare clay-based PET nanocomposites, their morphological and structural interpretation, improvements in mechanical and material characteristics, melt-state rheological and crystallization behavioural patterns, and, subsequently, current PET-based clay nanocomposites materials applications are all discussed.

Keywords

Poly (Ethylene Terephthalate), Clay, Nanocomposites, Characterization, Properties

Abbreviations

PET, Poly (ethylene terephthalate)
PE, Polyethylene
PVA, Polyvinyl alcohol
MMT, Montmorillonite
OMMT, Organic MMT
DCM, Dichloromethane
WXRD, Wide-angle X-ray diffraction
TEM, Transmission Electron Microscopy
DSC, Differential Scanning Calorimetry
AFM, Atomic Force Microscopy
EG, Ethylene Glycol

Adv. App. of Micro and Nano Clay II – Synthetic Polymer Composites   Materials Research Forum LLC
Materials Research Foundations 129 (2022) 183-202   https://doi.org/10.21741/9781644902035-8

SAXS, Small-Angle X-ray Scattering
TGA, Thermo Gravimetric Analysis
MAH, Maleic Anhydride
PENTA, Pentaerythritol
Mag-Q16, Magadiite-quinolinium
Hect-Q16, Hectorite- quinolinium
MMT-Q16, Montmorillonite- quinolinium
MMT-L, Montmorillonite-lauryl

Contents

Adv. App. of Micro and Nano Clay II – Synthetic Polymer Composites   Materials Research Forum LLC
Materials Research Foundations 129 (2022) 183-202   https://doi.org/10.21741/9781644902035-8

1.   Introduction

PET is a lower price thermoplastic polymer commonly applied in beverage bottles, coatings and textile fibers owing to its high physico-chemical characteristics. PET polymers are reinforced and given certain new qualities by the addition of some inorganic ingredients at the nanoscale or micro level [1]. The utilization of nanoscale agents has been increasingly popular, owing to the dramatic increases in composite characteristics that may be obtained however at lower additive loadings. Nanoclay is a prominent nanoscale ingredient used in polymer composites. Ca or Na montmorillonites are used to make nanoclay, which is a 2-D silicate with nanoparticles that have a height of 1 nm and length of several hundred nm. Nanoclay is made by modifying native Na or Ca montmorillonites with quaternary ammonium compounds. In dried form, commercialized nanoclay has particles with sizes ranging from 8-20 microns. Clay nanocomposites have homogenous dispersions and highly porous clay platelets within the polymer matrix [2]. It has been thoroughly characterized how nanoclay aggregation in polymer melt exfoliates [3]. Exfoliation occurred only when the micronized aggregation broke down into tiny tactoids and the polymer chain was intercalated and diffused into clay channels. Along with many other polymers, the global demand for PET, including its use as a matrix polymer in the preparation of nanocomposites, has grown rapidly over the last decade [3]. About 18% of the world's polymer production comprises polyester, which ranks 3rd after polypropylene (PP) and polyethylene (PE) [4].

PET is utilized in a variety of applications due to its lower price and high performance, including fibres, bottles, films, and engineering plastics for automobiles and electronics. PET has high impact resistance and is strong and naturally colorless with high transparency [5]. It has good gas barrier properties. For this reason, PET is the most popular material for soft drinks bottles. Sometimes polyvinyl alcohol (PVA) is sandwiched between PET films to give a further improvement in the gas barrier properties to meet the most challenging applications for sensitive beverages.

It is clear that PET is a very useful and important plastic from an industrial point of view. For this reason, worldwide research is being conducted on PET to make it more useful. One approach is to prepare nanocomposites of PET by using different types of nanofillers. Among the diverse nanofillers, like earth, carbon nanotubes, and metal-oxide nanoparticles, the dirt-based nanocomposites are vital since they frequently show upgraded warm and mechanical properties, heat opposition, gas porousness, and combustibility qualities. Thus, most of this part will focus on the readiness, portrayal, and properties of PET mud-based nanocomposites. The essential objective of creating PET-mud nanocomposites is to further develop the lattice's gas hindrance property for drink and food bundling [6].

2.   Types of PET based nanocomposites

PET-based composites may be categorized into three types based on nanofiller used: clay-based, carbon-nanotube based, and metal oxide nanoparticle based PET composites. Generally, majority of clays are ecologically benign and have extremely high modulus as well as aspect ratios, which seem to be the fundamental conditions for improving physical properties following nanocomposite synthesis. Carbon nanotubes and nanofibers are another two types of nanofillers that are commonly used, in addition to clay. They have received a great deal of attention due to their superior Young's modulus and higher aspect ratio, both of which are advantageous features for the production of multifunctional polymer nanocomposites. Moreover, substantial agglomeration, complex surface modification and relatively expensive cost, carbon-based nanofillers are not widely used as effective nanofillers in many applications [7]. For this reason, the most widely used nanofillers are clay. Compositions of PET based clay nanocomposites are illustrated in Table 1.

Table 1. Compositions of PET/clay nanocomposites

Materials	PET (wt%)	Modified clay (wt%)	Inorganic (wt%)	Ref.
PET	100	-	-	[8]
PET/ Montmorillonite -quinolinium	96.5	3.5 (MMT-Q16)	3	[8]
PET/ Magadiite -quinolinium	96	4.0 (Mag-Q16)	3	[9]
PET/ Montmorillonite -lauryl	88	12.0 (MMT-L)	3	[10]
PET/Hectorite- quinolinium	96.5	3.5 (Hect-Q16)	3	[11]

3.   Preparative methods

There are three major types of methods for creating nanocomposite, which can be classified as per the initial materials utilized as well as the manufacturing processes used [12].

3.1   Intercalation of PET from solution

This technique employs a soluble polymer or pre-polymer (solvent system) and swellable silicate layers. An appropriate solvent, including chloroform, water, or toluene, is used to swell the clay for the first time before use. The combination of polymer chains and clay solutions results intercalation of polymer chains into the silicate interlayers, displacing the solvent. When the solvent is removed, the intercalated structure is retained, forming a nanocomposite.

The three principal groups of layered silicate minerals are kaolinite, smectite, and illite or mica. The most frequent layered silicates employed in polymer nanocomposite technologies are smectite kinds, specifically montmorillonite (MMT), hectorite, and saponite. MMT seems to be the most widely utilized layered silicate for polymer nanocomposites due to its abundance and low cost [13]. The fundamental structure of MMT is characterized by the attraction of nearby hydrophilic platelets to one another through the use of several anionic charges and interchangeable metal counterions. In MMT's multilayered structure, there is a massive strength of ionic attraction, making it tough to split and distribute uniformly in hydrophobic polymeric matrixes [14]. This difficulty is typically resolved by altering MMT with particular organic chemicals, which has been shown to be an efficient method of enhancing the solubility between the clay solutions and polymer matrixes.

3.2   *In situ* intercalative polymerization

In this process, the clay is inflated in the presence of a liquid monomer, allowing polymer synthesis in between intercalated sheets. It is possible to commence polymerization by means of heat, by diffusion of an appropriate initiator, or by incorporation of an organic catalyst into the interlayer prior to the swelling process, depending on the circumstances.

For preparing PET nanocomposites, there are various ways of *in situ* intercalative polymerization. The first technique involves intercalating and ring-opening polymerizing ETC (ethylene terephthalate cyclic oligomers) in pre-swollen Organic MMT (OMMT) galleries. OMMT is typically swollen with dichloromethane (DCM) to achieve this effect. It is necessary to dissolve the ETC in DCM separately. An intercalated nanocomposite is created by vigorously mixing OMMT solution with ETC solution as well by drying and solvent extraction processes [15]. Multiple sessions of substantial nitrogen purging and extreme vacuum pretreatment are required throughout the polymerization process.

A PET/organically enhanced mica hybrid was created via transesterification reaction of ethylene glycol (EG) and dimethyl terephthalate (DMT) in existence of isopropyl titanate, as reported by Chang et al. (2005) [16]. Immediately after transesterification reactions, they cooled it and cleaned with distilled water and then finally the sample is vacuum dried in order to achieve PET hybrid. The sample was extruded via a capillary rheometer die following compression molding process [17].

Adv. App. of Micro and Nano Clay II – Synthetic Polymer Composites   Materials Research Forum LLC
Materials Research Foundations 129 (2022) 183-202   https://doi.org/10.21741/9781644902035-8

3.3   Melt intercalation

This process entails annealing a polymer-organically modified clay mixture statically or even under shear at the polymers softening point. Compared to in situ intercalative polymerization, this approach has several significant advantages. First and foremost, this technology is environmentally friendly due to the lack of use of organic solvents in its construction. Secondly, it can be used in conjunction with existing industrial processes including injection molding and extrusion processes.

Extremely thermally stabilized surfactants like Hexadecyl-quinolinium and L-surfactants were used to modify natural clays including MMT, magadiite, and hectorite as reported by Costache et al. (2006) [17]. In a Brabender plastic order at 280°C for 7 minutes, they created PET/clay nanocomposites via melt blending. They discovered that, while the modified clay exhibits improved thermal stability, the nanocomposites exhibit no increase in stability, and in some cases, a reduction. They felt that during high-temperature preparation, the clay begins to disintegrate and, as a result, the thermal stability of the product declines.

Hence, when producing a PET/clay nanocomposite, it is necessary to consider the clay surface's increased thermal stability, larger d-spacing, and compatibility with the PET matrix. The polar solubility component can be used to determine compatibility via group contribution approaches namely the Fedors strategy. Increased clay platelet dispersion in nanocomposite can be achieved with the introduction of a PET ionomer via melt blending process [18]. It is preferable to prepare the clay using a catalyst prior to polymerization while conducting in situ intercalation [19].

4.   Structural characterization

Wide-angle X-ray diffraction (WXRD) and transmission electron microscopy (TEM) are well-known technologies for characterizing the structural composition of clay-containing composites. Ordinarily, WXRD configurations begin with a diffraction angle of (2θ) ~2°. As a result, if the clay signal is lacking beyond the 2° diffraction angle in nanocomposite material, it is difficult to think that the clay levels are gently exfoliate in polymer matrix. Therefore, WXRD results can be in conflict with TEM results in some cases [19, 20]. The distribution of nanoparticles can also be observed directly in TEM pictures. However, it is preferable not to utilize the images of most densely packed area; rather, it is desirable to utilize the mean dispersion properties of low TEM images. To improve image quality in TEM, the clay might be stained with osmium tetroxide. It is required to conduct small-angle X-ray scattering (SAXS) investigations in order to determine the full extent of exfoliate. When using SAXS, it is possible to estimate both the chance of detecting neighbouring clay stacks within a given distance and the density profile of the electrons. This kind of study offers a far better grasp of the mean dispersion properties than other types of analysis.

It is important to incorporate 3 to 5 weight percent of organically modified clays into a polymer nanocomposite (clay based) in order to obtain a nanocomposite with better characteristics [20]. Geometric restrictions on the deformation of clay platelets are exacerbated by a rise in the quantity of clay loaded in the nanocomposite. As a result, it is extremely challenging task to obtain complete delamination of clay platelets using this method. The images acquired by SEM can also be used to visualize the huge clay clusters (0.5–1 μm in size) scattered in a matrix of clay particles. In contrast, as comparison to SEM, atomic force microscopy (AFM) images offer a considerably significantly more accurate view of the distributed clay strata in a polymer matrix [20].

5. Properties of PET based clay nanocomposites

In current days, polymer-based clay nanocomposites have considerable interest owing to its significant advantages over other nanocomposites. One of the primary goals of researchers is to enhance the different features of PET by incorporating relatively small amounts of nanoclay into the polymer matrix viz. thermal stability, mechanical property, flame retardancy, degree of crystallization etc. However, improvements of such properties are mainly depending on the extent of dispersion of clay into the polymer matrix. In the field of nanocomposite research, pristine montmorillonite (MMT) is commonly used clay [21]. The low compatibility of pristine clay, today's research mainly focused on uses of clay modified with organic surfactants in order to enhance the physical and chemical properties. Most of the cases, quaternary ammonium surfactant have served the purpose. Besides, thermally stable surfactant such as imidazolium, phosphonium and pyridinium surfactants have been found to be used [21, 22].

5.1 Thermal property and degree of crystallization

Thermal analysis is defined as a technique used to analysis the properties of a sample against time or temperature changes when the sample is heated or cooled. Among the different other thermal analysis technique, Differential scanning calorimetry (DSC) is one of the most frequently used tool to analyze polymers, developed by Ardakani et al. (2017) [23]. DSC measures quantitatively amount of heat flow either absorbed or released with time or temperature changes. It is also allowed us to study phase changes, melting point, heat of fusion, kinetics of crystallization, stability etc.

Glass transition temperature (T_g):- The temperature at which substance undergoes changes from hard brittle to soft rubbery state [24].

Crystallization temperature (T_c):-The temperature at which crystallization process takes place called Crystallization temperature.

Half time of crystallization ($t_{1/2}$):- It denotes the rate of crystallization that is proportional to the combined impact of nucleation and crystal growth, as determined by Eq. 1 [25].

$$t_{\frac{1}{2}} = \frac{T^c_{on} - T^c_c}{X} \qquad (1)$$

$$Where, \qquad T^c_{on} = onset\ of\ crystallisation$$

$$T^c_c = crystallization\ at\ the\ exothermic\ peak$$

$$X = cooling\ rate$$

Degree of crystallinity (X_c): Crystallization is a process by which atom/molecules arranged themselves in a definite ordered manner. From the thermodynamic point of view, crystallization is an exothermic process. It takes place in two steps viz. nucleation and crystal growth. The degree of crystallinity is defined by the following Eq. 2, where ΔH_m is second melting enthalpy of the sample (J/g), $\emptyset$ is the weight fraction of filler and ΔH^o_m is the enthalpy value of 100% crystalline form [26].

$$X_c = \frac{\Delta H_m}{(1 - \emptyset)\Delta H^o_m} \times 100 \qquad (2)$$

From the DSC study conducted by Chang and Moon (2007) [27] with recycled PET(r-PET) and Cloisite10A nanocomposite, it has been observed that addition of clay content (1-6 %wt) into polymer matrix, T_c and X_c slightly increases. The author suggested this increase in T_c and X_c could be owing to the nucleation impact of clay platelets in PET matrix. In addition, reduction of T_g on introducing clay in polymer because of higher mobility of polymer chain in presence of clay. Choi et al. (2006) [28] reported that, $t_{1/2}$ value for nanocomposites are lower than those of extruded PET, though Cloisite 30B shows higher rate of crystallization. Moreover, modification of Cloisite 30B with oleic acid exhibit smaller effect on rate of crystallization due to improvement of dispersion of nanoparticles into polymer matrix. The smallest $t_{1/2}$ value is 1.35 min and 1.45 min found for 90/10 PET/8 K-M organoclay and 90/10 PET/8 K-P organoclay nanocomposites respectively [29]. Similar way studied on polymer-based nanocomposites described that nano-sized clay particles affect the crystallization behavior of polymer matrix. In all cases, the crystallization process of clay/PET nanocomposite is much more times higher than that of pure PET [30]. This indicates that clay nano-particle dispersed in matrix acts as nucleating agent. Particles enhance the rate of crystallization by decreasing the interfacial surface energy. However higher clay content of clay nano-particles in nanocomposites slows down crystallization rate owing to the agglomeration of the organoclay in the PET matrix [31].

5.2   Mechanical properties

The different mechanical properties of nanocomposite are largely depending on the modification of clay surface, chemical treatment as well as processing conditions. To enhance mechanical properties of nanocomposite compared to polymer, it is necessary for the nanoclay to be dispersed well over the polymer matrix.

In case of phosphonium modified PET nanocomposites, there is a significant enhancement of modulus when clay incorporates into the polymer matrix because of higher modulus of clay compared to polymer [32]. Tensile modulus is useful mechanical property of a material that actually measures the stiffness *i.e.,* how much material deform when subjected to a particular load. It is defined as the ratio of tensile stress to its strain. Experimentally, it is determined from the slope of stress strain curve. However, the strength as well as elongation at break decreased as mechanical property mainly depends on the extent of delamination of nanoparticles into polymer matrix [33].

Tensile strength is one of the mechanical properties which are defined as maximum amount of load that a material experienced before it stretches or break. The effect of clay content with tensile strength investigated thoroughly and it has been found that the tensile strength and tensile module increases with the increasing clay content. The increasing percentage is about 10% and 28% for tensile strength and module respectively with 6% weight percentage of clay [34]. This happens due to homogeneous dispersion of clay in the PET matrix. However, a sharp reduction of elongation at break from 3.7% to 3.3% observed which is common in clay reinforcement PET matrix system [35]. The epoxy group directly bound with polymer chain results in enhancement of delamination (Fig. 1B). The synthetic scheme for preparation of r-PET modified organoclay (MMT-IM organoclay) is demonstrated in Fig. 1A. According to tensile test, incorporating MMT-IM to the recycled PET (2170 MPa) as well as virgin PET (2286 MPa) leads to the increase of Young modulus value 2628 MPa and 2898 MPa respectively (Table 2). In some cases, silicates-based montmorillonite clay PET nanocomposite showed significant improvement of mechanical properties with the addition of additives like maleic anhydride (MAH) and pentaerythritol (PENTA).

Adv. App. of Micro and Nano Clay II – Synthetic Polymer Composites   Materials Research Forum LLC
Materials Research Foundations 129 (2022) 183-202   https://doi.org/10.21741/9781644902035-8

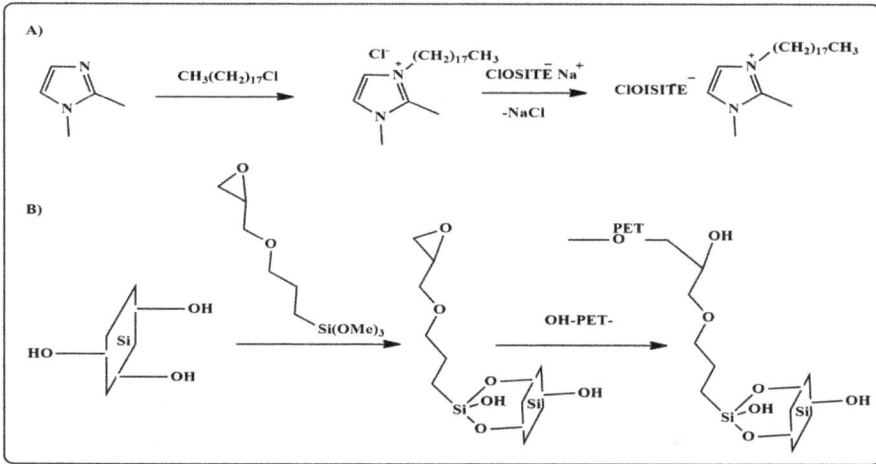

Figure 1. A) Preparation of MMT-IM organoclay; B) Polymer chain epoxi-silanization and bonding to the silicate layer

Table 2. Mechanical property of PET-clay based nanocomposites

Material	Tensile Strength [MPa]	Young's Modulus [MPa]	Elongation at Break [%]	Ref.
PET-R(IM)	20.6	2628	210.4	[36]
PET-IM	30.4	2898	227	[36]
PET-R	54.7	2170	316.5	[37]
PET	57.1	2286	327.2	[38]

5.3   Rheological properties

Rheological study of a polymer nanocomposite is an effective tool to indirectly probe the structures along with intermolecular interactions of nanocomposites. It is also used for characterizing the state of clay dispersion and exfoliation of nanoplatelets in polymer matrices [39]. The prime advantage of rheological study over other technique is that measurements are taken in molten state and method is useful for both linear as well as non-linear deformation. Besides, it also provides insights for improving the processability of polymer nanocomposites [40]. Till today, rheological behavior of numerous numbers of polymeric nanocomposites system have been reported and related processes have been modeled [41]. The preparation and rheological property of PET/o-MMT nanocomposites by direct melt compounding in presence of polyester ionomer [42]. Study revealed that complex viscosity was much higher than that of neat PET at low frequencies. Increase in polyester ionomer (PETi), improved the dispersion of O-MMT into PET matrix [43]. Giraldi monitored how screw rotation speed alters the rheological behavior of recycled PET clay nanocomposites prepared by Twin-Screw Extruder. In absence of antioxidant, the intrinsic viscosity reduced to 22% after the extrusion. However, reduction was decreased to 16%, upon the addition of Irganox B561 (antioxidant). This result confirms that antioxidant had some serious influence on avoiding degradation.

5.4   Thermal stability

Polyester based nanocomposites have wide range of applications ranging from textile to packaging. PET based nanocomposites was prepared at elevated temperature (above 200°C), thus there is a chance of degradation. To avoid such degradation, it is better to kept it in vacuum overnight before processing. Therefore. researcher mainly focused on development of organoclay that are thermally stable at high temperature. In addition, the processing time is also reduced by preparing sample in extruder as these method are responsible for degradation of commonly used ammonium cation, which cause decomposition of the polymer matrix [44]. In a study conducted by Roy et al. (2012) [44], three different organoclay viz. Ammonium-modified silicate clay (Cloisite 30B), phosphonium and imidazolium-modified montmorillonites melt compounded with PET in a co-rotating twin-screw extruder. Thermo gravimetric analysis (TGA) is a technique for determining thermal stability of any compound in which mass changes is measured with temperature changes. If there is no mass change over a temperature range, then the species is considered as thermally stable [45].   An investigation of the thermal stability characteristics of Recycled PET (r-PET) modified Cloisite 10A nanocomposites was carried out using TGA under oxidative as well as pyrolytic environment. The compounds exhibited two steps degradation in TGA curve, which is concomitant with the previously reported systems [46]. Initial decomposition was mostly caused by degrading polymer chains begun by end-groups, which was followed by breakdown of products generated during the initial degradation process. Because of the destruction of cross-linked carbon - containing components, the second degradation of PET occurs between 520 and 640°C. For PET nanocomposites, the TGA curve is in similar pattern than that of PET [47]. Moreover, the second decomposition, which takes place at 520-720°C becomes broader and broadness increases with increase in clay content (1-6 wt%) [48]. Hu et al. (2012) [49] concluded that these types of broadening observed may be due to the large number of cross linked structure formation in nanocomposites resulting slow decomposition process in wide range of temperature. In the pyrolytic environment, one sharp breakdown was observed as the carbonaceous crosslink structures formed in the first step, remain intact in presence of nitrogen atmosphere. Due to decomposition of alkyl ammonium enhancer which exist in Cloisite 10A, temperatures at 1% and 5% weight loss as well as the onset of decomposition is lower for the nano-composites compared to PET was observed [50]. On the other hand, phosphonium salts like (4-carboxybutyl) triphenylphosphonium bromide and dodecyltriphenyl phosphonium chloride, 2-carboxyethyl (phenylphosphinic) acid (HPPPA) was used in MMT based PET/clay nanocomposites to enhance thermal stability characteristics. Cui et al. (2015) [51] suggested that, alkyl ammonium modified clay degraded at 200-300°C by Hoffmann elimination reaction mechanism results in the formation of Bronsted acidic sites along with amines and alpha-olefines which subsequently facilitated the process of degradation (Fig. 2). In addition, Cui et al. (2015) [51] examined thoroughly how hydroxyl group (–OH) on the edge of clay pellets and ammonium linker influence the degradation process.

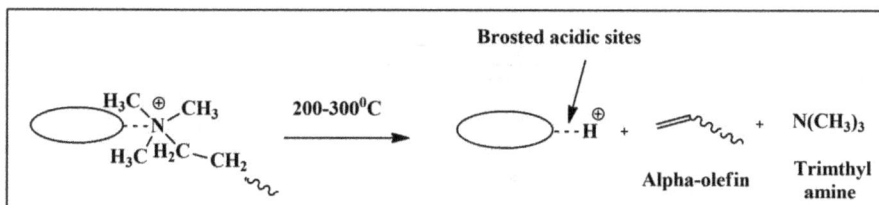

Figure 2. Hoffmann elimination in alkyl ammonium modified clay

5.5   Gas-barrier properties

Throughout the most recent years, polymers are generally utilized in bundling applications attributable to its minimal expense, lightweight and simplicity of preparing. Late examinations showed that about 40% of polymers have applications in bundling materials and a big part of them are utilized in food bundling as films, sheets, and bottles and so forth [52]. In spite of being flexible applications, the burdens of polymeric materials in food packeting are their innate penetrability to gases and fumes, including oxygen, carbon dioxide, and natural fumes. Improvement of gas obstruction property in nanocomposites to make the materials more attractive is one of the principle worries by the analyst [53]. The obstruction properties of nanocomposites are upgraded by joining of impermeable lamellar fillers. For the most part Graphene and Montmorillonite (MMT) served the purpose which adjust the gas entrance way. These nanofillers impeded gas particles dispersion and increment the convolution brings about a lengthy and convoluted voyaging pathway of the diffusing gas to further develop the gas-obstruction properties of the composites [54].

Reichert et al. (2020) [55] reported, the in situ blend of Poly (ethylene terephthalate) (PET)/mud nanocomposites (PCNs) with N-methyl diethanol amine (MDEA) - based organoclay and investigated gas boundary property by estimating oxygen penetrability through the PCN sheet. Oxygen obstruction property of MEDA-III PCN is found to show two overlay upgrades with just 1% wt earth composite contrasted with unadulterated PET. Also, in some mud upheld impetus, O_2 porousness decrease is practically 11.3-15.6 with 1-5%wt mud. The instrument of intercalation of impetus into dirt interlayer's by metathesis response is summed up in Fig. 3.

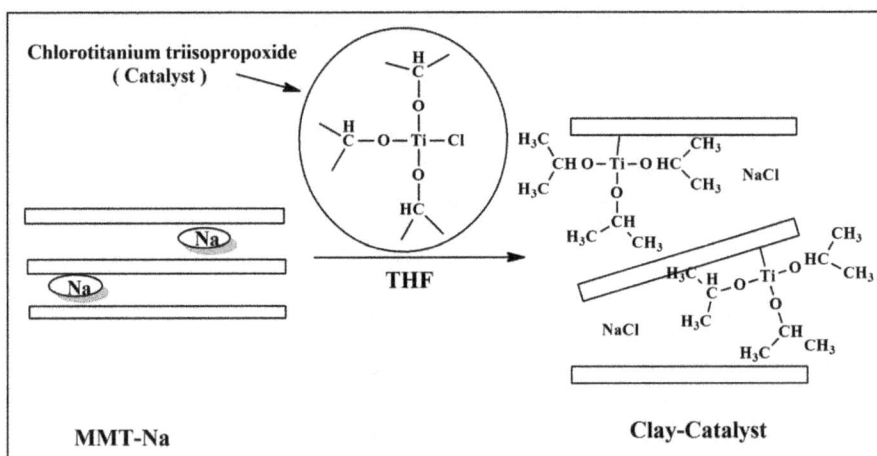

Figure 3. Intercalation of catalyst into clay interlayer's by metathesis reaction

6.   Applications of PET based clay nanocomposites

PET is a synthetic fiber that is extensively utilized in textile industries and contains the highest percentage of synthetic fibers on the marketplace. It can be utilized as a stand-alone substance or in combination with cellulose. Fiber made of PET is by far the most crystalline and compact of all fibres, and it has a distinct hydrophobic characteristic. Its hydrophobic characteristic makes it difficult to remove moisture out from body and difficult to cleanse. The fiber industry has long sought to improve PET's passive and lethargic behaviour through surface modification despite losing bulk qualities [56]. Alkaline hydrolysis of PET fibers is often used to enhance soil release characteristics and moisture regain. Recently, PET/SiO$_2$ nanocomposites have been developed via *in situ* polymerization and melt spinning to produce fibers. Such nanocomposites exhibit tougher superfine structures, which facilitate certain applications like deep dyeing. An experiment was conducted on a

Adv. App. of Micro and Nano Clay II – Synthetic Polymer Composites   Materials Research Forum LLC
Materials Research Foundations 129 (2022) 183-202   https://doi.org/10.21741/9781644902035-8

PET/cotton material sheet in order to enhance the mechanical qualities of the fabric by depositing several resin/clay mixes with varying clay percentages. Furthermore, a variety of mechanical tests, including tear tests, tensile, abrasion, and breakout were carried out. Except for polyvinyl acetate, which exhibited a drop in fabric tear resistance while utilizing greater than 20% clay, other resins seemed to improve mechanical strength when using additional clay. The coated cloth without clay had much lower mechanical properties than the reference, but improved while clay was introduced to resin. As a consequence, it is evident that the kind of resin used is significant, as the outcomes vary through one resin to another [57, 58].

Conclusion

The super downstream enterprises dependent on PET are the creation of polyester filaments, representing around 65% of worldwide utilization, and PET container gums devouring around 30%. Different applications are for polyester film and polyester designing tars. Albeit much exploration has been done to work on the mechanical, obstruction, and warm properties of PET tar, it is as yet important to work on these properties to satisfy the high needs on this sap. This section gives an extensive outline of clay based PET nanocomposites. The various strategies applied to get ready clay-based PET nanocomposites, their primary and morphological portrayal, worked on mechanical and physical characteristics, soften state rheological and crystallization bahavioural patterns, and, lastly current PET-based clay nanocomposites materials applications are completely examined.

References

[1] A.K. Singh, R. Bedi, B.S. Kaith, Composite materials based on recycled polyethylene terephthalate and their properties - A comprehensive review, Compos. Part B Eng. 219 (2021) 108928. https://doi.org/10.1016/j.compositesb.2021.108928

[2] S. Fu, Z. Sun, P. Huang, Y. Li, N. Hu, Some basic aspects of polymer nanocomposites: A critical review, Nano Mater. Sci. 1 (2019) 2-30. https://doi.org/10.1016/j.nanoms.2019.02.006

[3] D.R. Paul, L.M. Robeson, Polymer nanotechnology: Nanocomposites, Polymer (Guildf). 49 (2008) 3187-3204. https://doi.org/10.1016/j.polymer.2008.04.017

[4] S. Sinha Ray, M. Okamoto, Polymer/layered silicate nanocomposites: a review from preparation to processing, Prog. Polym. Sci. 28 (2003) 1539-1641. https://doi.org/10.1016/j.progpolymsci.2003.08.002

[5] V. Mittal, Polymer Layered Silicate Nanocomposites: A Review, Materials (Basel). 2 (2009) 992-1057. https://doi.org/10.3390/ma2030992

[6] S.-S. Lee, Y.T. Ma, H.-W. Rhee, J. Kim, Exfoliation of layered silicate facilitated by ring-opening reaction of cyclic oligomers in PET-clay nanocomposites, Polymer (Guildf). 46 (2005) 2201-2210. https://doi.org/10.1016/j.polymer.2005.01.006

[7] S.Y. Hwang, W.D. Lee, J.S. Lim, K.H. Park, S.S. Im, Dispersibility of clay and crystallization kinetics forin situ polymerized PET/pristine and modified montmorillonite nanocomposites, J. Polym. Sci. Part B Polym. Phys. 46 (2008) 1022-1035. https://doi.org/10.1002/polb.21435

[8] M.C. Costache, M.J. Heidecker, E. Manias, C.A. Wilkie, Preparation and characterization of poly(ethylene terephthalate)/clay nanocomposites by melt blending using thermally stable surfactants, Polym. Adv. Technol. 17 (2006) 764-771. https://doi.org/10.1002/pat.752

[9] T. Mariappan, D. Yi, A. Chakraborty, N.K. Singha, C.A. Wilkie, Thermal stability and fire retardancy of polyurea and epoxy nanocomposites using organically modified magadiite, J. Fire Sci. 32 (2014) 346-361. https://doi.org/10.1177/0734904113516268

[10] A. Sanchez-Solis, A. Garcia-Rejon, O. Manero, Production of nanocomposites of PET-montmorillonite clay by an extrusion process, Macromol. Symp. 192 (2003) 281-292. https://doi.org/10.1002/masy.200390038

[11] D. Bikiaris, Can nanoparticles really enhance thermal stability of polymers? Part II: An overview on thermal decomposition of polycondensation polymers, Thermochim. Acta. 523 (2011) 25-45. https://doi.org/10.1016/j.tca.2011.06.012

[12] S. Sinha Ray, M. Okamoto, Polymer/layered silicate nanocomposites: a review from preparation to processing, Prog. Polym. Sci. 28 (2003) 1539-1641. https://doi.org/10.1016/j.progpolymsci.2003.08.002

[13] A. Amin, E.H. Ahmed, M.W. Sabaa, M.M.H. Ayoub, I.K. Battisha, Dielectric Behavior of Some Vinyl Polymers/Montmorillonite Nanocomposites on the Way to Apply Them as Semiconducting Materials, Open J. Org. Polym. Mater. 03 (2013) 73-80. https://doi.org/10.4236/ojopm.2013.33012

[14] M. Yin, C. Li, G. Guan, X. Yuan, D. Zhang, Y. Xiao, In-situ synthesis of poly(ethylene terephthalate)/clay nanocomposites using TiO 2 /SiO 2 sol-intercalated montmorillonite as polycondensation catalyst, Polym. Eng. Sci. 49 (2009) 1562-1572. https://doi.org/10.1002/pen.21388

[15] S.-S. Lee, Y.T. Ma, H.-W. Rhee, J. Kim, Exfoliation of layered silicate facilitated by ring-opening reaction of cyclic oligomers in PET-clay nanocomposites, Polymer (Guildf). 46 (2005) 2201-2210. https://doi.org/10.1016/j.polymer.2005.01.006

[16] J.-H. Chang, M.K. Mun, I.C. Lee, Poly(ethylene terephthalate) nanocomposite fibers by in situ polymerization: The thermomechanical properties and morphology, J. Appl. Polym. Sci. 98 (2005) 2009-2016. https://doi.org/10.1002/app.22382

[17] M.C. Costache, M.J. Heidecker, E. Manias, C.A. Wilkie, Preparation and characterization of poly(ethylene terephthalate)/clay nanocomposites by melt blending using thermally stable surfactants, Polym. Adv. Technol. 17 (2006) 764-771. https://doi.org/10.1002/pat.752

[18] A. Ammala, C. Bell, K. Dean, Poly(ethylene terephthalate) clay nanocomposites: Improved dispersion based on an aqueous ionomer, Compos. Sci. Technol. 68 (2008) 1328-1337. https://doi.org/10.1016/j.compscitech.2007.12.012

[19] T.-Y. Tsai, C.-H. Li, C.-H. Chang, W.-H. Cheng, C.-L. Hwang, R.-J. Wu, Preparation of Exfoliated Polyester/Clay Nanocomposites, Adv. Mater. 17 (2005) 1769-1773. https://doi.org/10.1002/adma.200401260

[20] M. Kráčalík, M. Studenovský, J. Mikešová, J. Kovářová, A. Sikora, R. Thomann, C. Friedrich, Recycled PET-organoclay nanocomposites with enhanced processing properties and thermal stability, Journal of Applied Polymer Science. 106 (2007) 2092-2100. https://doi.org/10.1002/app.26858

[21] A. Sánchez-Solís, I. Romero-Ibarra, M.R. Estrada, F. Calderas, O. Manero, Mechanical and rheological studies on polyethylene terephthalate-montmorillonite nanocomposites, Polymer Engineering & Science. 44 (2004) 1094-1102. https://doi.org/10.1002/pen.20102

[22] A.L.F. de M. Giraldi, M.T.M. Bizarria, A.A. Silva, J.I. Velasco, M.A. D'Ávila, L.H.I. Mei, Effects of extrusion conditions on the properties of recycled poly(ethylene terephthalate)/nanoclay nanocomposites prepared by a twin-screw extruder, Journal of Applied Polymer Science. 108 (2008) 2252-2259. https://doi.org/10.1002/app.27280

[23] K. Majdzadeh-Ardakani, S. Zekriardehani, M.R. Coleman, S.A. Jabarin, A Novel Approach to Improve the Barrier Properties of PET/Clay Nanocomposites, Int. J. Polym. Sci. 2017 (2017) 1-10. https://doi.org/10.1155/2017/7625906

[24] S.E. Vidotti, A.C. Chinellato, G.-H. Hu, L.A. Pessan, Preparation of poly (ethylene terephthalate)/organoclay nanocomposites using a polyester ionomer as a

compatibilizer, Journal of Polymer Science Part B: Polymer Physics. 45 (2007) 3084-3091. https://doi.org/10.1002/polb.21311

[25] S.H. Kim, S.C. Kim, Synthesis and properties of poly(ethylene terephthalate)/clay nanocomposites byin situ polymerization, Journal of Applied Polymer Science. 103 (2007) 1262-1271. https://doi.org/10.1002/app.25120

[26] X.-G. Ge, D.-Y. Wang, C. Wang, M.-H. Qu, J.-S. Wang, C.-S. Zhao, X.-K. Jing, Y.-Z. Wang, A novel phosphorus-containing copolyester/montmorillonite nanocomposites with improved flame retardancy, European Polymer Journal. 43 (2007) 2882-2890. https://doi.org/10.1016/j.eurpolymj.2007.03.040

[27] J.-H. Chang, M.K. Mun, Nanocomposite fibers of poly(ethylene terephthalate) with montmorillonite and mica: thermomechanical properties and morphology, Polymer International. 56 (2007) 57-66. https://doi.org/10.1002/pi.2110

[28] W. Joon Choi, H.-J. Kim, K. Han Yoon, O. Hyeong Kwon, C. Ik Hwang, Preparation and barrier property of poly(ethylene terephthalate)/clay nanocomposite using clay-supported catalyst, Journal of Applied Polymer Science. 100 (2006) 4875-4879. https://doi.org/10.1002/app.23268

[29] X. Xu, Y. Ding, Z. Qian, F. Wang, B. Wen, H. Zhou, S. Zhang, M. Yang, Degradation of poly(ethylene terephthalate)/clay nanocomposites during melt extrusion: Effect of clay catalysis and chain extension, Polymer Degradation and Stability. 94 (2009) 113-123. https://doi.org/10.1016/j.polymdegradstab.2008.09.009

[30] R.R. Chowreddy, K. Nord-Varhaug, F. Rapp, Recycled Poly(Ethylene Terephthalate)/Clay Nanocomposites: Rheology, Thermal and Mechanical Properties, Journal of Polymers and the Environment. 27 (2019) 37-49. https://doi.org/10.1007/s10924-018-1320-6

[31] B. Lecouvet, M. Sclavons, S. Bourbigot, C. Bailly, Towards scalable production of polyamide 12/halloysite nanocomposites via water-assisted extrusion: mechanical modeling, thermal and fire properties, Polym. Adv. Technol. 25 (2014) 137-151. https://doi.org/10.1002/pat.3215

[32] C.N. Barbosa, F. Chabert, V. Nassiet, J.C. Viana, P. Pereira, Effect of Clay Amounts on Morphology and Mechanical Performances in Multiscale PET Composites, in: 2011: pp. 779-784. https://doi.org/10.1063/1.3589610

[33] C. Gao, S. Zhang, F. Wang, B. Wen, C. Han, Y. Ding, M. Yang, Graphene Networks with Low Percolation Threshold in ABS Nanocomposites: Selective Localization and

Electrical and Rheological Properties, ACS Appl. Mater. Interfaces. 6 (2014) 12252-12260. https://doi.org/10.1021/am501843s

[34] Y.-J. Ma, C.-W. Xia, H.-Y. Yang, R.J. Zeng, A rheological approach to analyze aerobic granular sludge, Water Res. 50 (2014) 171-178. https://doi.org/10.1016/j.watres.2013.11.049

[35] R. Krishnamoorti, E.P. Giannelis, Rheology of End-Tethered Polymer Layered Silicate Nanocomposites, Macromolecules. 30 (1997) 4097-4102. https://doi.org/10.1021/ma960550a

[36] M. Kráčalík, M. Studenovský, J. Mikešová, J. Kovářová, A. Sikora, R. Thomann, C. Friedrich, Recycled PET-organoclay nanocomposites with enhanced processing properties and thermal stability, J. Appl. Polym. Sci. 106 (2007) 2092-2100. https://doi.org/10.1002/app.26858

[37] J. Bandyopadhyay, S. Sinha Ray, Clay-containing poly(ethylene terephthalate) (PET)-based polymer nanocomposites, in: Adv. Polym. Nanocomposites, Elsevier, 2012: pp. 277-320. https://doi.org/10.1533/9780857096241.2.277

[38] F. Awaja, D. Pavel, Recycling of PET, Eur. Polym. J. 41 (2005) 1453-1477. https://doi.org/10.1016/j.eurpolymj.2005.02.005

[39] S.E. Vidotti, A.C. Chinellato, G.-H. Hu, L.A. Pessan, Preparation of poly(ethylene terephthalate)/organoclay nanocomposites using a polyester ionomer as a compatibilizer, J. Polym. Sci. Part B Polym. Phys. 45 (2007) 3084-3091. https://doi.org/10.1002/polb.21311

[40] J. Bandyopadhyay, S. Sinha Ray, Clay-containing poly(ethylene terephthalate) (PET)-based polymer nanocomposites, in: Adv. Polym. Nanocomposites, Elsevier, 2012: pp. 277-320. https://doi.org/10.1533/9780857096241.2.277

[41] S.S. Ray, Recent Trends and Future Outlooks in the Field of Clay-Containing Polymer Nanocomposites, Macromol. Chem. Phys. 215 (2014) 1162-1179. https://doi.org/10.1002/macp.201400069

[42] A.M. Elbaz, A. Gani, N. Hourani, A.-H. Emwas, S.M. Sarathy, W.L. Roberts, TG/DTG, FT-ICR Mass Spectrometry, and NMR Spectroscopy Study of Heavy Fuel Oil, Energy & Fuels. 29 (2015) 7825-7835. https://doi.org/10.1021/acs.energyfuels.5b01739

[43] A.K. Singh, R. Bedi, B.S. Kaith, Composite materials based on recycled polyethylene terephthalate and their properties - A comprehensive review, Compos. Part B Eng. 219 (2021) 108928. https://doi.org/10.1016/j.compositesb.2021.108928

[44] N. Roy, R. Sengupta, A.K. Bhowmick, Modifications of carbon for polymer composites and nanocomposites, Prog. Polym. Sci. 37 (2012) 781-819. https://doi.org/10.1016/j.progpolymsci.2012.02.002

[45] C.I. Idumah, C.M. Obele, Understanding interfacial influence on properties of polymer nanocomposites, Surfaces and Interfaces. 22 (2021) 100879. https://doi.org/10.1016/j.surfin.2020.100879

[46] K. Cao, C.P. Siepermann, M. Yang, A.M. Waas, N.A. Kotov, M.D. Thouless, E.M. Arruda, Reactive Aramid Nanostructures as High-Performance Polymeric Building Blocks for Advanced Composites, Adv. Funct. Mater. 23 (2013) 2072-2080. https://doi.org/10.1002/adfm.201202466

[47] H. Lu, S. Nutt, Restricted Relaxation in Polymer Nanocomposites near the Glass Transition, Macromolecules. 36 (2003) 4010-4016. https://doi.org/10.1021/ma034049b

[48] X. Xu, Y. Ding, Z. Qian, F. Wang, B. Wen, H. Zhou, S. Zhang, M. Yang, Degradation of poly(ethylene terephthalate)/clay nanocomposites during melt extrusion: Effect of clay catalysis and chain extension, Polym. Degrad. Stab. 94 (2009) 113-123. https://doi.org/10.1016/j.polymdegradstab.2008.09.009

[49] J. Hu, Y. Zhu, H. Huang, J. Lu, Recent advances in shape-memory polymers: Structure, mechanism, functionality, modeling and applications, Prog. Polym. Sci. 37 (2012) 1720-1763. https://doi.org/10.1016/j.progpolymsci.2012.06.001

[50] A.P. Kumar, D. Depan, N. Singh Tomer, R.P. Singh, Nanoscale particles for polymer degradation and stabilization-Trends and future perspectives, Prog. Polym. Sci. 34 (2009) 479-515. https://doi.org/10.1016/j.progpolymsci.2009.01.002

[51] Y. Cui, S. Kumar, B. Rao Kona, D. van Houcke, Gas barrier properties of polymer/clay nanocomposites, RSC Adv. 5 (2015) 63669-63690. https://doi.org/10.1039/C5RA10333A

[52] M. Dong, H. Zhang, L. Tzounis, G. Santagiuliana, E. Bilotti, D.G. Papageorgiou, Multifunctional epoxy nanocomposites reinforced by two-dimensional materials: A review, Carbon N. Y. 185 (2021) 57-81. https://doi.org/10.1016/j.carbon.2021.09.009

[53] X.-G. Ge, D.-Y. Wang, C. Wang, M.-H. Qu, J.-S. Wang, C.-S. Zhao, X.-K. Jing, Y.-Z. Wang, A novel phosphorus-containing copolyester/montmorillonite nanocomposites with improved flame retardancy, Eur. Polym. J. 43 (2007) 2882-2890. https://doi.org/10.1016/j.eurpolymj.2007.03.040

[54] Y. Gao, J. Wu, Q. Wang, C.A. Wilkie, D. O'Hare, Flame retardant polymer/layered double hydroxide nanocomposites, J. Mater. Chem. A. 2 (2014) 10996. https://doi.org/10.1039/c4ta01030b

[55] S. Michałowski, K. Pielichowski, Nanoparticles as flame retardants in polymer materials: mode of action, synergy effects, and health/environmental risks, in: Heal. Environ. Saf. Nanomater., Elsevier, 2021: pp. 375-415. https://doi.org/10.1016/B978-0-12-820505-1.00017-1

[56] C.L. Reichert, E. Bugnicourt, M.-B. Coltelli, P. Cinelli, A. Lazzeri, I. Canesi, F. Braca, B.M. Martínez, R. Alonso, L. Agostinis, S. Verstichel, L. Six, S. De Mets, E.C. Gómez, C. Ißbrücker, R. Geerinck, D.F. Nettleton, I. Campos, E. Sauter, P. Pieczyk, M. Schmid, Bio-Based Packaging: Materials, Modifications, Industrial Applications and Sustainability, Polymers (Basel). 12 (2020) 1558. https://doi.org/10.3390/polym12071558

[57] M. Maiti, M. Bhattacharya, A.K. Bhowmick, Elastomer Nanocomposites, Rubber Chem. Technol. 81 (2008) 384-469. https://doi.org/10.5254/1.3548215

[58] F. Rault, S. Giraud, F. Salaün, Flame Retardant/Resistant Based Nanocomposites in Textile, in: 2015: pp. 131-165. https://doi.org/10.1007/978-3-319-03467-6_6

Chapter 9

Polypropylene/Clay Nanocomposites

M. Ramesh*, M. Tamil Selvan, P. Hariprasad, D. Ashok Kumar

Department of Mechanical Engineering, KIT-Kalaignarkarunanidhi Institute of Technology, Coimbatore, Tamil Nadu, India

* mramesh97@gmail.com

Abstract

Polymers play a vital role in material science; they have been utilised for more than a century. Cosmetics, geotextiles, aircraft, and automobiles are just a few of the sectors that employ polypropylene. The addition of appropriate clay, can improve the mechanical and barrier characteristics of the material. This chapter deals with various types of fabrication as extrusion process by using twin screw extruder, solution blending method, melt-blending method, in-situ polymerization method, direct melt compounding, ultra sound-aided extrusion and master batch dilution. To characterize the polypropylene/clay nano composites by different property such as mechanical, thermal, tribological, optical, viscoelasticity, viscoplasticity, creep failure, hygrothermal, rheological and morphological.

Keywords

Polypropylene, Clay, Nanocomposites, Properties, Characterization

Contents

1.   Introduction

Polymers have a wide range of applications in the packaging, car, aircraft, and infrastructure materials due to their significant characteristics. By using nano-fillers into the polymer, we can produce polypropylene, nylon, polyethylene, polystyrene, epoxy resins etc. In that polypropylene is a thermoplastic polymer belonging to the poly family (a-olefins). Because of its great characteristics, such as cheap price and reduced densities, this polymer is regarded one of the viable options to substitute industrial plastic products. Nanocomposites offer great barrier properties that are beneficial in product packaging. The gas barriers characteristic is improved by adding tiny amounts of nano clay. Nano clay improves clarity and lifespan while also reducing haze. Isotactic polypropylene is a common commodities polymer that has been widely utilised in a spectrum of uses, includes vehicle internal and external decorating materials, electrical distribution products, and so on. Although from a research and commercial standpoint, polypropylene has a basic molecular structure $[CH_2\text{-}CH(CH_3)]_n$ and is one of the most fundamental and significant polymers. Polypropylene is one of the most cohesive looks of the polyolefin group from an industrial standpoint because of its exceptional advent of social cost, lighter volume, mild thermal distort temperature of around 100 °C, and excellent utility in regards of characteristics usage and recycle. Because of its uniform structural behaviour, polypropylene has a greater crystalline structure. It has a crystallite size that is halfway between higher and lower density polyethylene. The thin polymeric polypropylene has excellent mechanical, electromechanical, and physiological characteristics. It has a good durability and resists stress breaking well. Insulating characteristics are beneficial in a variety of industrial purposes. Hazard, acids, de-icing chemicals, and galvanic assaults have no effect on it [1, 2].

Polypropylene is utilised in a variety of household items in our everyday routines. Polypropylene packaging materials are high-quality and may be securely cleaned in the washer. It's utilised in the production of grocery bags and aerosol can. Coloured polypropylene fibres in textiles create attractive and long-lasting carpets. The usage of polypropylene in the automobile industry has grown to include bumper and batteries boxes. Its exterior doesn't really support microbial growth, making it suitable for use in a variety of medical devices. Compressor and various sorts of pipelines are manufactured with it in the building industry. Package, labelling, stationary, scientific instruments, and polymers are all key uses. Because of their better mechanical strength and heat tolerance, nanomaterials founded on stacked ionic substances such as clay particles are widely used. The kind of clay and the technique of pre-processing, the polymeric material chosen, and the technique of clay infiltration all have an impact on the nanoparticle characteristics. Clay quality and distribution have an impact on the characteristics of polymer nanocomposites. Because of their polar differences, polymer and clays are not mixable [3, 4]. Nanoclay crystal lattice structure is presented in Fig. 1 [3].

The clay of class montmorillonite, whose belongs to the phyllosilicates group, is used to make these customised clays. Each leaf is made up of an octahedral layer of aluminium sandwiched between two exterior tetrahedral layers of silicon, with the octahedral layer sharing its oxygen molecules with the tetrahedral layer. From a pragmatic perspective, it is critical to be able to estimate certain features. It is critical to contrast research data with conventional micromechanical models offered for such systems in order to enhance nanocomposite properties. A range of simulations for predicting mechanical characteristics, and porosity in nanoparticle-containing systems have been suggested. In the next part, we'll go through several of these approaches [5].

Adv. App. of Micro and Nano Clay II – Synthetic Polymer Composites   Materials Research Forum LLC
Materials Research Foundations 129 (2022) 203-232   https://doi.org/10.21741/9781644902035-9

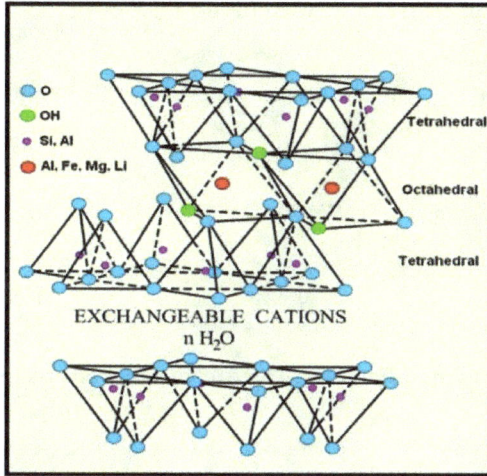

Fig. 1 Nanoclay crystal lattice structure [3].

2.   Fabrication of polypropylene/clay nanocomposites

One of the most extensively used polyolefin polymers is polypropylene. Polypropylene–clay nanomaterials have increasingly got a lot of interest due to their impressive characteristic's enrichments at much lower mass loading conditions than traditional fillers. polypropylene/clay nanocomposites have showed significantly tensile strengths, as well as enhanced barriers resilience, flame retardance, and thermal properties after only a minimum amount of clay was added. The use of a minuscule portion of inorganic filler, such as clay to improve the physical characteristics of polypropylene has been widely described in the research papers. Because polypropylene backbone lacks polar functional groups, it was anticipated that homogenous diffusion of the interlayer space would be problematic. By combining polypropylene and clay using maleic anhydride-grafted polypropylene to increase the interface among polypropylene and clay, significant progress has been made in the manufacturing of polypropylene nano-composites.

Even though organic clay is hydrophilic in nature it leads to uptake in water in lager amount, it must be surface modified to become counterproductive and hence miscible with a wide range of polypropylene fabrication. Chemically surface modification to be done by pre-treatment by using alkaline, silane surfactants are commonly used to modify surfaces.

Adv. App. of Micro and Nano Clay II – Synthetic Polymer Composites   Materials Research Forum LLC
Materials Research Foundations 129 (2022) 203-232   https://doi.org/10.21741/9781644902035-9

(a) preparation of HMDA modified PP-g-MA

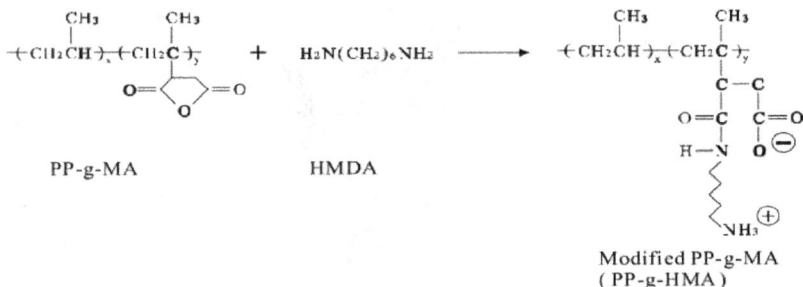

(b) preparation of modified PP-g-HMA/clay nanocomposites

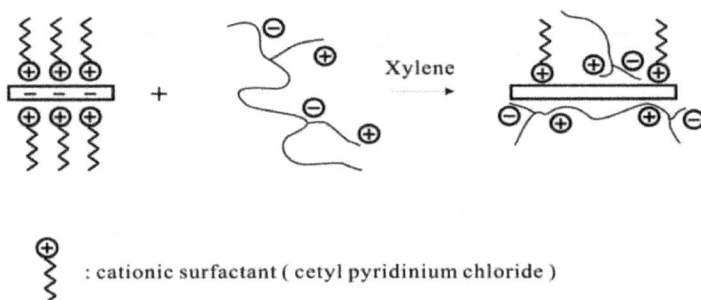

Fig. 2 Fabrication of (a) Hexamethylenediamine (HMDA) modified polypropylene-grafted- maleic-anhydride and (b) Modified polypropylene-grafted-HMA/clay nanocomposites [8].

As a result, the basal spacing has been increased, leading to greater polymeric impregnation in the integrally and enhanced interactive relationship between both the clay and polypropylene. Impregnation is energetically unfavourable for strongly non-polar polymers (e.g., polypropylene, polyethylene), hence a compatibilizer must be included in the nanoparticle meld. A remobilization is a polymer containing side- and end compounds, such as maleic anhydride, that have better thermal relationships with the clay surface while maintaining polymeric compatibility. Despite the enormous community of scholars researching on the fabrication of polypropylene nanomaterials, due to a polarity of the

Adv. App. of Micro and Nano Clay II – Synthetic Polymer Composites   Materials Research Forum LLC
Materials Research Foundations 129 (2022) 203-232   https://doi.org/10.21741/9781644902035-9

matrix material, no straight intercalation of polymerization in conventional organically modified colloidal silica has been seen. To solve this challenge, two alternative approaches are applied. The first involves the addition of maleic anhydride or hydrophobic interactions to the Polypropylene backbone [6, 7].

For the fabrication of polypropylene/clay nanocomposites used several techniques to optimize the characteristics of materials. To achieve a better outcome for the materials required specified addition of fillers, grafted treatment of the matrix materials, equipment selection plays a vital role to obtain the quality of polypropylene/clay nanocomposites. To selection of specified equipment based on the application of materials, feasibility of manufacturing polypropylene/clay nanocomposites and the quantity of materials to be used to synthesis of the polypropylene/clay nanocomposites. In this chapter gives a detailed study on the fabrication methods such as extrusion method, solution-blending method, melt-blending method, in-situ polymerization method, direct melt compounding, ultra sound-aided extrusion and master batch dilution. Fabrication of polypropylene-clay nanocomposites is explained in Fig. 2 [8].

2.1   Extrusion method

Extrusion can be done by using twin-screw extruders. Twin-screw extruders are primarily used to fabricate polypropylene-clay nanomaterials. The ultimate distribution levels reached are acknowledged to be influenced by processing parameters such as screw speed and feed rate. The influence of these factors on the progression of dispersal along the extruder for a polymer, polypropylene patched with maleic anhydride, and a natural clay mixture is examined. Relatively close readings were done simultaneously at several axial positions as well as the die exit. Local average body temperatures and lowest residence duration were also measured at the same locations. Furthermore, materials were rapidly taken from the extruder for extracting features. Distribution grew quickly upon melts, irrespective of the operating conditions, though at a reduced rate downstream. Using uniquely engineered specimen gathering equipment, substance specimens were tested from the sensation at the L/D guidelines developed in Fig. 3. Extrusion trials undertaken at the greatest screw speed (300 rpm) and the highest feed rates (6 and 9 kgh^{-1}) were still unfinished at L/D = 10. An accurate estimation of the median body heat was obtained by inserting a rapid deployment thermometer into the melted specimen just removed from the extruder. Furthermore, monitoring the duration between introducing a colouring marker into the screw and detecting the start of a colour change in the melted channel might be used to estimate the locally optimal retention time [9, 10].

Adv. App. of Micro and Nano Clay II – Synthetic Polymer Composites   Materials Research Forum LLC
Materials Research Foundations 129 (2022) 203-232   https://doi.org/10.21741/9781644902035-9

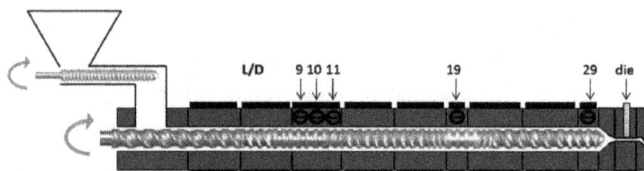

Fig. 3 The twin-screw extruder's construction and screws pattern designation from feeder to nozzle [9]

The goal of that research is to improve the twin-screw equipment to synthesis of polypropylene-clay nanomaterials. Using a Leistritz Micro 27 twin-screw equipment, proportionate quantities is in the ratio 3:1 of polypropylene grafted maleic anhydride and natural-clay loadings of 1, 3, and 5 wt. % were heat up with a polypropylene chitosan. (i) A polypropylene grafted maleic anhydride and natural-clay was mixed in first step, proceeded by diluting with the polypropylene resin in a followed step; (ii) all three aspects were treated in a single cycle; and (iii) natural mixes polypropylene and organoclay were treated repeatedly. The influence of screws respective direction and series on organoclay flaking and diffusion was investigated using both clockwise direction and anti-clockwise direction techniques [11]. Furthermore, nanomaterials are used to estimate individual melt-blended room temperature value. The findings show unambiguously that perhaps the twin-screw technique's large lateral outperforms the single-screw operation with in-line supercritical carbon dioxide introduction in producing excellently, poured polypropylene-clay nanomaterials. Extended retention time between clay and supercritical carbon dioxide is most likely required for supercritical carbon dioxide to have a favourable effect on exfoliate. For the novel polypropylene/clay nanocomposite manufacturing technique, a twin-screw extruder was employed. The equipment was a Japan Steel Works, Ltd. TEX30 77BW-20V both direction rotation intermeshing extrusion. The screws sizes were 32 mm, with a 77:1 length-to-diameter proportion. The twin extruder's large screw length was a distinguishing feature. The cumulative averages were limited to 6 kg/h by supplying polypropylene and polypropylene grafted maleic anhydride. Using a downturn hopper, Na-type montmorillonite and octadecyl trimethyl ammonium chloride were fed in amounts proportional to the polypropylene and polypropylene grafted maleic anhydride. By pumping, water was fed into the crusher at a speed of 2.5 kg/h. The screws rotated at 300 revolutions per minute. Due to the extreme thermal energy by shearing, the extruder temperature was determined at 180 °C—200 °C, although the average temperature of the resins was 200 °C—220 °C. A pressure of 1 MPa was recorded.

Adv. App. of Micro and Nano Clay II – Synthetic Polymer Composites   Materials Research Forum LLC
Materials Research Foundations 129 (2022) 203-232   https://doi.org/10.21741/9781644902035-9

The water vapour pressure in the twin-screw extrusion was assumed to be sustained that's because the saturate vapour pressure of water at 200 °C is predicted to be 1.7 MPa. A two-step mixing technique was used to make polypropylene/clay nanomaterials in that research. Polypropylene grafted maleic anhydride /clay reference samples comprising 30 and 40 wt. % clay were produced prior to the production of final polypropylene/clay nanomaterials. In a tumble blender, polypropylene grafted maleic anhydride powder or pellets were combined with clay and 0.2 wt. % stabiliser. Around 150 °C –190 °C and 150 rpm, the solution was melt-blended in a discrimination based on race twin-screw extruder (Werner and Pfleiderer; ZSK 25). Ten segmentation chambers with 3 churning regions made up the screws. Screw speed, chamber thermal efficiency, and include at were the most important processing factors. For the first compound phase, the screw speed was set at 150 rpm and the barrel temperature ranged from 150 °C at the first chamber to 190 °C at the very last cylinder after numerous attempts. The layers were pelletized and dried for 12 hours at 80 °C [12-15].

2.2   Solution-blending method

In the solution-blending method, the multi-layered silica or chemically treated silicate is swelled in a solution that also dissolves the copolymer in this process. By substituting the water molecule in the chambers, the expanded clay allows the intermolecular forces to diffuse in between the polymer matrix. Amorphous nanomaterials are formed when the solution is withdrawn under suction and the amorphous structure remains. Numerous publications on the production of polypropylene nanocomposites using the solution blending technique have been reported [16]. By disintegrating polypropylene in O-dichlorobenzene at 180 °C in a specifically constructed round barrel, nanomaterials were synthesized. Supersonic combination was used to distribute the natural clay in O-dichlorobenzene, proceeded by blending of the polypropylene solvent. Precipitation in cold ethanol segregated the polypropylene /organoclay nanomaterials, and the supernatant was evaporated in a separatory funnel. It was discovered that in nanomaterials with high clay concentrations (5%), the clay tactoids persist as cluster in some places, causing discontinuity, poor adhesion, and, as a result, tensile characteristics degradation. Polypropylene and clay were dissolved in 1,2,4-trichlorobenzene to make nanomaterials. The mixtures were maintained at an elevated temperature until systematic risks produced, then poured onto steel plates that were maintained at a heat of about 140 °C to vaporize the 1,2,4-trichlorobenzene solvent. Although without the need of a surfactant, X-ray diffraction and scanning electron microscopy outcome indicated that the clay granules were finely distributed. Revealed a more complicated method for making hydrophobic polypropylene /clay nanomaterials. Diacetone acrylamide was crosslinked prepared by the addition of polypropylene solvent in toluene, in contrast to the optimization technique.

Although the desired clay distribution was not accomplished, the resulting nanomaterials exhibited outstanding mechanical characteristics [17-19].

2.3   Melt-blending method

Since its inception in 1993, this technique has grown in popularity. Due to its versatility with current fabrication methods like as extrusion, blow moulding, and additive manufacturing, it has become the most preferred route to create polypropylene nanomaterials. Melt blending is the process of combining clay or customized clay with polypropylene at a heat above the polypropylene melting point in an extrusion or blender, ideally during shearing [1]. Melt blending mixes comprising a polypropylene and the described earlier naturally modified clay were used to synthesize melt-blend polypropylene/clay nanoparticles. Blending was done with a twin-screw extrusion with co-rotating intermeshing screws diameter should be 250 mm, length to diameter ratio must be 40 mm, set at 170 °C processing temperatures, 190 °C die temperature, 200 rpm screw speed, and 1 kg/h feed rate. After numerous attempts at obtaining the optimal physiological characteristics, the blending settings are chosen. Extrusion nanocomposites were conditioned before being injection moulded into tensile test rods (ASTM D638). Izod test rods (ASTM D256) in an Eckert & Ziegler GmbH injection moulding machine at 190–200 °C [20, 21]. Process of melt-blending is depicted in Fig. 4 [20].

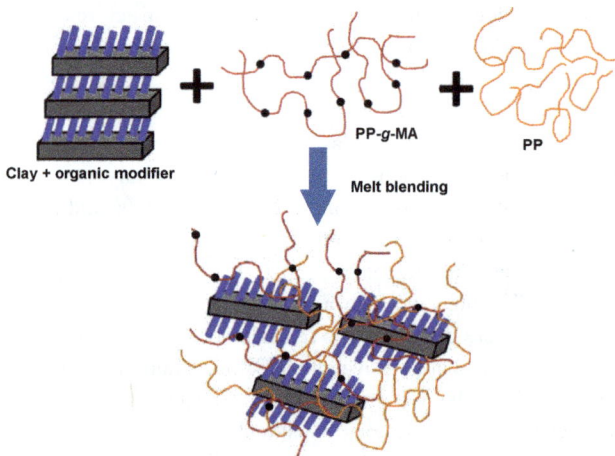

Fig. 4 Process of melt-blending [20]

Adv. App. of Micro and Nano Clay II – Synthetic Polymer Composites   Materials Research Forum LLC
Materials Research Foundations 129 (2022) 203-232   https://doi.org/10.21741/9781644902035-9

2.4   In-situ polymerization method

Preparation of polypropylene/clay nanocomposites by in situ polymerization method, the funnel type furnace was utilised to polymerize propylene in hexane using a slurry polymerization technique at a pressure of 4 bar. The boiling point of hexane, which was used as a solution, limited the processing temperature. As a result, the polymerization heat was limited to no more than 60 °C. As a co-catalyst and an external do-nor, trieisobutylaluminium and dimethoxymethylcyclohexyl silane were utilised, accordingly. The enzyme was then introduced into the reactor, with hydrogen serving as a chain transfer agent. The polymers were washed multiple times with methanol at the end of the polymerization phase, then filtered and dried in a vacuum oven at 70 °C for 24 h. For membrane preparation in situ polymerized polypropylene clay particle nano-material, a twin-screw extrusion having diameter of 250 mm, length to diameter ratio as 40 mm, blade velocity of 120 rpm was utilised. The composites granules were then injection moulded into an ASTM D638 tensile test bar. Moulding was carried out at processing temperatures of 175 °C and a pressure of 70 bars, correspondingly [22]. The thermal-distortion rate of the synthesised nylon-clay nanomaterials was dramatically raised by inserting a few volumes fraction of customized clay nanoparticles into the nylon matrices via in-situ polymerization from 65 to 152 °C. Tensile strength, and elongation the yield stress was also considerably reduced. Within that research, they describe a unique fabrication method for producing elevated   polypropylene nanocomposite using in-situ polymerization under moderate experimental circumstances, which yields a marked enhancement in mechanical characteristics and recyclability over early stuff. Additional agonists such as perfluoroaryl borates at elevated-pressure or elevated temperature processing parameters, are not essential in this unique technique to initiate olefin polymerization. Various polyolefin–clay nanomaterials can be easily made using this method [23]. Polypropylene nanomaterials were made from conventional polypropylene mixtures in the liquid phase and participate in specific of polypropylene/organophilic clays customised salt - affected clay with varying proportions of fillers. The master batches were made utilising 4th generation Ziegler-Natta accelerators based on $MgCl_2/TiCl_4$ including clay and in situ polymerization. The analysis indicates that the blended materials showed mechanical characteristics were improved when compared to a conventional polypropylene, and that the clay basal separation in the matrix material was increased, particularly when maleated polypropylene was used in the mixture. The resulting nanocomposites' first thermally deterioration enhanced by 80 °C [24, 25].

2.5   Ultra sound-aided extrusion

In ultra sound-aided extrusion numerous researches of the impact of ultrasonic on polymer have been conducted and published during the last few years. Long-chain polymer monomers have been found to be broken by elevated ultrasonic during melt extruded. Ultrasonic causes the C-C bond to rupture, resulting in the creation of lengthy radical. Polymer-based radical may terminated on the clay surface or interact with the interface identify the suitable natural-clay to establish a covalent linkage in polymer–filler complexes. A single-stage and a two-stage procedure were established. A double side rotating twin-screw extrusion process was used to make the nanomaterials, accompanied by a single-screw extrusion process with an ultrasonic die assembly. There were two types of continuing to feed regiments developed: starvation and flood giving. The ultrasound treatments zone's spacing dimension was adjusted. The tension on the die and the amount of electricity consumed were also assessed [8, 26, 27]. Die pressure vs. flow rate for the polypropylene nanomaterials is presented in Fig. 5 [8].

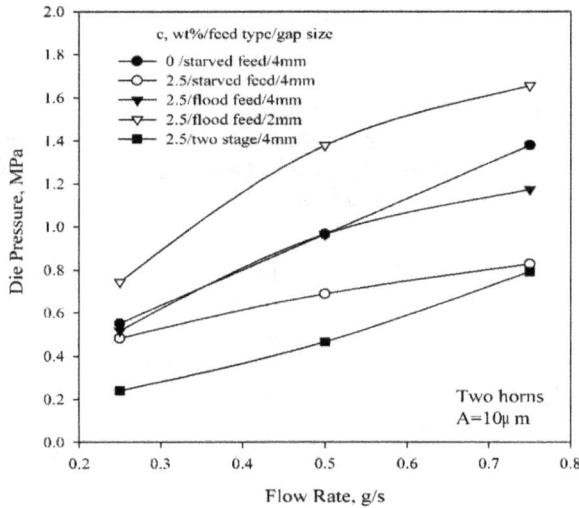

Fig. 5 Graph plot die pressure vs. flow rate for the polypropylene nanomaterials [8]

Adv. App. of Micro and Nano Clay II – Synthetic Polymer Composites   Materials Research Forum LLC
Materials Research Foundations 129 (2022) 203-232   https://doi.org/10.21741/9781644902035-9

During the single-stage process, different feeder regimens and gap sizes were investigated. Intercalation under ultrasonography was higher for flood feeding and a narrower gap, according to the findings. Under the same conditions, the deterioration of the polymer matrix was significant. The single-stage procedure significantly increased the elongation at break and toughness of nanocomposites. This enhancement did not outperform the materials produced by the two-stage method. Further exploration of the processing parameters is required to improve additional mechanical qualities, such as raising processing temperature and increasing die pressure to facilitate filler dispersion [28].

3. Properties of polypropylene/clay nanocomposites

Polypropylene/clay nanocomposites were synthesized by many ways like extrusion method, solution-blending method, melt-blending method, in-situ polymerization method and ultra sound-aided extrusion methods. But we have to characterize the materials to optimize the fabrication technique and added to that study the materials behaviour under several conditions by using mechanical, thermal, tribological, optical, viscoelasticity, viscoplasticity, creep failure, hygrothermal, rheological and morphological.

3.1 Mechanical properties

Polypropylene/clay nanocomposites to be studied mechanical properties, by intense stirring at 50 rpm for 10 min, polypropylene/clay nanomaterials with 0.5, 1, 3, and 5 wt. % montmorillonite were synthesized. The original components or process parameters were altered for the greatest clay content (5 wt. %) to see if they had an indirect impact. On the one hand, 10% polypropylene-grafted-Maleic anhydride was included was substituted with organo-modified. The mechanical properties are explained in Fig. 6 [29].

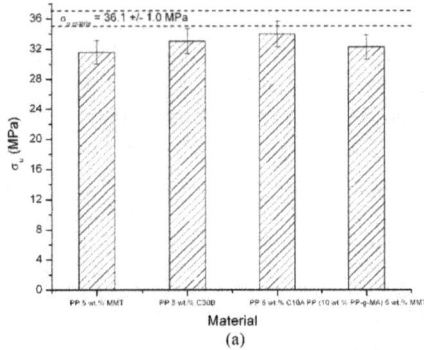

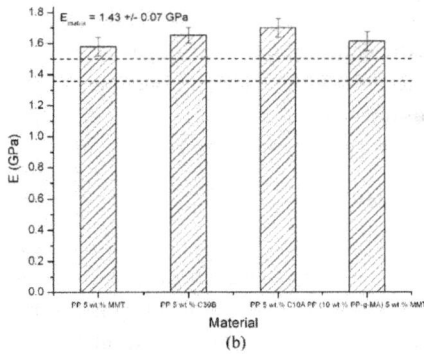

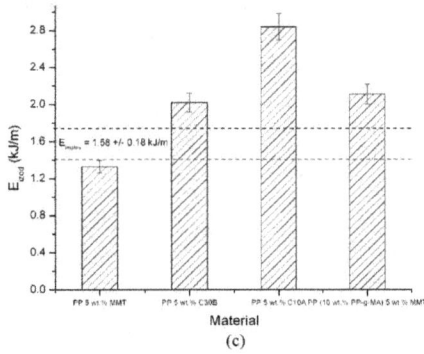

Fig. 6 (a) Tensile strength (b) Young's modulus and (c) impact energy of nanomaterials with 5 wt. % of montmorillonites, (organo-modified clays) C10A, and C30B and influence polypropylene-grafted-maleic anhydride incorporation [29].

Adv. App. of Micro and Nano Clay II – Synthetic Polymer Composites   Materials Research Forum LLC
Materials Research Foundations 129 (2022) 203-232   https://doi.org/10.21741/9781644902035-9

The elastic modulus was shown to be enhanced by clay inclusion and increased as a proportion of clay amount. The tensile strength, impact strength and toughness, on the other hand, were not improved, which could be attributable to inadequate inter-face adherence between the polypropylene matrix and unaffected clay. When comparing the suitability of clay and polymer showing distinctive clay minerals or polypropylene-grafted-maleic anhydride, it was obvious that the most hydrophobic clay had the greatest influence on higher mechanical characteristics and enhanced dispersal as measured by microscopy (C10A). However, when compared to standard variances, the variations in material behavior were not significant, with the exception of impact strength and toughness, which showed a significant rise for the nanoparticles with C10A. The proportion of dispersal and mechanical characteristics were not affected by the operating conditions [29].

Melt extrusion was used to create exfoliated nanomaterials of polypropylene-clay nanoparticles utilizing maleic anhydride customized polypropylene and natural clay. The clay minerals behave as a nucleation site for the framework of polypropylene-maleic anhydride, although their existence has no effect on the sequential rate of growth or total crystallisation rate. Nucleation is one of the most successful methods for controlling the amount of polymeric chains complexation into silicate chambers, and thus the mechanical properties of polypropylene/clay nanomaterials [30, 31]. The melt blended of polypropylene/clay nanomaterials was addressed. Unmodified, by using octadecyl ammonium modification, and by using dimethyl dialkyl ammonium modification commercial montmorillonite has all been utilized. Only nanomaterials made from naturally modified montmorillonites exhibited enhancements in mechanical characteristics. Nanoparticles strengthened with 5 wt. % naturally synthesized montmorillonites–polypropylene exhibited young's modulus and tensile strength values of 871 and 29.3 MPa, correspondingly, which were greater, by roughly 23 % and 4.6 %, than young's modulus and tensile strength values of pure polypropylene.

When montmorillonites were added to the polypropylene matrix, the crystallite size rate increased when compared to pure polypropylene. By adopting plastic order, maleic anhydride/grafted polypropylene and natural modification done on clay mixtures were synthesized. Dynamic mechanical analysis was used to evaluate mechanical characteristics. Although maleic anhydride/grafted polypropylene with a lower molecular weight and a larger maleic anhydride concentration could improve clay dispersal in polypropylene-clay combinations, the mechanical characteristics of polypropylene/maleic anhydride-grafted polypropylene/clay composites were deteriorated. The hybrids were studied to determine the primary parameters that influence the materials' ultimate appearance and mechanical properties. The complexation potential of the plasticizers in the clay particles, as well as the content of the surfactant in the Polypropylene/clay composites, were shown to be two critical parameters that aided exfoliating and homogenous dispersal of the clay layers [32-34].

3.2   Thermal properties

The current detailed information about the thermal characterization of polypropylene and clay nanomaterials is reported in this section. A study of thermal behaviour of several polypropylene/clay nanomaterials has also been developed. To produce a polypropylene/clay nanomaterials with appropriate thermal behaviour, we must achieve a best solution of several factors, including clay proportion and type, maleic anhydride concentration, polypropylene/maleic anhydride-grafted molecular weight, and polymer type. The XRD graph nanocomposites is depicted in Fig. 7 [3].

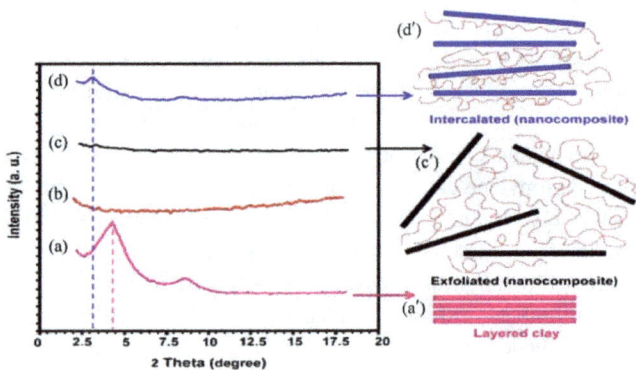

Fig. 7 XRD graph for (a) organo-montmorillonite, (b) polyester, (c) polyester/4 wt.% organo-montmorillonite, (d) polyester /10 wt.% organo-montmorillonite and pictorial view for (a') layered clay, (c') exfoliated nanocomposite, (d') intercalated nanocomposite [3].

Because of the intercalated topology, the maxima correlated with the baseline position are displaced towards angle with shorter durations, resulting to a range between the most essential lamina. The lack of a peak intensity is caused by exfoliated morphologies. Polymer nanocomposites based on polymer and naturally changed nano clays are of specific importance since they have significantly better material characteristics than raw polypropylene, such as water solubility, thermal expansion rigidity, flame retardancy, and chemical stability, when compared to raw polypropylene. In contrast to processes equipped with traditional fillers, these advantages are realised with a reinforcing element clay amount of less than 5%. As a result, the polymeric strengthened with clay nanostructures are much lighter than typical nanocomposite and can compete with other polymers in some uses [6]. The heat sustainability of short alkyl long chain imidazolium-oriented, ionic liquid as a montmorillonite enhancer was studied in comparison to the typical anionic sodium agent. The impact of these two distinct raw-clay modifications on polypropylene-nanomaterial was studied using integrated method degradation temperature. The Flynn–Wall–Ozawa technique was used to determine the activation energy for these specimens. The activation energy was also calculated using multiple linear regression in addition to assessing the efficiency of this approach when used to nanomaterials. In this study, thermal properties of polypropylene nanomaterials with imidazolium-oriented ionic liquid treated montmorillonite was investigated. The increase in thermal resilience, as measured by the temperature of the as integral process disintegration, was considerable. Imidazolium-based enhancers had integral process breakdown temperatures that ranged from 47 °C to 91 °C higher than peracetic acid solutions. The findings suggest that using imidazolium-based ionic liquids with short alkyl chain lengths as organic modifiers for lamellar clays rather than traditional quaternary ammonium salts is promising [35]. Thermal behaviour of polypropylene-grafted- vinyltriethoxysilane and polypropylene-grafted-maleic anhydride fabricated by melt processing were examined. According to thermal study, the introduction of polypropylene-grafted-vinyltriethoxysilane increased the crystallisation temperatures and crystalline structure of nanomaterials after clay inclusion compared to polypropylene-grafted-maleic anhydride. Thermal properties were also improved in both compatibilized nanomaterials. In 5 wt. % clay with polypropylene-grafted-maleic anhydride, a higher value of 98 °C was achieved [36, 37].

3.3   Tribological properties

If the clay granules are evenly distributed, it is thought that the introduction of a tiny quantity of clay can greatly enhance the characteristics of polymer. Because of their extremely high surface area, silica films are useful for quickly dispersing energies if the silica films and parent matrix contain nano-level thickness interphases. It was hypothesised that infusing polypropylene with clay increased thermal performance, resulting in increased thermal properties in an oxidising environment. Thermogravimetric analysis was widely used to examine the thermo-oxidative deterioration of polypropylene nanocomposites. The thermogravimetric analysis results in an oxidising environment revealed that the weight reduction profiles of nanomaterials shifted towards elevated temperature as compared to pure polymer. These findings are ascribed to nanoparticles significantly improving the thermal properties of the matrix phase in an oxidising environment. The dynamic mechanical behaviours of polypropylene nanocomposites were

Adv. App. of Micro and Nano Clay II – Synthetic Polymer Composites   Materials Research Forum LLC
Materials Research Foundations 129 (2022) 203-232   https://doi.org/10.21741/9781644902035-9

investigated to determine the influence of clay on the apparent viscosity of the matrix phase. It is expected that as the concentration of polymer bonding develops, so will the system's viscosity and mechanical [1].

Several distinctive characteristics, including as clay concentration, empirical way, and refinement method, impact the thermal behaviour of polypropylene-clay nanomaterials. In this part, we compared the findings of many researches to get a sense of the characteristics that impact the thermal characteristic of polypropylene-clay nanomaterials. As the clay concentration increases, the crystallisation temperature decline from 169 °C to 163 °C. Because the use of methylethylketone as a solution improves the available space between the polymer matrix, the glass transition temperature is reduced from 117 °C to 12 °C. The drop in glass transition temperature from 117 °C to 30 °C suggests that at lower temperatures, the heterogeneous molecules become more mobility. The variance in glass transition temperature is due to the change in customising intermediary of clay surfaces: 15A clay was altered by a lengthy alkyl chain with a quaternary ammonium group, whereas montmorillonite was altered by an alkyl chain with an octadecyl amine. The reagent allows for oxidation and deterioration, resulting in a reduction in thermal characteristics [6].

Table 1 Thermal characteristics of polypropylene/clay nanomaterials [3]

T_m (°c)	169	165	166	165	165	163	166
T_g (°c)	117	6	30	-16	12	10	30
T_c (°c)	127	116.4	116	116	116	116	127
X_c (%)	46.1	41.3	34.95	56	46	-	-

T_m represent that melting temperature, T_g represent that glass transition temperature, T_c represent that crystallization temperature and X_c represent that crystallinity rate [6].

3.4 Optical properties

Melt mixing was used to synthesize polypropylene-clay nanomaterials. To enhance the distribution of conventional nanoclays, two types of admixtures were used: polyolefin elastomer functionalized maleic anhydride and polypropylene functionalized maleic anhydride. The optical characterise of the nanomaterials increased with the addition of polypropylene functionalized maleic anhydride compared to the polyolefin elastomer functionalized maleic anhydride compatibilized instance. Moreover, as contrasted to polypropylene functionalized maleic anhydride, polyolefin elastomer functionalized maleic anhydride significantly enhanced the interlayer spacing of the clay. This intriguing finding is relevant to the complicated topology of nano clay nanomaterials [38-40].

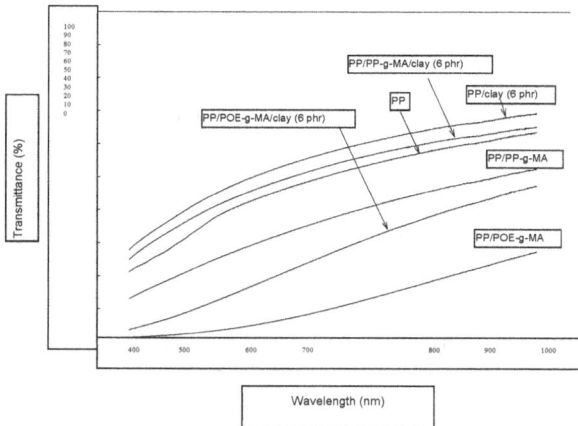

Fig. 8 Optical transmission of polypropylene-clay nanomaterials that have been compatibilized [38]

As shown in Fig. 8, the optical transmission of the nanoparticles without aggregates was typically greater than that of the transparent polypropylene material. This was mostly owing to the decreased austenite grain size observed due to the nucleating action of additional organic-clay, which was not evident in the traditional nucleating agent used to create the clear polypropylene. Based on the matrix characteristics, the included quasi polypropylene functionalized maleic anhydride proceeded to lower polypropylene optical transmission via the creation of scattered maleic anhydride compound. Furthermore, this consequence was more noticeable for the polypropylene/polyolefin elastomer functionalized maleic anhydride mixture due to the greater interoperability of polypropylene and polyolefin elastomer functionalized maleic anhydride via predominantly adsorption polyolefin elastomer functionalized maleic anhydride into the surface and edge of clay molecule as seen in the Transmission electron microscopy and prepare observe move analyses results. This effect was also seen in research using polyvinyl chloride-clay treated with epoxy. In this example, the transmission of polyvinyl chloride/clay nanomaterials rose with increasing clay content after they were treated with epoxy as an intercalant and the epoxy dispersed towards the clay chamber [36].

3.5   Viscoelasticity properties

Polypropylene-organoclay nanomaterials have been studied. Mathematical expressions for the viscoplastic reactions at 3-D deformations with modest stresses are derived. A combination of nanocomposite is modelled as an analogous inviscid, transient, non-affine value chain. Visco-plasticity is related to displacement connection moving in the analogous system. Visco-elasticity reflects network reorganisation. The goal of this article is to analyze the influence of modified organoclay and compatibilizer on viscoelastic properties by preparing nanomaterials using modified organoclay and polypropylene, with and without compatibilizer modified-Isopropenyl-a-a-dimethylbenzyl-isocyanate grafted Polypropylene. The incorporation of organoclay to both nanocomposites, either with or without compatibilizer, increased the amount of the mechanical properties, according to dynamic mechanical investigation. The strengthening action of organoclay is credited with the rise in volume fraction. In comparison to polypropylene and other nanostructured materials, 9 % organoclay strengthening without compatibilizer led to a tremendous storage modulus at all temperatures. At 0 °C, the storage modulus of a composite containing % organoclay was approximately 41 % greater than that of virgin polypropylene.

The incorporation of modified-Isopropenyl-a-a-dimethyl benzyl-isocyanate grafted polypropylene as the compatibilizer had a detrimental effect on the composite's dynamic modulus. All nanomaterials without a compatibilizer had a higher dynamic modulus than their equivalents with a compatibilizer. This is due to the consistent dispersion of organoclay, as evidenced by atomic force microscopy pictures. At increased organoclay loading, the complex modulus increases, resulting in more dissipation of energy in the system under viscoelastic distortion. Although frequencies had no influence on dynamic modulus, they were capable of changing the glass transition temperature of nanomaterials with greater clay addition. According to the findings of this investigation, compatibilizers do not significantly improve viscoelastic characteristics [41-44]. The prominent lower frequencies modulus peak, the residual stresses reported in steady shear flow, and the outcomes of stiffness degradation after the reduction of steady flow evidenced by polypropylene-natural clay specimens were determined in accordance with the establishment of a 3-D physical network in steady-state conditions that is easily ruptured when low shear rates are applied. The expanded Cox-Merz analogy was demonstrated. To create a connection between viscoelastic characteristics and filler content in virgin polypropylene and its clay-nanomaterials. To get the centreline main stress differential during intergranular flow, flow birefringence is conducted through a small hole. Even at low silicate ratios, the midline stress profile of clay-nanomaterials exhibited extra viscoelastic character, but no remarkable strain hardening was observed.

The impacts of increased filler amount were investigated extensively during the basic shearing flow in order to address non-linear viscoelastic behaviour in terms of damping factor. The fact that the damping ratio increased with clay content indicated that polymer-nanomaterials were more strain responsive. At all clay percentages investigated, Wagner's exponential damping coefficient properly represented the time-strain separability. Both studies demonstrated that the polymer were time-strain detachable at all clay percentages investigated during deformation and basic shearing processes, although the filler positions differed depending on the melt flow behaviour. The non-linear viscoelastic characteristics of virgin polypropylene and clay nanomaterials are investigated in order to develop a framework connection in respect to clay percentages. Even at low silicate percentages, the midline stress profile of clay nanomaterials exhibits extra viscoelastic character, as well as comparable short-time chain breakdown. The impacts of greater clay content are investigated subsequently during the basic shearing process to take into account the damping characteristics of the clay nanomaterials [45-47].

3.6   Visco-plasticity properties

At 3-D displacements with modest stresses, constitutive equations for visco-plastic and visco-plastic reactions are derived. An analogous inviscid, transient, non-affine network of links is regarded as a hybrid nanomaterial. Deflection sliding of junctions in the analogous network is linked to visco-plasticity [37]. The viscoelastic and visco-plastic behaviors of nanomaterials with semi-crystalline matrices are developed from constitutive equations. A two different continuity is often used to represent a nanomaterial. Both stages are considered as visco-elasto-plastic media, but with unique plastic flow kinetic equations and considerations about the impact of plastic deformation on viscoelastic behaviour. The stress–strain relationships are composed of nine variable components that are identified by fitting the data [42, 43]. Tensile experiments on double end notched samples were being used to evaluate fracture toughness of nanocomposites, as for tensile strain assessed far away from the connection. Experiments were observed in the range of temperatures and with different strain rates. The information was used to create stress-strain illustrations under loading/unloading (controlled by sample rupture), assess the energy contained in fractured samples owing to visco-plastic deformation, and accommodate for it in the energy balance law. The result indicated a noticeable rise (by double) in specific vital work of fracture increasing clay content added to that considerable impact of the rate of crack initiation on and limited impact of connection width on the fracture process [40, 42, 45].

3.7   Creep failure

The use of nano-clay to reinforce polypropylene has led to a significant increase in creep resistance. The duration to rupture of nanomaterials was demonstrated to be longer even than polypropylene with all stresses examined. The numerical solutions have been used to evaluate the time-dependent reaction in long-term creep experiments analytically. Creep studies revealed when the cumulative strain goes high, that is regardless of stress level inside the heating rate studied, polypropylene and its blends experience fast mechanical failure. This critical failure strain is used with the time-strain combination approach to estimate the creep lifespan of Polypropylene and its nanomaterials in the next sections [48, 49].

3.8   Hygrothermal properties

Hygrothermal of the blended material improved with soak period until it reached a concentration stage, at which stage the water content of the composite stayed stable. Hygrothermal reduced as the amount of maleated polypropylene and organoclay in the composite increased [46, 49].

3.9   Rheological properties

Polypropylene clay nano-composites get a considerably lower strain-sensitive linear behaviour area than with the polymer matrix in linear viscoelastic tests. Some rheological behaviours of these composites are greatly influenced by silicate crystallites/tactoids [47]. The isolation of polymer molecules inside the modified montmorillonite structures would be the cause of rising abilities. The viscosities of restricted polymer materials are shown to be higher even than mass strings [48]. The initial normal stress variance in nanomaterials includes multiple sources: the polymer matrix as well as the clay granules. When relative to empty polymer resin, the clay granules just aren't properly aligned and linked with each other to make a system pattern in lower frequency, it able to deliver extra constant force [49]. The sequential rheological behavior of nano-materials implies the creation of a dense as well as durable padding underlying network with nanomaterials, emphasising the vital role performed by clay in this structure's production. A rheological behavior of filled polymers has recently been described for being directly related to the amount, size, but also scatter of filled particles. Like a result, rheological examinations may be used to characterise the level of clay scattering with in polymer as an adequate passive diagnostic of microstructure of polypropylene clay nano-materials [51]. Only one small number of strains showed rheological behaviour. The blended elements' storage modulus was substantially higher than that of the matrix polymer, below certain frequency. Desquamation was identified theologically, as well as the findings and for modified clay specimen can be cited as fact of desquamation. The unmodified clay, on the other hand, exhibited no signs of desquamation [50-53].

3.10 Morphological properties

The composites dissolution and softening temperatures were enhanced by adding clay, cross-linking agent and wood powder. In relation to clean polypropylene the composites crystals maximum temperature is raised when clay and wood powder were added [54]. Polarized optical microscopy was used to examine the morphology of polypropylene and polypropylene clay nano-materials at a cooling rate of 20 °C. The spherulites of polypropylene clay nano-materials are significantly smaller than those of clean polypropylene, indicating that organo-clay served as a nucleating agent and that the addition of organoclay hastened the crystallisation of polypropylene. The study of non-isothermal kinetic parameters corroborated this observation [55, 56].

4.   Merits, demerits and applications

4.1   Merits

The growing interest in these nanomaterials is predicted to have the following merits:

- The qualitative case study of modest quantities of fillers relative to other fillers can be anticipated to occur in lightweight materials.
- It is simple to reuse since no filler has to be degraded before remanufacturing [3].

4.2   Demerits

Moreover, there are several potential issues with these nano-composites:

- A little volume of clay can be used to enhance the structure; but, if a significant amount is applied, the impact strength may drop.
- Synthesized clays that could outperform organic clays in terms of its functionality and price have yet to be increased [3].

4.3   Applications

These nanomaterials might be utilised in a variety of applications, including:

- Resin compounds for moulding, particularly automobile parts requiring increased strength.
- Use in sheet metal application and mainly food packaging films.
- Being used in elastomers that need carrying out a detailed, such as automobile hoses.
- Use in resin elements for flame-resistant home electrical equipment [3].

5.   Summary

This chapter paves the way for future research into the development of polypropylene/clay nanocomposites. The general effect that polypropylene has on clay nanoparticles is discussed. It is concluded that clays with a larger particle size behave differently to those with smaller particle size. Point properties were found to be improved in this study, indicating that these materials may find potential use as nanocomposites for high-performance applications such as construction and engineering plastics [57]. A wide range of polymer nanocomposites are constantly being produced, with many of them having potential uses. Clay nanocomposites with strong compensatory impacts when tiny amounts of ingredients are added have garnered international attention, as has the impact of gas boundaries, and large biochemical conversion firms have been equipped with the knowledge into these nanocomposites. Due to the different properties of clay and polypropylene, these two materials are not miscible when formed into nano-composites. However, since their polarity is different, it is necessary to alter the clay's organophilic properties to enable the formation of these composite materials [58]. Creating better clay dispersion in polypropylene is possible through the introduction of various functional groups or blending a modified anhydride. The addition of clays also enhances the mechanical properties and the thermal properties of the polypropylene. Whereas significant effort has been expended in developing polypropylene nanomaterials for diverse applications by attaining uniform distribution of clay particles, it is still far from resolving issues such as filler–polymer relationships [59].

Conclusion

In order to comprehend architecture connections in polypropylene nanomaterials, it is critical to first study the link between composite material. The enormous surface- to-volume ratio of nanomaterials causes significant changes in their characteristics. Point properties of polypropylene/clay nano composites were investigated and found to depend on the particle size and crystallinity of the clay. Nano-clay was added to polypropylene in order to increase strength and stiffness at low filler concentrations, while nanoparticle diameter was increased to decrease thermal conductivity. The results showed that by increasing the loading level (volume fraction) of clay, there is an increase in modulus, strength, toughness and impact strength. The same trends were also shown with polypropylene nano-clay. This would support those clays with larger particle sizes behave differently than those with smaller particles. The crystallite size of clay was calculated by XRD, both with and without the addition of polypropylene. Clay size decreased with increasing loading level, which was to be expected since there are fewer clay particles in polypropylene /clay nanocomposites. Since the crystallite size of clay is related to the conformation of the macromolecule, it is reasonable to suggest that these crystals are more likely to have ordered structures. Changes in point properties between composites with different particle sizes were investigated by means of compression testing under tension at strain rates for between 0.1 and 10 μm/s.

References

[1] K.R. Reddy, Polypropylene clay nanocomposites. In: Handbook of Polymer nanocomposites. Processing, Performance and Application, Springer, Berlin, Heidelberg, 2014, pp. 153-175. https://doi.org/10.1007/978-3-642-38649-7_2

[2] A. Saravana Kumar, P. Maivizhi Selvi, L. Rajeshkumar, Delamination in drilling of sisal/banana reinforced composites produced by hand lay-up process. Appl. Mech. Mater. 867 (2017) 29-33. https://doi.org/10.4028/www.scientific.net/AMM.867.29

[3] K. Zdiri, A. Elamri, M. Hamdaoui, Advances in thermal and mechanical behaviors of PP/clay nanocomposites. Polym. Plast. Technol. Eng. 56(8) (2017) 824-840. https://doi.org/10.1080/03602559.2016.1233282

[4] M. Ramesh, L. Rajeshkumar, Wood flour filled thermoset composites. In: Abdullah M. Asiri, Anish Khan, Imran Khan, Showkat Ahmad Bhawani (eds.) Thermoset composites: preparation, properties and applications, Materials Research Foundations, 38 (2018) 33-65. https://doi.org/10.21741/9781945291876-2

[5] V. Kumar, A. Singh, Polypropylene clay nanocomposites. Rev. Chem. Eng. 29(6), (2013) 439-448. https://doi.org/10.1515/revce-2013-0014

[6] A. Usuki, N. Hasegawa, M. Kato, S. Kobayashi, Polymer-clay nanocomposites. Inorganic polymeric nanocomposites and membranes, 2005, pp. 135-195. https://doi.org/10.1007/b104481

[7] M. Ramesh, L.R. Kumar, A. Khan, A.M. Asiri, Self-healing polymer composites and its chemistry. In: Anish Khan, Mohammad Jawaid, Abdullah Mohammed Ahmed Asiri (eds.) Self-Healing Composite Materials, Elsevier, 2020, pp. 415-427. https://doi.org/10.1016/B978-0-12-817354-1.00022-3

[8] S. Lapshin, A. I. Isayev, Ultrasound-aided extrusion process for preparation of polypropylene-clay nanocomposites. J. Vinyl Addit. Technol. 13(1) (2007) 40-45. https://doi.org/10.1002/vnl.20095

[9] J. M. Barbas, A. V. Machado, J.A. Covas, Processing conditions effect on dispersion evolution in a twin-screw extruder: polypropylene-clay nanocomposites. Chem. Eng. Technol. 37(2) (2014) 257-266. https://doi.org/10.1002/ceat.201300303

[10] M. Ramesh, L. Rajeshkumar, V. Bhuvaneshwari, Bamboo Fiber Reinforced Composites. In: Jawaid M., Mavinkere Rangappa S., Siengchin S. (eds) Bamboo Fiber Composites. Composites Science and Technology. Springer, Singapore, 2021. https://doi.org/10.1007/978-981-15-8489-3_1

[11] J. Ma, Z. Qi, Y. Hu, Synthesis and characterization of polypropylene/clay nanocomposites. J. Appl. Polym. Sci. 82(14) (2001) 3611-3617. https://doi.org/10.1002/app.2223

[12] J.Y. Wu, T.M. Wu, W.Y. Chen, S.J. Tsai, W.F. Kuo, G.Y. Chang, Preparation and characterization of PP/clay nanocomposites based on modified polypropylene and clay. J. Polym. Sci. Part B: Polym. Phy. 43(22) (2005) 3242-3254. https://doi.org/10.1002/polb.20605

[13] M. Ramesh, L.R. Kumar, Bioadhesives. In: Inamuddin R, Boddula MI, Ahamed, Asiri AM (eds) Green adhesives. 2020, pp. 145-167. https://doi.org/10.1002/9781119655053

[14] N. Rezaiean, H. Ebadi-Dehaghani, H.A. Khonakdar, P. Jafary, S.M.A. Jafari, R. Ghorbani, Microstructure and properties of polypropylene/clay nanocomposites. J. Macromol. Sci. Part B 55(10) (2016) 1022-1038. https://doi.org/10.1080/00222348.2016.1230462

[15] M. Ramesh, C. Deepa, L.R. Kumar, M.R. Sanjay, S. Siengchin, Life-cycle and environmental impact assessments on processing of plant fibres and its bio-composites: a critical review. Journal of Industrial Textiles, (2020). https://doi.org/10.1177/1528083720924730

[16] M.A. Treece, J.P. Oberhauser, Processing of polypropylene-clay nanocomposites: Single-screw extrusion with in-line supercritical carbon dioxide feed versus twin-screw extrusion. J. Appl. Polym. Sci. 103(2) (2007) 884-892. https://doi.org/10.1002/app.25226

[17] M.A. Treece, W. Zhang, R.D. Moffitt, J.P. Oberhauser, Twin-screw extrusion of polypropylene-clay nanocomposites: Influence of masterbatch processing, screw rotation mode, and sequence. Polym. Eng. Sci. 47(6) (2007) 898-911. https://doi.org/10.1002/pen.20774

[18] D. Balaji, M. Ramesh, T. Kannan, S. Deepan, V. Bhuvaneswari, L. Rajeshkumar, Experimental investigation on mechanical properties of banana/snake grass fiber reinforced hybrid composites. Mater Today: Proc. 42 (2021) 350-355. https://doi.org/10.1016/j.matpr.2020.09.548

[19] M. Ramesh, C. Deepa, M. Tamil Selvan, L. Rajeshkumar, D. Balaji, V. Bhuvaneswari, Mechanical and water absorption properties of calotropis gigantea plant fibers reinforced polymer composites, Mater. Today: Proc. 46 (2020) 3367-3372. https://doi.org/10.1016/j.matpr.2020.11.480

[20] H. Baniasadi, S.A. Ramazani, S.J. Nikkhah, Investigation of in situ prepared polypropylene/clay nanocomposites properties and comparing to melt blending method. Mater. Des. 31(1) (2010) 76-84. https://doi.org/10.1016/j.matdes.2009.07.014

[21] V. Bhuvaneswari, M. Priyadharshini, C. Deepa. D. Balaji, L. Rajeshkumar, M. Ramesh, Deep learning for material synthesis and manufacturing systems: A review, Mater. Today: Proc. 46(9) (2021) 3263-3269. https://doi.org/10.1016/j.matpr.2020.11.351

[22] M. Kato, M. Matsushita, K. Fukumori, Development of a new production method for a polypropylene-clay nanocomposite. Polym. Eng. Sci. 44(7) (2004) 1205-1211. https://doi.org/10.1002/pen.20115

[23] C.H. Hong, Y.B. Lee, J.W. Bae, J.Y. Jho, B.U. Nam, T.W. Hwang, Preparation and mechanical properties of polypropylene/clay nanocomposites for automotive parts application. J. Appl. Polym. Sci.98(1) (2005) 427-433. https://doi.org/10.1002/app.21800

[24] M. Pannirselvam, A. Genovese, M. C. Jollands, S.N. Bhattacharya, R.A. Shanks, Oxygen barrier property of polypropylene-polyether treated clay nanocomposite. Exp. Polym. Lett. 2(6) (2008) 429-439. https://doi.org/10.3144/expresspolymlett.2008.52

[25] M. Ramesh, J. Maniraj, L. Rajesh Kumar, Biocomposites for Energy Storage. Biobased Composites: Processing, Characterization, Properties, and Applications. In: Anish Khan, Sanjay M. Rangappa, Suchart Siengchin, Abdullah M. Asiri (eds) Biobased Composites: Processing, Characterization, Properties, and Applications, Wiley Online Library, 2021 pp.123-142. https://doi.org/10.1002/9781119641803.ch9

[26] M. Ramesh, L. Rajeshkumar, D. Balaji, V. Bhuvaneswari, Green composite using agricultural waste reinforcement. In: Thomas S., Balakrishnan P. (eds) Green Composites. Materials Horizons: From Nature to Nanomaterials. Springer, Singapore, 2021, pp 21-34. https://doi.org/10.1007/978-981-15-9643-8_2

[27] A.I. Isayev, Rishi Kumar, T.M. Lewis, Ultrasound assisted twin screw extrusion of polymer-nanocomposites containing carbon nanotubes. Polym. 50 (2009) 250-260. https://doi.org/10.1016/j.polymer.2008.10.052

[28] D.S. Dlamini, S.B. Mishra, A.K. Mishra, B.B. Mamba, Comparative studies of the morphological and thermal properties of clay/polymer nanocomposites synthesized via melt blending and modified solution blending methods. J. Compos. Mater. 45(21) (2011) 2211-2216. https://doi.org/10.1177/0021998311401074

[29] J.A. Tarapow, C.R. Bernal, V.A. Alvarez, Mechanical properties of polypropylene/clay nanocomposites: effect of clay content, polymer/clay compatibility, and processing conditions. J. Appl. Polym. Sci. 111(2) (2009) 768-778. https://doi.org/10.1002/app.29066

[30] M.S. Kim, J. Yan, K.M. Kang, K.H. Joo, Y.J. Kang, S.H. Ahn, Soundproofing ability and mechanical properties of polypropylene/exfoliated graphite nanoplatelet/carbon nanotube (PP/xGnP/CNT) composite. Int. J. Prec. Eng. Manuf. 14(6) (2013) 1087-1092. https://doi.org/10.1007/s12541-013-0146-3

[31] M. Ramesh, L. Rajeshkumar, Technological advances in analyzing of soil chemistry. In: Inamuddin, Mohd Imran Ahamed Rajender Boddula, Tariq Altalhi (eds) Applied Soil Chemistry, Wiley- Scrivener Publishing LLC, USA, 2021, pp 61-78. https://doi.org/10.1002/9781119711520.ch4

[32] M. Avella, S. Cosco, G.D. Volpe, M.E. Errico, Crystallization behavior and properties of exfoliated isotactic polypropylene/organoclay nanocomposites. Adv. Polym. Technol. J. Polym. Proc. Inst. 24(2) (2005) 132-144. https://doi.org/10.1002/adv.20036

[33] M. Ramesh, L. Rajeshkumar, D. Balaji, Aerogels for insulation applications. In: Inamuddin (ed) Aerogels II: Preparation, Properties and Applications. Materials Research Foundations, United states. 2021, pp. 57-76. https://doi.org/10.21741/9781644901298-4

[34] V. Bhuvaneswari, L. Rajeshkumar, K.N.S. Ross, Influence of bioceramic reinforcement on tribological behaviour of aluminium alloy metal matrix composites: experimental study and analysis. J. Mater. Res. Technol. 15 (2021) 2802-2819. https://doi.org/10.1016/j.jmrt.2021.09.090

[35] F.C. Chiu, P.H. Chu, Characterization of solution-mixed polypropylene/clay nanocomposites without compatibilizers. J. Polym. Res. 13(1) (2006) 73-78. https://doi.org/10.1007/s10965-005-9009-7

[36] Y. Wang, S.W. Huang, Solution intercalation and relaxation properties of maleated polypropylene/organoclay nanocomposites. Polym. Plast. Technol. Eng. 46(11) (2007) 1039-1047. https://doi.org/10.1080/03602550701522377

[37] M. Ramesh, L. Rajeshkumar, D. Balaji, Mechanical and dynamic properties of ramie fiber reinforced composites. In: Rajini Nagarajan, Senthil Muthu Kumar Thiagamani, Senthilkumar Krishnasamy, Suchart Siengchin (eds) Mechanical and Dynamic Properties of Biocomposites. Wiley, Germany, 2021, pp 275-322. https://doi.org/10.1002/9783527822331.ch15

[38] S.M. Lai, W.C. Chen, X.S. Zhu, Melt mixed compatibilized polypropylene/clay nanocomposites: Part 1-The effect of compatibilizers on optical transmittance and mechanical properties. Compos. Part A: Appl. Sci. Manuf. 40(6-7) (2009) 754-765. https://doi.org/10.1016/j.compositesa.2009.03.006

[39] M, Ramesh, L. Rajeshkumar, D. Balaji, V. Bhuvaneswari, S. Sivalingam, Self-healable conductive materials. In: Inamuddin, Mohd Imran Ahamed, Rajender Boddula, Tariq A. Altalhi (eds) Self-Healing Smart Materials, Wiley, United States, 2021, pp. 297-320. https://doi.org/10.1002/9781119710219.ch11

[40] M. Ramesh, L. Rajeshkumar, R. Saravanakumar, Mechanically-induced self-healable materials. In: Inamuddin, Mohd Imran Ahamed, Rajender Boddula, Tariq A. Altalhi (eds) Self-Healing Smart Materials, Wiley, United States, 2021, pp. 379-404. https://doi.org/10.1002/9781119710219.ch15

[41] A. Oya, Y. Kurokawa, H. Yasuda, Factors controlling mechanical properties of clay mineral/polypropylene nanocomposites. J. Mater. Sci. 35(5) (2000) 1045-1050. https://doi.org/10.1023/A:1004773222849

[42] M. Ramesh, L. Rajeshkumar, D. Balaji, Influence of process parameters on the properties of additively manufactured fiber-reinforced polymer composite materials: A review. J. Mater. Eng. Perform. 30 (7) (2021) 4792-4807. https://doi.org/10.1007/s11665-021-05832-y

[43] M. Ramesh, C. Deepa, K. Niranjana, L. Rajeshkumar, R. Bhoopathi, D. Balaji, Influence of Haritaki (Terminalia chebula) nano-powder on thermo-mechanical, water absorption and morphological properties of Tindora (Coccinia grandis) tendrils fiber reinforced epoxy composites. J. Nat. Fib. 2021. https://doi.org/10.1080/15440478.2021.1921660

[44] D. Mohankumar, V. Amarnath, V. Bhuvaneswari, S.P. Saran, K. Saravanaraj, M.S. Gogul, S. Sridhar, G. Kathiresan, L. Rajeshkumar, Extraction of plant based natural fibers - A mini review. IOP Conf. Ser.: Mater. Sci. Eng. 1145 (2021) 012023. https://doi.org/10.1088/1757-899X/1145/1/012023

[45] A. Chafidz, I. Ali, M.A. Mohsin, R. Elleithy, S. Al-Zahrani, Nanoindentation and dynamic mechanical properties of PP/clay nanocomposites. J. Polym. Res. 19(7) (2012) 1-12. https://doi.org/10.1007/s10965-012-9906-5

[46] M. Ramesh, C. Deepa, L. Rajeshkumar, K. Tamilselvan, D. Balaji, Influence of fiber surface treatment on the tribological properties of Calotropis gigantea plant fiber reinforced polymer composites. Polym. Compos. (2021). https://doi.org/10.1002/pc.26149. https://doi.org/10.1002/pc.26149

[47] M. Ramesh, L. Rajeshkumar. Case-studies on green corrosion inhibitors. Sustainable Corrosion Inhibitors, Materials Research Foundations, 107, 2021, pp. 204-221. https://doi.org/10.21741/9781644901496-9

[48] R.S. Chen, S. Ahmad, S. Gan, Characterization of recycled thermoplastics-based nanocomposites: Polymer-clay compatibility, blending procedure, processing condition, and clay content effects. Compos. Part B: Eng. 131 (2017) 91-99. https://doi.org/10.1016/j.compositesb.2017.07.057

[49] A.F. Sahayaraj, M. Muthukrishnan, M. Ramesh, L. Rajeshkumar, Effect of hybridization on properties of tamarind (Tamarindus indica L.) seed nano-powder incorporated jute-hemp fibers reinforced epoxy composites. Polym. Compos. (2021), https://doi.org/10.1002/pc.26326. https://doi.org/10.1002/pc.26326

[50] M. Ramesh, L. Rajeshkumar, R. Bhoopathi. Carbon substrates: A review on fabrication, properties and applications. Carbon Lett. 31 (2021) 557-580. https://doi.org/10.1007/s42823-021-00264-z

[51] B. Devarajan, R. Saravanakumar, S. Sivalingam, V. Bhuvaneswari, K. Fatemeh, L. Rajeshkumar, Catalyst derived from wastes for biofuel production: a critical review and patent landscape analysis. Appl. Nanosci. (2021), https://doi.org/10.1007/s13204-021-01948-8. https://doi.org/10.1007/s13204-021-01948-8

[52] M. Ramesh, D. Balaji, L. Rajeshkumar, V. Bhuvaneswari, R. Saravanakumar, Anish Khan, A.M. Asiri, tribological behavior of glass/sisal fiber reinforced polyester composites. In: Jawaid M., Khan A. (eds) Vegetable Fiber Composites and their Technological Applications. Composites Science and Technology. Springer, Singapore. 2021, pp.445-459. https://doi.org/10.1007/978-981-16-1854-3_20

[53] T. Sun, J.M. Garces, High-performance polypropylene-clay nanocomposites by in-situ polymerization with metallocene/clay catalysts. Adv. Mater. 14(2) (2002) 128-130. https://doi.org/10.1002/1521-4095(20020116)14:2<128::AID-ADMA128>3.0.CO;2-7

[54] K. Hariprasad, K. Ravichandran, V. Jayaseelan, T. Muthuramalingam. Acoustic and mechanical characterisation of polypropylene composites reinforced by natural fibres for automotive applications. J. Mater. Res. Technol. 9 (2020) 14029-14035. https://doi.org/10.1016/j.jmrt.2020.09.112

[55] K. Shirvanimoghaddam, K.V. Balaji, R. Yadav, O. Zabihi, M. Ahmadi, P. Adetunji, M. Naebe, Balancing the toughness and strength in polypropylene composites. Compos. Part B: Eng. (2021) 109121. https://doi.org/10.1016/j.compositesb.2021.109121

[56] M. Ramesh, L. Rajeshkumar, C. Deepa, M. Tamil Selvan, V. Kushvaha, M. Asrofi, Impact of silane treatment on characterization of ipomoea staphylina plant fiber reinforced epoxy composites. J. Nat. Fib. (2021). https://doi.org/10.1080/15440478.2021.1902896

[57] I.O. Oladele, M.O. Oladejo, A.A. Adediran, B.A. Makinde-Isola, A.F. Owa, E.T. Akinlabi, Influence of designated properties on the characteristics of dombeya buettneri fiber/graphite hybrid reinforced polypropylene composites. Sci. Rep. 10 (2020) 1-13. https://doi.org/10.1038/s41598-020-68033-y

[58] L. Lei, Z. Yao, J. Zhou, B. Wei, H. Fan. 3D printing of carbon black/polypropylene composites with excellent microwave absorption performance. Compos. Sci. Technol. 200 (2020) 108479. https://doi.org/10.1016/j.compscitech.2020.108479

[59] H. Hadiji, M. Assarar, W. Zouari, F. Pierre, K. Behlouli, B. Zouari, R. Ayad, Damping analysis of nonwoven natural fibre-reinforced polypropylene composites used in automotive interior parts. Polym. Test. 89 (2020) 106692. https://doi.org/10.1016/j.polymertesting.2020.106692

Adv. App. of Micro and Nano Clay II – Synthetic Polymer Composites   Materials Research Forum LLC
Materials Research Foundations 129 (2022) 233-281   https://doi.org/10.21741/9781644902035-10

Chapter 10

Sonochemical Synthesis of Polymer Nanocomposites

M.K. Poddar [1,*], V.S. Moholkar [2], S. Chakma [3,*]

[1]Department of Chemical Engineering, National Institute of Technology Karnataka, Srinivasnagar, Mangalore – 575 025, Karnataka, India

[2]Department of Chemical Engineering, Indian Institute of Technology Guwahati, Guwahati – 781039, Assam, India

[3]Department of Chemical Engineering, Indian Institute of Science Education and Research Bhopal, Bhopal – 462066, Madhya Pradesh, India

* maneesh.poddar@nitk.edu.in (M.P.), schakma@iiserb.ac.in (S.C.)

Abstract

Polymer nanocomposite materials have drawn the attention of scientists due to their outstanding properties as compared to native polymers. The nanocomposites of polymer are widely used in packaging, aerospace, nanocoatings, solar energy, automotive, electronics, semiconductors, cosmetic, and construction. The modifications in the properties are done according to their applications in various fields. The properties of such nanocomposite materials depend on several variables such as nanoparticles dispersion and distribution, flame retardancy, mechanical and tensile strength, thermal properties, electrical and magnetic properties. In addition to these, the nanofillers play a vital role in nanocomposites to enhance or modify their mechanical and physical properties. In this chapter, various aspects of the nanocomposites of polymer have been discussed, including ultrasound-assisted *in-situ* polymerization intercalation and nanoparticles dispersion. The processing of nanocomposites using different filler materials such as ZnO, reduced graphene oxide (RGO), Cloisite-30B, etc. as well as the percentage of filler materials has been discussed in detailed. Finally, the nanocomposites have been characterized using different analytical techniques to elucidate its properties.

Keywords

Polymer Nanocomposite, Copolymerization, PMMA/Cloisite 30B, PMMA/ZnO, PMMA/RGO, PMMA-Magnetic Particle, Polymer Matrix, Nanofiller, Ultrasound

Contents

1.   Introduction

The synthesis of polymer nanocomposites has attracted significant attention as they possess enhanced physical and chemical properties than pristine polymer. It has been observed that addition of various nanofillers such as nanoclay, metal nanoparticles, metal oxides, carbon-nanotubes, and carbon-nanofibers, etc. can enhance the physical characteristics of the pristine polymer via transferring its inherent unique properties into long-chain polymer matrix [1–7]. For example, the addition of montmorillonite (MMT) nanoclay into the matrix of polymer can remarkably increase the thermal-stability of pristine polymer. While addition of metal nanofillers like silver and gold nanoparticles, semiconductor zinc oxide (ZnO), titanium oxide (TiO_2) or combination of metals nanoparticles increase the optical and electrical properties of nanocomposites which are widely used in light-emitting diode (LED), gas sensors, photoluminescence solar cells, and shielding against UV light radiation. Recently, the modification of polymer nanocomposites with carbon-based nanofillers (*viz.* carbon nanotubes of single and multi-wall, graphene and carbon nanofibers, etc.) are fascinating as it improves the electrical conductivity of the composites. These materials are identified as one of the ideal materials for fabrication of various modern electrical and electronic appliances in the form of sensors, supercapacitors, shielding resistance against electromagnetic radiation. Other than carbon-based nanofiller, nanocomposites synthesized with iron nanoparticles (Fe_3O_4/Fe_2O_3) have also attracted exceptional attention due to its immense application in biomedical and biotechnological fields such as magnetic biosensors, magnetic storage application, targeted drug delivery, magnetic resonance, etc. Most recently, the synthesis of hybrid polymer nanocomposites with a combination of magnetic and carbon-based nanofillers (Fe_3O_4/CNT) has been started for shielding resistance against harmful electromagnetic interference (EMI). These polymer nanocomposites are being extensively used in electronic devices due to their high electrical conductivity and magnetic permeability achieved when the polymer matrix contents hybrid nanofillers [8–9]. The conventional methods for synthesis of polymer nanocomposites induced by mechanical agitation are unable to provide homogenous and orderly distribution of nanofillers into the matrix of polymer due to generation of shear force. It resulted in non-uniform dispersion of nanofiller into the polymer matrix leading to the agglomeration of nanoparticles in the bulk polymer matrices. Although the limitation of agglomeration can be intensified with increasing the nanofiller loading; however, it also adversely affects by lowering the physical properties of the nanocomposites through the formation of interfacial restriction between the base polymer matrix and the filler materials. Therefore, synthesis of polymer nanocomposites with a uniform dispersion of nanofillers and no agglomeration is a challenge for the scientific community in the area of polymer nanocomposites.

2.   Processing techniques of nanocomposite

The nanocomposite materials are the artificially synthesized materials consisting of a polymer matrix and inorganic nanoparticles/nanofillers. The composite materials have multiple phases within it which are chemically different from each other with a distinct interface between these chemical phases. A few techniques are reported for processing nanocomposites of polymer. The most commonly used methods for synthesis of polymer nano-composites are: (1) Intercalation of polymers, (2) *In-situ* intercalative polymerization, (3) Melt intercalation, (4) Direct mixing polymerization, (5) Template polymerization, (6) Sol-gel method polymerization, (7) In-situ polymerization, (8) Emulsion polymerization. The intercalation technique is a polymer nanocomposites preparation technique in which silicate layers are formed using a solvent where the polymer is dissolved and silicate layers are swelled. Depending upon the degree of polymer chains' penetration, differently structured ranging from intercalation to exfoliated nanocomposites can be achieved. In this method, the polymer matrix is entered between the silicate layers to form a crystallographically regular fashion intercalated nanocomposite with highly ordered multi-layer morphology where polymeric and inorganic layers are placed alternately. When the nanocomposites are synthesized through *in-situ* polymerization technique, the process is involved in swelling of nanofillers into the monomer or the low-molecular weight monomer solution with drips into the layers [10–11]. Then the polymerization reaction occurs in between the inter-layers and forms intercalated nanocomposites. Also, it allows grafting of polymers on the filler surfaces and improves the properties of the nanocomposites significantly. Using this technique, partially exfoliated composite structures can be achieved as the inorganic nanoparticles are dispersed in the solution of monomer and proper intercalation of nanofillers/nano-reinforcements in the polymer matrix [12]. The template method or matrix polymerization is another technique for the synthesis of nanocomposites where the interaction is involved between the preformed macromolecule (template), *e.g.* layered material, hexagonal, nanotubes etc., and the growing chain. The technique is popular for the preparation of mesoporous or double-layered hydroxide based composites. In this method, the polymer acts as a nucleating agent and upholds the inorganic fillers crystal growth. This method is favorable for developing controlled polymerization, including tacticity, molecular distribution, and kinetics [13]. However, this method is not favorable for the formation of layered silicates due to the aggregation tendency resulting from the polymer degradation caused by the high temperature during the synthesis process. In the direct mixing, monomer or polymer and the reinforcing phase are premixed before the polymerization reaction by applying temperature, catalyst or radical initiators.

The melt intercalation is a nanocomposite processing technique under specific processing conditions. This processing method is highly favorable for immiscible polymer and the reinforcements. In this process, the choice of the polymer matrix, surface modification of the filler materials, compatibility of fillers are essential parameters to achieve a well-dispersed nano-reinforcement in the polymer matrix. The process requires high shear mixing as the polymer solution viscosity is significantly high due to the presence of nano-reinforcements [12, 14].

3.   The basic mechanism of ultrasound in synthesizing of polymer nanocomposites

The ultrasound-assisted synthesis of polymer nanocomposites has been reported to be beneficial due to high monomer conversion, faster polymerization rate, high molecular weight polymer; and more importantly uniform and homogenous distribution of nanofillers into the matrix of polymer [15–17]. During the application of ultrasound, its physical effect generates intense microturbulence in the medium. On the other hand, the sonochemical effect due to cavitation is responsible for the formation of reactive species such as $^\bullet$H, $^\bullet$OH etc. Cavitation is basically the nucleation, enlargement and intense collapse of the micro-bubbles induced by the pressure variation in the medium due to the passage of

Adv. App. of Micro and Nano Clay II – Synthetic Polymer Composites   Materials Research Forum LLC
Materials Research Foundations 129 (2022) 233-281   https://doi.org/10.21741/9781644902035-10

ultrasound waves. This leads to the generation of strong concentric energy on extremely short span of time [18–21]. The tiny cavitation bubbles collapse adiabatically as shown in the Fig. 1 [22]. During the collapse, hot spots are generated resulting in extreme peaks of temperature & pressure within the bubble (~5000 K, ~500 bar) [22–23] and generates free radicals in the medium that are responsible for initiating the *in-situ* polymerization reaction and enhance the monomer to polymer propagation rate [24–25].

During the application of ultrasound, it induces micro-turbulence, micro-convection, and shockwaves along with the generation of reactive species. When the ultrasonic longitudinal wave passes through the liquid medium comprising compression and rarefaction cycles alternatively, it generates oscillatory motion of the fluid elements around their mean position. The amplitude of this oscillatory motion is typically ~0.1–1 micron; and the time scale is of few micro-seconds (*e.g.,* 50 µs for 20 kHz wave). Additionally, the radial motion of the cavitation bubbles also induces micro-convection in the form of the oscillatory motion of fluid elements and shock waves (or acoustic waves). The shock waves are induced due to a sudden reflection of the fluid elements from the gas-liquid interface – when the radial motion of the cavitation bubble comes to a sudden halt. This effect is resulted from the back-pressure exerted by the non-condensable gas present in the bubble during adiabatic compression. Thus, the microstreaming generated in the reaction mixture by both ultrasound and cavitation creates fine emulsification of the monomer/water solution. This not only leads to a higher encapsulation of nanofiller material in the polymer matrix but also help in uniform dispersion in the matrix. Both the effects are highly beneficial towards enhancing the properties of nanocomposites over that of a pristine polymer.

Moreover, the fine emulsification of aqueous/organic phases generates large interfacial area in a reaction mixture that boosts the kinetics of the polymerization. In the last decade, nanocomposites preparation of polymer has been performed in the presence of ultrasound using several methods such as *in-situ* emulsion polymerization, solution casting and melt intercalation [26–29]. Out of these techniques, *in-situ* emulsion polymerization is the most preferable and widely used for the synthesis of polymer nanocomposites due to its simplicity, low processing cost, controlled polymerization and its environmental benign [16].

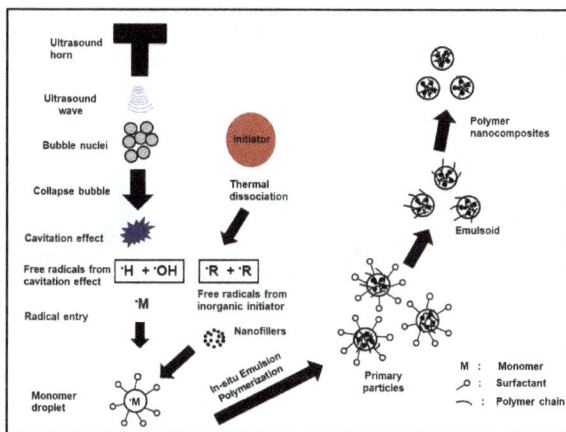

Figure 1: A mechanism of ultrasound-stimulated in-situ emulsion polymerization for preparation of polymer nanocomposites [22].

4.   Case studies of sonochemically synthesized nanocomposites

4.1   Clay supported polymer nanocomposites

In the clay supported polymer nanocomposites, the dispersion of organically modified nanoclay into the matrix of polymer tremendously boosts the thermal, mechanical, and flame retardance properties of original polymer [30–32]. Cloisite-30B clay, which has a layered silicate structure belongs to the family of smectites group is widely used as a filler material in nanocomposites. The characteristic properties of clays are defined in terms of d (001) spacing and gallery spacing, also called as '*interlayer spacing*'. It is noteworthy to mention that a mere interaction of nanoclay with polymer does not enhance the properties of nanocomposites. During the polymer/clay interaction, final nanocomposites structure can be enlightened based on the formation of three diverse phases based on the degree of separation of clay-platelets into the matrix of polymer. The phase separation can be defined as follows: (1) phase-separated or unmixed construction, (2) intercalated structure, and (3) exfoliated/ desquamated structure. A schematic representation of these structures is shown in Fig. 2 [33]. During the phase separation or unmixed, the tightly bound clay layers do not allow the polymer chain to penetrate into its clay galleries and lead to high agglomerated clay structure in final nanocomposites. In intercalated structure, polymer and clay layers are alternately attracted in a repeated form via injection between the polymer chains. In exfoliated clay structure, the most desirable structure is attained when almost all the clay layers are individually separated from each other and dispersed into the matrix of polymer. The exfoliated clay structure yields maximum interactions with polymer and enhances physical and chemical properties significantly in the final nanocomposites.

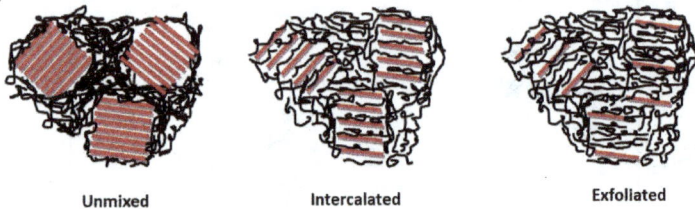

Unmixed   Intercalated   Exfoliated

Figure 2: Different types of polymer/clay nanocomposites structures obtained during clay dispersion in the polymer during the synthesis of polymer/clay nanocomposite [33].

Synthesis procedure of Cloisite-30B based PMMA nanocomposites: The nanocomposite synthesis of PMMA-Cloisite-30B has been accomplished using the optimum condition (*viz.* 15 g MMA, 0.75 g KPS and 0.87 g SDS) obtained from statistical design of experiments [34]. Further, nanocomposites of PMMA/Cloisite-30B with different clay loading (*viz.* 1, 2, 4 and 5 wt.%) were prepared using ultrasound-aided *in-situ* emulsion polymerization technique. The effect of ultrasound has significantly increased the *d*-spacing in the nanoclay platelets, including its homogenous dispersion into the pristine PMMA matrix. The results also revealed an increase in mechanical strength as well as thermal properties of the nanocomposites. The salient features of this investigation are discussed below.

Structural morphology analysis: To determine the intercalated or exfoliated structure of polymer/clay nanocomposites, the powder X-ray diffractometer analysis was performed for original Cloisite-30B clay and its nanocomposites PMMA-Cloisite-30B with various clay loading and the results are depicted in Fig. 3A [34]. The nanocomposites of PMMA-Cloisite-30B for all loadings (1- 5 wt.%) showed no peak, while the pristine clay exhibits a peak at $2\theta = 4.85°$ corresponds to d (001). The disappearance of this peak substantially confirms the exfoliation or homogenous dispersion of individual clay platelets into PMMA matrix-induced under high-intensity ultrasound wave. To study the effect of ultrasound on the exfoliation of clay structure, sonication of pristine clay with similar ultrasound power and frequency as used during nanocomposites synthesis was performed. The XRD results of the as-received pristine Cloisite-30B clay and treated pristine clay under sonication are shown in Fig. 3B [35]. The result of ultrasound treated clay revealed to have increased *d*-

spacing between the clay platelets as shown in Fig. 3B – which could be attributed to the effect of ultrasound and cavitation [35]. From the Fig. 3B, one can easily observe the shifting of 2θ value of as receive Cloisite 30B clay from 4.85° to 4.10° after 15 min of sonication. The change in interlayer d (001) spacing of sonicated clay was found using Bragg's law and it showed a substantial increase of 2.16 nm as compared to pristine clay (1.82 nm). The increase in d (001) spacing highlights and supports the role of micro-turbulence induced by the ultrasound as well as cavitation in the clay exfoliation.

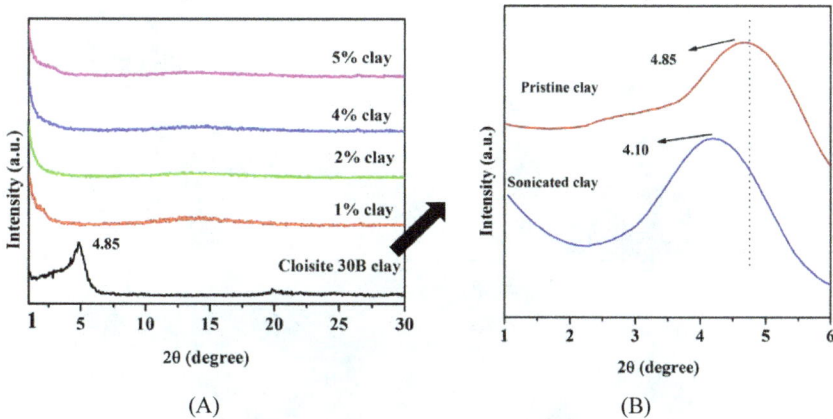

Figure 3: XRD diffractogram of (A) pristine Cloisite 30B clay and different loadings of PMMA/Cloisite 30B nanocomposites [34] (B) Pristine clay and clay after sonication [35].

The change in d (001) spacing is due to the microturbulence and shock-waves produced at the time of cavitation bubbles collapse are responsible for exfoliation and homogeneous dispersion of nanoclay into the PMMA matrix. However, mere results obtained from XRD cannot guarantee the complete exfoliation of clay platelets during nanocomposites synthesis. The intercalated/exfoliated clay structure into polymer nanocomposites can be confirmed using TEM analysis. The results of the inner structure of nanocomposites analyzed with TEM micrograph are depicted in Fig. 4 [36]. The TEM images clearly showed the presence of clay structures as dark layers which are almost uniform in nanocomposites with loading rate of clay 1.0 and 2.0 wt.% - which confirmed the exfoliated structure. However, for 4.0 wt.%, all the clay layers are not entirely separated, and few are partially stacked and termed as semi exfoliated structure. Further increasing the clay loading to 5.0 wt.%, the results showed highly clustered and crowded clay structure attributing the low *d*- spacing with intercalated structure of nanoclay inside the PMMA matrix. Similar observation of intercalated and exfoliated clay structure with varying clay loading of 1 and 3 wt% was also reported by Borthakur et al. [37]. In their studied, Borthakur et al. [37] found a partial exfoliation of final nanocomposites structure with the clay loading of 1.0 and 2.0 wt.% and a complete exfoliation of nanocomposites was achieved with 0.5 wt% clay loading.

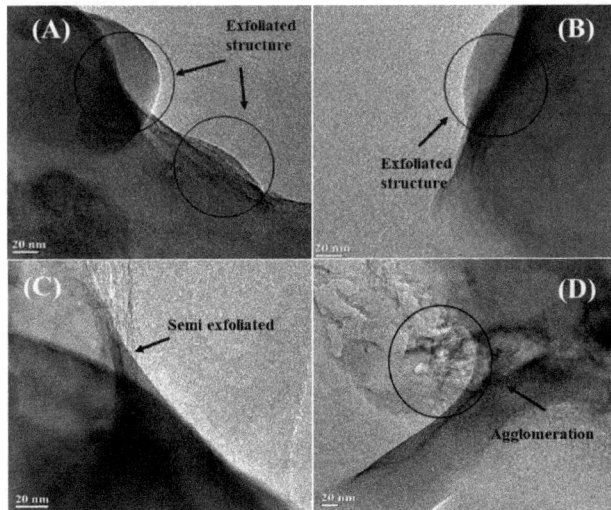

Figure 4: TEM micrograph of Cloisite-30B based PMMA nanocomposites prepared at different clay loading (A) 1.0 wt.%, (B) 2.0 wt.%, (C) 4.0 wt.%, and (D) 5.0 wt.%. [36].

TGA analysis: Thermogravimetric analysis (TGA) measures the thermal stability of solid mass with variation in sample weight over time with the change in temperature. Fig. 5 [38] shows the TGA curve of pure PMMA and Cloisite-30B based PMMA nanocomposites at different loading rate which clearly shows the changes in thermal stability. It can be seen that the thermal decomposition occurs in two steps for both neat PMMA and its nanocomposites with various clay loadings. The first step of thermal decomposition occurred up to 240°C, which correspond to the loss of residual moisture present in the nanocomposite samples and cleavage of head to head linkage of long PMMA chains. The thermal decomposition of PMMA and its nanocomposites prepared at different clay loadings is same for all the samples up to 240°C. However, a substantial weight loss was seen above 240°C which are reported in terms of $T_{5\%}$, $T_{10\%}$ and $T_{50\%}$ that correspond to 5, 10 and 50% weight loss of the sample mass at their corresponding temperatures, respectively. A summary of the thermal decomposition of $T_{5\%}$, $T_{10\%}$ and $T_{50\%}$ of pristine PMMA and its nanocomposites are reported in Table 1 [38]. For neat PMMA the thermal decomposition temperature $T_5\%$ occurred at 267.4°C. However, with 1.0 and 2.0 wt% clay loading, $T_{5wt\%}$ was increased with a value of 7.3 and 9.2°C respectively as compared to the pristine PMMA. For 4.0 and 5.0 wt% clay loading, reduction in thermal stability was observed. The reduction of thermal stability for $T_{5wt\%}$ with increasing clay loading is due to the rapid thermal degradation of cationic surfactants present in Cloisite-30B clay. The thermal decomposition temperature of neat PMMA corresponding to $T_{10}\%$ exhibits at 277.2°C; whereas the value was increased up to 12.5, 19.5, 25.1 and 30.9°C for the clay loading of 1.0, 2.0, 4.0, and 5.0 wt%, respectively. The marked rise in thermal stability of nanocomposites at $T_{10}\%$ weight-loss is observed due to the loss of thermally unstable molecules of cationic surfactants present in the clay and successful encapsulation of nanoclay into the PMMA matrix. The differential increase in decomposition temperature at $T_{50\%}$ showed marked improvement of 14.1, 30.7, 43.3 and 55.9 °C when clay loading was 1.0, 2.0, 4.0, and 5.0 wt% which is much higher than the neat PMMA.

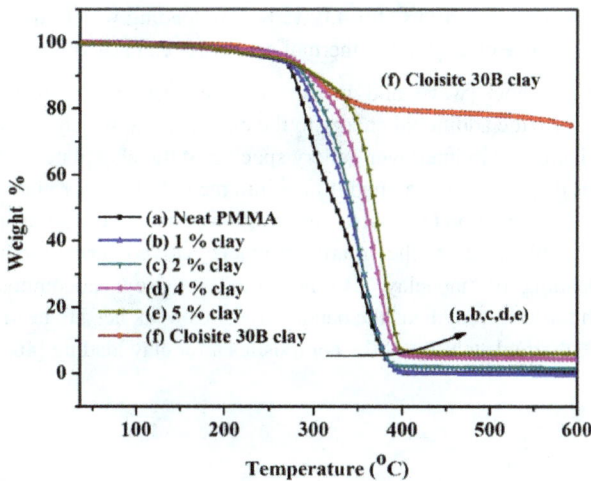

Figure 5: Thermogravimetric analysis of neat PMMA and PMMA/Clay composites (B) synthesized with various clay loadings [38].

Table 1: The TGA results of PMMA and Cloisite-30B based PMMA composites [38].

Name of Sample	$T_{5\%}$	$T_{10\%}$	$T_{50\%}$	$\Delta T_{10\%}$	$\Delta T_{50\%}$
Neat PMMA	267.4°C	277.2°C	326.0°C	–	–
PMMA/clay (1 wt%)	274.7°C	289.7°C	340.1°C	12.5°C	14.1°C
PMMA/clay (2 wt%)	276.6°C	296.7°C	356.7°C	19.5°C	30.7°C
PMMA/clay (4 wt%)	267.3°C	302.3°C	369.3°C	25.1°C	43.3°C
PMMA/clay (5 wt%)	240.4°C	308.1°C	381.9°C	30.9°C	55.9°C

The results of differential thermometric analysis of pristine PMMA and the nanocomposites with Cloisite-30B, which corresponds to the maximum thermal degradation rate, are depicted in Fig. 6A [38]. The DTG curve of neat PMMA for maximum thermal degradation was observed at 376 °C. With an increase in clay loading, DTG curves moved from left to right and showed marked enhancement of 4, 9, and 12 and 17 °C with clay addition of 1, 2, 4 and 5 wt%, respectively. The glass transition temperature (T_g) has been determine in order to investigate the molecular mobility of polymer chains and the results are depicted in Fig. 6B [38]. It clearly showed that the addition of clay enhances the T_g value up to 125°C for 4.0 wt.% clay loading which the T_g value of neat PMMA was 116°C. The change in the thermal properties can be explained as follows:

The high-intensity shock wave and microturbulence generated during irradiation of ultrasound wave provide additional energy to the clay material which widens its interlayer *d*-spacing. The increased in interlayer gallery spacing of the clay platelets allows uniform and homogenous dispersion of clay nanofillers into the bulk polymer matrix and enhances the polymer/clay combination [39]. Also, the addition of nanoclay into the polymer matrix provides a thermal hindrance to the volatile products evolved during the nanocomposites heating. The loading of nanoclay also helps to increase nanocomposites cross-link structure, which further strengthens the nanocomposites weak head to head carbon linkage and augment the thermal stability of the composites after clay loading [40–41].

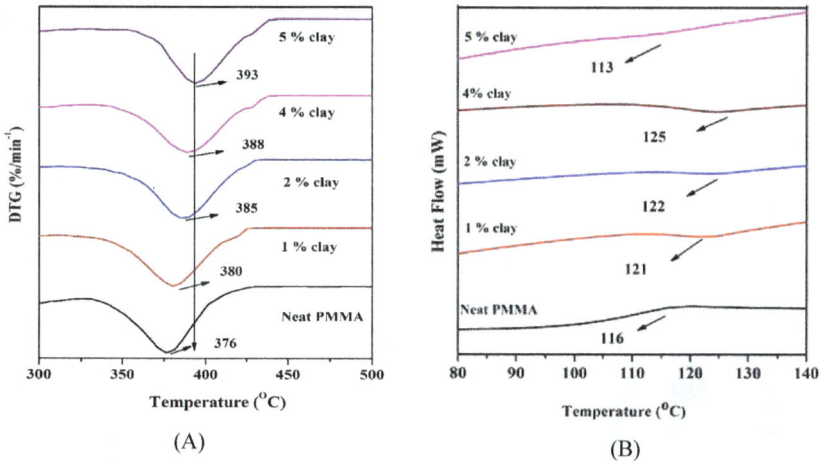

Figure 6: (A) Analysis of differential thermogravimetric (DTG) curve and (B) glass transition temperature (T_g) of neat PMMA and Cloisite-30B based PMMA composites synthesized at different loadings [38].

Mechanical Properties: Fig. 7A-C shows the results of the Young's modulus, tensile strength, and elongation percentage of pure PMMA and Cloisite-30B based PMMA composites loaded with different % [38]. Results of tensile strength of pristine PMMA as depicted in Fig. 7A shows a maximum value of 32.74 MPa. Conversely, clay loading affected the mechanical properties of the composite, *e.g.*, the tensile strength of composites continuously decreased and reported a minimum value of 22.1 MPa for 5.0 wt.% of clay loading. This is because of the highly amorphous and brittle structure of PMMA, which is responsible for the decline in tensile strength when clay loading increases [42].

On the other hand, the Young's modulus was increased as the clay loading increases up to 4.0 wt.% as shown in Fig. 7B and was reported a maximum of 2.2 GPa as compared to that of 1.12 GPa for original PMMA. Encapsulation of clay loading provides additional neighboring hydrogen sites into PMMA matrix that increase polymer's stiffness and its resistance against elastic deformation which increases Young's modulus of final nanocomposites structure. However, further increase in clay loading of 5.0 wt.%, Young's modulus value decreased to 1.9 GPa which anticipated due to the formation of clay agglomerates at higher clay loading and yielded intercalated structure as reported in TEM analysis. The results of percentage elongation as shown in Fig. 7C illustrates the reduction in its value with increase in clay loading and shows a similar trend with results as obtained during tensile strength. The magnitude of percentage elongation of neat PMMA shows the maximum value of 3.92%. Addition of clay into the PMMA matrix showed a constant reduction in percentage elongation and was a minimum of 0.7% elongation for clay loading of 5 wt%. The possible reason for the reduction in percentage elongation is attributed to the marked enhancement in stiffness and brittle behavior of nanocomposites as also observed in tensile strength.

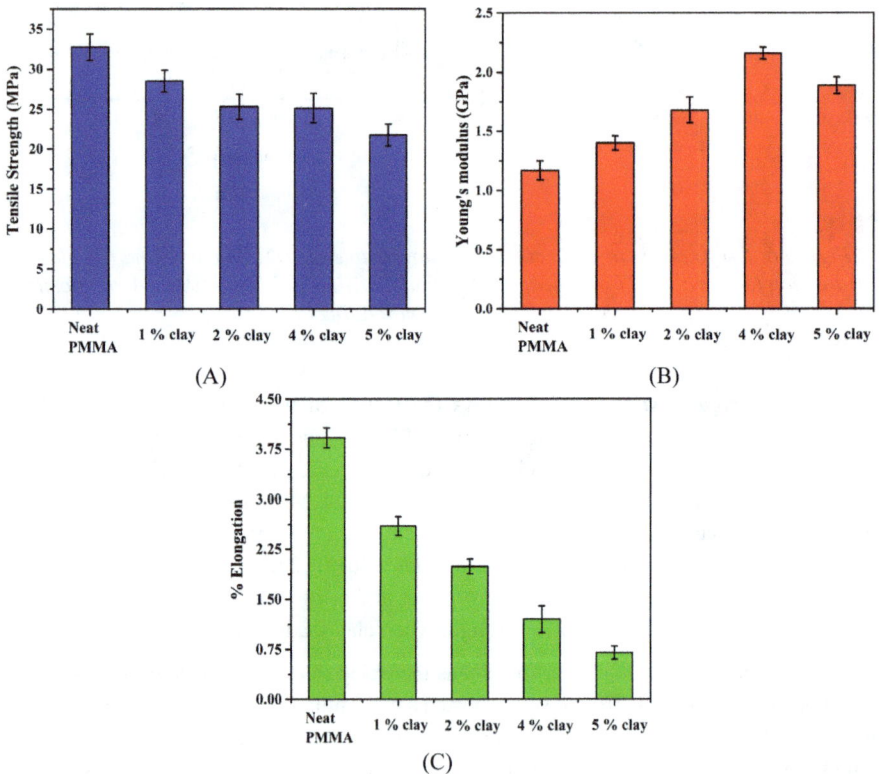

Figure 7: (A) Tensile strength, (B) Young's modulus, and (C) percentage elongation results of PMMA and Cloisite-30B based PMMA composites prepared at different loading of clay, [38].

Adv. App. of Micro and Nano Clay II – Synthetic Polymer Composites   Materials Research Forum LLC
Materials Research Foundations 129 (2022) 233-281   https://doi.org/10.21741/9781644902035-10

4.2   PMMA-ZnO nanocomposites

In the recent past synthesis of polymer nanocomposites with ZnO encapsulation nanofiller into polymer matrix has attracted significant attention of research interest owing to marked enhancement in nanocomposite optical, electrical and thermal properties. In modern technology, polymer/ZnO nanocomposites device in the form of thin-film are widely used in optoelectronic industries as protection against harmful UV-radiation, antireflective coating, synthesis of light-emitting diode, transport barrier [43–47]. However, an inherent hydrophilic character of ZnO cannot allow its better compatibility with hydrophobic polymer and results in large agglomerates formation during nanocomposites synthesis. Surface modification of ZnO surface via grafting of hydrophobic ligands polymer onto ZnO surface are alternate chemical techniques and is commonly used in controlling nanoparticles agglomerates formed during ZnO synthesis. However, the chemicals methods used during surface modification are highly expensive, time-consuming, environmentally unfriendly and is unpreferable due to its high potential toxicological effect.

However, the application of high-intensity ultrasound wave can provide an advantageous effect of dispersion of ZnO nanofillers during the synthesis of polymer/ZnO nanocomposites even without surface modification of ZnO. The following sections discuss the change in nanocomposites properties with loading of different amount of zinc oxide (ZnO) nanofiller into the PMMA polymer matrix using ultrasound-aided *in-situ* emulsion polymerization. The preparation of PMMA/ZnO composites was comprised in two steps: (i) sonochemical preparation of ZnO NPs from zinc acetate as a precursor followed by calcination at 500°C, (ii) ultrasound-assisted *in-situ* synthesis of ZnO-based PMMA composite loaded with calcined ZnO nanoparticles (1.0, 2.0, 4.0 and 5.0 wt.%). This can necessarily yield narrow size distribution and uniform dispersion of ZnO into the polymer matrix.

TEM analysis: The morphological structure of nanocomposites of PMMA/ZnO obtained from TEM analysis is depicted in Fig. 8 [48]. The ZnO nanoparticles encapsulation in the PMMA matrix has been shown with the presence of small black dots. These black dots of ZnO nanoparticles are almost homogenous and uniformly dispersion into PMMA matrix for 1 wt% ZnO loading even without nanoparticles agglomeration. However, with further increase in ZnO loading, the size of black dots increases, which illustrate the agglomeration of nanoparticles into a polymer matrix at high ZnO loading.

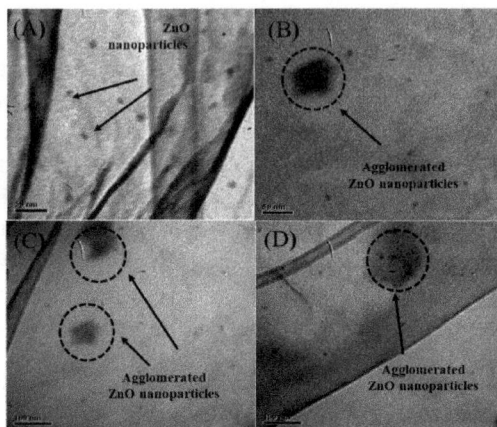

Figure 8: TEM micrographs of ZnO-based PMMA composites prepared at different loadings of ZnO: (A) 1.0 wt.%, (B) 2.0 wt.%, (C) 4.0 wt.%, (D) 5.0 wt% [48].

TGA analysis: The thermal stability analysis results of PMMA-ZnO nanocomposites with various loadings of ZnO into neat PMMA are shown in Fig. 9 [48]. A summary of the TGA analysis of nanocomposites weight loss *vs.* decomposition temperature is also mentioned in Table 2 [48]. It showed that the ZnO nanoparticles addition in neat PMMA increased the thermal degradation temperature of nanocomposites for $T_{5\%}$ and $T_{50\%}$ weight loss. The enhancement in thermal stability of nanocomposites for 5% weight loss ($T_{5\%}$) was seen in between 4°C to 12°C, and the maximum was 12°C for 4.0 wt.% ZnO loading as compared to neat PMMA. Successful encapsulation of ZnO nanoparticles can offer significant thermal resistance to volatile material present in neat PMMA matrix and its consequent increases the thermal stability of nanocomposites corresponding to 5% weight loss.

On the other hand, thermal decomposition temperature for 50% loss in weight ($T_{50\%}$), a sharp increase of 40°C is observed for nanocomposites with 1.0 wt.%. The possible reason for the remarkable rise in thermal stability for $T_{50\%}$ is due to the thermal hindrance induced by ZnO-nanoparticles which further restricted the hemolytic breakage of the weak vinyl group present in the PMMA molecules. The addition of ZnO-nanoparticles also induced the cross-linking of the PMMA chains and enhanced its thermal stability for higher temperature range [49]. However, with further increase in ZnO loading of 2-5 wt%, no significant changes in thermal decomposition temperature were observed. The possible reason could be an agglomeration of ZnO nanoparticles at higher loading.

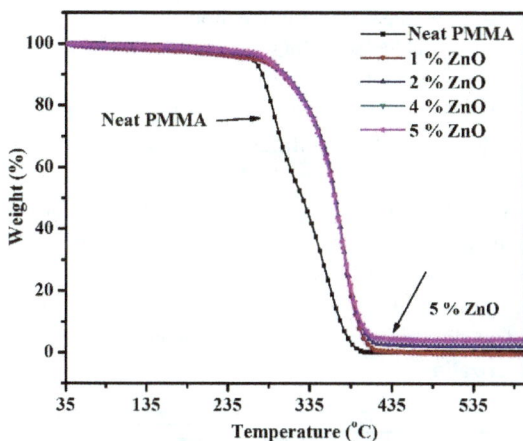

Figure 9: TGA analysis of neat PMMA and ZnO-based PMMA composites prepared with different loadings of ZnO [48].

Table 2: The TGA results of PMMA and ZnO-based PMMA composites [48]

Name of Sample	$T_{5\%}$	$T_{50\%}$	Char Value (at 600°C)	Encapsulation of ZnO
Neat PMMA	267.4°C	326.0°C	0.28%	–
PMMA+ 1 wt.% ZnO	271.3°C	366.0°C	0.30%	0.32%
PMMA+ 2 wt.% ZnO	279.3°C	367.5°C	2.28%	2.46%
PMMA + 4 wt.% ZnO	279.9°C	368.1°C	4.26%	4.82%
PMMA+ 5 wt.% ZnO	278.6°C	367.2°C	6.85%	7.76%

The maximum degradation temperature obtained from the DTG curve is shown in Fig. 10A [48]. The result shows that the maximum degradation temperature of neat PMMA was 376°C. However, with the incorporation of ZnO nanofillers, enhancement in maximum degradation temperatures have been observed and reported maximum of 10.3°C for 5.0 wt.% ZnO loading. This increased in thermal stability was attributed to the intense microturbulence generated via ultrasound wave which provides uniform and homogenous dispersion of nanosize ZnO-particles into the monomer MMA during in-situ emulsion polymerization. This successful encapsulation of ZnO-nanoparticles into PMMA matrix increases the thermal stability of resulting PMMA/ZnO nanocomposites via restricting the thermal scission of PMMA chains.

Addition of ZnO nanofillers into PMMA matrix for change in its molecular mobility can be accessed via measuring glass transition temperature (T_g). In a similar facet glass transition temperature (T_g) of PMMA-ZnO nanocomposites, as shown in Fig. 10B [48] also increased with ZnO loading. The T_g with the addition of 2 wt% ZnO loading has been found to have the highest value of 122.5°C. This mark increased in T_g value confirms the successful encapsulation of ZnO nanofillers into PMMA matrix via restricting the segmental motion of the PMMA chain matrix. In a similar work, with variation in T_g value with the incorporation of ZnO nanofillers into polystyrene matrix polymer has been reported by Hong et al. [47]. However, further increasing loading of ZnO with 4.0 and 5.0 wt.%, the T_g values decrease up to 116 and 112.5°C, respectively. The possible reason for the reduction in T_g value at high ZnO loading may be due to agglomerates formation induced by intermittent dispersion of ZnO-nanoparticles in the PMMA matrix during *in-situ* emulsion polymerization.

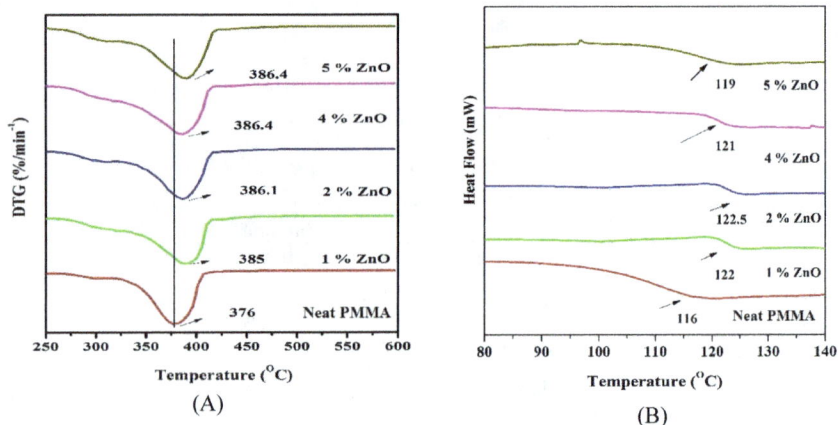

Figure 10: (A) Analysis of differential thermogravimetric (DTG) curve, and (B) glass transition temperature (T_g) of ZnO-based PMMA composites prepared at different loadings of ZnO and neat PMMA [48].

UV–Visible spectroscopy analysis: The optical characteristics of the thin film of PMMA/ZnO nanocomposites to confirm its UV absorption capacity was measured using UV-vis absorption, and the results are presented in Fig. 11 [48]. Pristine PMMA lacks an absorption peak in UV-region at 372 nm and characterizes as a poor material for UV-absorption. However, with the loading of ZnO nanofillers into PMMA matrix, a sharp

Adv. App. of Micro and Nano Clay II – Synthetic Polymer Composites   Materials Research Forum LLC
Materials Research Foundations 129 (2022) 233-281   https://doi.org/10.21741/9781644902035-10

absorption peak at 372 nm is observed, which indicates the successful loading of ZnO-nanoparticles into the long PMMA chain [45]. Interestingly, with increasing in ZnO loading, the UV-absorption peak intensity at 372 nm increases and reported maximum for 5 wt% ZnO loading. Increase in UV-absorption capacity with the loading of ZnO into PMMA matrix attributes to the quantum behavior of nanosize ZnO-nanoparticles [50].

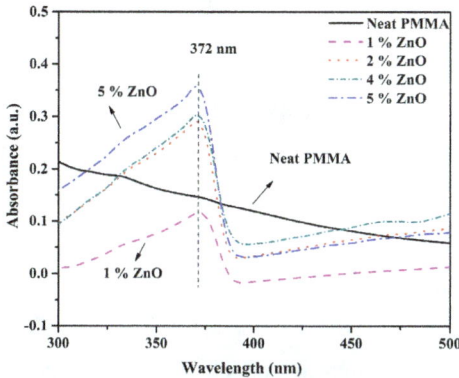

Figure 11: UV-visible absorption spectra of PMMA and ZnO-based PMMA composites prepared at different loadings of ZnO [48].

Electrical properties: Effect of ZnO loading into PMMA matrix for the variation in electrical conductivities is shown in Fig. 12 [48]. The results revealed that the conductivity of pristine PMMA is observed minimum at all radio frequencies. This low magnitude of conductivity of neat PMMA corresponds to the insulating character of pristine PMMA polymer. However, with the addition of ZnO nanofillers, a linear enhancement of electrical conductivity was seen in the nanocomposites. The maximum rise in electrical conductivity was 0.21 µS/cm for 5 wt% ZnO loading nanocomposites. This manifests that mobile charge carries present in ZnO can allow the flow of free electrons into the PMMA matrix and make it electrically conductive. A similar observation of electrical conductivity of ZnO nanoparticles into sago starch has also been reported by Rahman et al. [51]. In their analysis, the authors concluded that various factors viz. mobility of free charge, number of charge carriers, and accessibility of connections of the polar domains present in ZnO nanofillers provide a connecting network for the flow of free electrons and responsible for the augmentation in electrical conductivity of starch/ZnO nanocomposites.

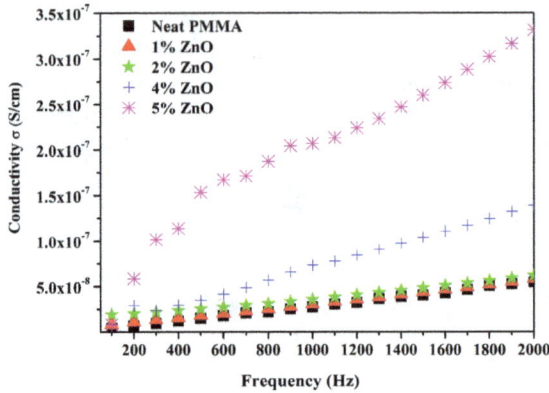

Figure 12: The electrical conductivity of PMMA and ZnO-based PMMA composites prepared at different loadings of ZnO [48].

Mechanical properties: The results of Young's modulus, tensile strength, and percentage elongation of PMMA and ZnO-based composites are shown in Fig. 13A-C [48]. The tensile strength results (Fig. 13A) indicated that PMMA with 1.0 wt.% ZnO loading have a maximum value 36.57 MPa as compared to that of the neat PMMA of 32.74 MPa. The increase in tensile strength of PMMA-ZnO nanocomposites with 1 wt% ZnO loading attributes to the uniform and homogenous dispersion of nanosize ZnO into the PMMA matrix due to strong and intense microturbulence generated by sonication. This intense mixing promotes a strong interfacial adhesion in between long polymer chain and ZnO nanoparticles with further increased in crystallinity of resulting nanocomposites. Another possible reason for the increase in tensile strength could be due to the restriction of a long polymer chain segment offered by ZnO nanoparticles as reported by Li et al. [52]. However, the tensile strength of nanocomposites was decreased to 27.77, 26.36, and 25.8 MPa when the ZnO loading was 2.0, 4.0, and 5.0 wt.%, respectively.

The variation of Young's modulus in the nanocomposites prepared with different loading of ZnO is shown in Fig. 13B. The Young's modulus of PMMA/ZnO nanocomposites with 1.0 wt.% ZnO nanocomposites was 1.47 GPa as compared to the PMMA with only 1.14 GPa. With increasing ZnO loading, a consistent contraction in Young's modulus has been observed. The mark dropped in Young's modulus of nanocomposites at higher ZnO loading is attributed to the ZnO nanoparticles agglomeration into the matrix of polymer.

The elongation percentage of ZnO-based PMMA nanocomposites with 1.0 wt.% loading is markedly high of 17.56% as compared to the percentage elongation of 3.92% for neat PMMA as shown in Fig. 13C. Further increase in ZnO loading the reduction in percentage elongation was observed and the minimum value was 10.98% for 5.0 wt.% ZnO loading. Although with the increase in ZnO loading, reduction in percentage elongation of nanocomposites was observed. However, its magnitude was still higher than the percentage elongation of neat PMMA. Addition of nanosize ZnO at low filler loading (1 wt%) enhances the interfacial-interaction between extended polymer matrix and nanoparticles in addition to an increase in nanocomposites crystallinity. This improves the ability to absorb maximum energy during deformation via restricting the segmental motion of the polymer chain. However, with an increase in ZnO loading, the high agglomerated structure reduces the ductility of the polymer matrix leading to a reduction in percentage elongations.

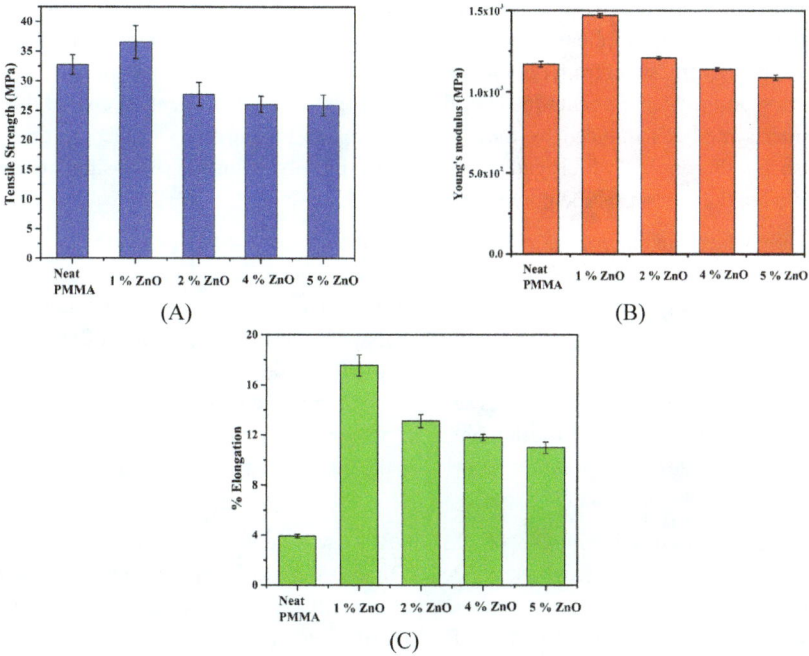

Figure 13: (A) Tensile strength, (B) Young's modulus, and (C) percentage elongation results of PMMA and ZnO-based PMMA composites prepared at different loadings of ZnO [48].

Adv. App. of Micro and Nano Clay II – Synthetic Polymer Composites   Materials Research Forum LLC
Materials Research Foundations 129 (2022) 233-281   https://doi.org/10.21741/9781644902035-10

4.3   PMMA/RGO nanocomposites

In the recent past, polymer nanocomposites synthesized with graphene-based nanofiller has drawn considerable interest due to significant enhancement in thermal, mechanical, and electrical properties of nanocomposites [53–57]. The high aspect ratio graphene nanosheets drastically increase the properties of the nanocomposites even at small loading% of nanosheets in the matrix of bulk polymer. However, polymer nanocomposites synthesized using graphene in pristine form is not suitable for interaction with the long polymer chains due to its poor dispersion ability and agglomerate formation into the polymer matrix. In order to achieve high compatibility and uniform dispersion of graphene nanosheets into a bulk polymer matrix, chemical functionalization of pristine graphene sheets is recommended [58]. The chemical functionalization of pristine graphene sheets using non-covalent technique is commonly preferred due to preserving the physical properties of nanosheets. A simpler and facile approach for chemical modification or functionalization of graphene sheets is carried out using graphene oxide. The graphene oxide can be synthesized from popular Hummer's methods which involve the use of strong oxidizing and reducing agents [59–60]. However, graphene nanosheets prepared through Hummer's methods induced by mechanical force are unable to exfoliate the graphene nanosheets thoroughly which causes a reduction in polymer nanocomposites properties after its loadings into pristine polymer chains.

The graphenes can be prepared by direct exfoliation of natural graphite flakes using sonication in the liquid medium of organic solvent, ionic liquids, or surfactants-water solution. The physical effects of ultrasound during the exfoliation of graphite in the liquid phase and chemical oxidation break the 3D graphite structure down to a 2D graphene structure. In-situ reduction plays an important role for achieving uniformly distributed and exfoliated graphene oxide nanosheets structure with improved physical properties. Among various factors which control the physical properties of GO nanosheets such as its dispersion abilities into polymer matrix is the ultrasound intensity, and is inversely proportional to the dispersants viscosity.

The ultrasound-assisted synthesis of PMMA/RGO was done in two steps. The first step involves the synthesis of graphene oxide (GO) according to modified Hummer's method using graphene powder as a precursor, and in the second step, simultaneous functionalization of GO into RGO and PMMA/RGO was accomplished using ultrasound-initiated *in-situ* polymerization technique. The advantages of simultaneous reduction of GO into RGO and PMMA/RGO synthesis is to increase the properties of the nanocomposites even at small loading of nanofillers via enhancing its compatibility with the long-chain polymer matrix. The physical properties of finally prepared nanocomposites with different RGO loadings were further investigated using various characterization techniques which are explained as follows:

Morphological analysis: TEM micrographs of the as synthesized graphene oxide (GO) nanosheets and its nanocomposites with PMMA polymer with varying RGO loading are shown in Fig. 14A-D [61]. The 0.2 and 0.4 wt.% RGO loading exhibit almost uniform dispersion into long PMMA chain as the results are depicted in Fig. 14B and 14C, respectively. Agglomeration of RGO nanosheets was observed when RGO loading increased to 1.0% which is due to the increase in the bulk density of RGO into the PMMA matrix. This can be clearly observed in Fig. 14D for the RGO-based PMMA nanocomposites preperaed with 1.0 wt.% RGO loading.

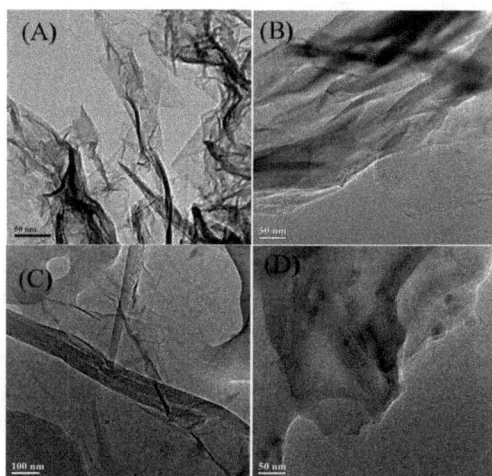

Figure 14: TEM micrographs of (A) graphene oxide nanosheet and PMMA/RGO nanocomposites: (B) PMMA+ 0.2 wt.% RGO loading, (C) PMMA+0.4 wt.% RGO loading, and (D) PMMA+ 1.0 wt% RGO loading [61].

Thermal stability: The thermal stability behavior of PMMA/RGO after incorporation of RGO into PMMA matrix has been measured using thermogravimetric analysis, and the results are presented in Fig. 15 [61]. A summary of this TGA curve in terms of nanocomposites weight loss against degradation temperature is also presented in Table 3 [61]. The results confirm that marked enhancement in thermal stability of pristine PMMA when loaded with RGO nanosheets. The augmentation in thermal stability of composites is reported highest for 1.0 wt.% RGO loading. This marked rise in thermal stability of nanocomposites for highest RGO loading (1.0 wt.%) is reported 23.85 and 49.7°C for $T_{5\%}$ and $T_{50\%}$ respectively.

Table 3: The TGA results of PMMA and RGO-based PMMA composites [61]

Name of Sample	$T_{5\%}$	$T_{50\%}$	$\Delta T_{5\%}$	$\Delta T_{50\%}$
Neat PMMA	267.4°C	326.1°C	–	–
PMMA + 0.2 wt.% RGO	283.2°C	373.0°C	15.8°C	46.9°C
PMMA + 0.4 wt.% RGO	288.2°C	375.2°C	20.8°C	49.1°C
PMMA + 1.0 wt.% RGO	291.3°C	375.8°C	23.9°C	49.7°C

The thermal stability of nanocomposites corresponds to maximum thermal degradation rate/inflection point also called DTG, which is illustrated in Fig. 16A [61]. The DTG peak of PMMA/RGO showed marked enhancement with an increase in RGO loading. The maximum peak value of nanocomposites was seen at 393°C for 1.0 wt% RGO loading. Increase in inflection point refers to exfoliation and uniform distribution of the nanosheets into the long polymer chain. This leads to better compatibility between PMMA matrix and exfoliated RGO nanosheets. Strong interaction between long polymer chain PMMA and exfoliated RGO nanosheets further inhibit the breakage of random polymer chains and therefore, increase the thermal stability of resulting nanocomposites. Another possible reason for the rise in thermal stability of nanocomposites could be due to the strengthening of a weak head to head linkage of long PMMA chain with the addition of exfoliated structure of RGO nanosheets induced during high energy sonication. The strong microturbulence and shock wave generated during *in-situ* emulsion polymerization derived from sonication exfoliate the stacked RGO nanosheets and reinforce the PMMA polymer chain into it and yielded a strong chemical interaction in the form of chemical bond. This chemical bonding also inhibits the thermal diffusion which evolves during heating of nanocomposites.

The RGO encapsulation into the PMMA matrix also influence the glass transition temperature (T_g) of resulting nanocomposites as shown in Fig. 16B [61]. It can be seen that with RGO with 0.2 and 0.4 wt% loading the T_g value increases up to 123.5 and 124.5°C as compared to minimal T_g value of neat PMMA of 116°C. The increase in T_g value attributes the restriction of segmental motion of long PMMA chain after addition of thermal stable RGO sheets into the PMMA matrix. However, RGO with 1 wt% shows a reduction in T_g

value. The possible consequence of intercalated structure or RGO nanosheets was obtained due to its non-homogenous dispersion and agglomerated structure into the PMMA matrix.

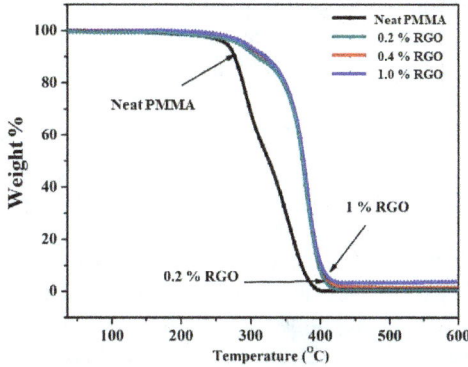

Figure 15: The TGA analysis results of PMMA and RGO-based PMMA nanocomposites prepared at different loadings of RGO [61].

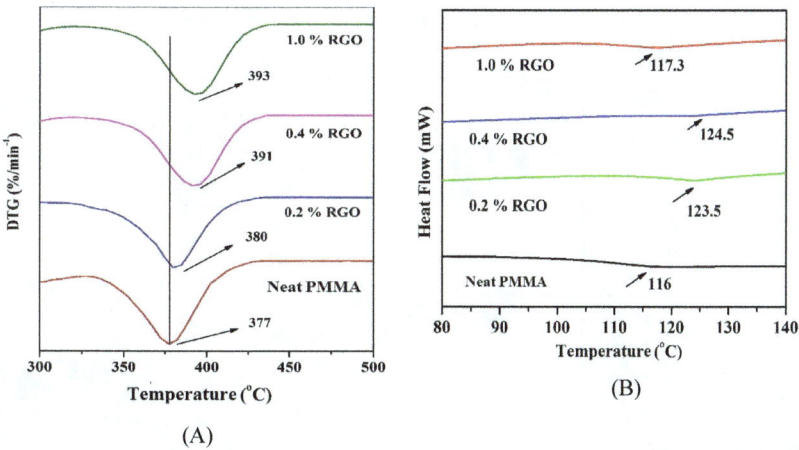

(A)   (B)

Figure 16: (A) Analysis of differential thermogravimetric (DTG) curve, and (B) glass transition temperature (T$_g$) of PMMA and RGO-based PMMA nanocomposites prepared at different loadings of RGO [61].

Electrical conductivity: The measurement of bulk electrical conductivity of pure PMMA and RGO-based PMMA with various RGO loading is depicted in Fig. 17 [61]. The results showed that the pristine PMMA exhibits a very small magnitude of electrical conductivity of 2.3×10^{-15} S/cm, which justify its electric insulating behavior. With the addition of RGO nanosheets into PMMA matrix, the electrical conductivity increases drastically and varies almost linearly with nanosheets loading. The maximum increase in electrical conductivity was reported to be 2×10^{-7} S/cm for RGO nanosheets loading for 1 wt%. The enhancement in electrical conductivity of nanocomposites can be explained based on two phenomena. First involves the nanocomposites synthesis method with the simultaneous reduction of GO into RGO and *in-situ* polymerization into PMMA/RGO nanocomposites. During this, the addition of RGO nanosheets into PMMA matrix facilitates the conductive network-like structure, which increases the transportation of free electron carrier through nanocomposites and makes it electrical conductive [62]. Second, the synthesis of the nanocomposites coupled with high-intensity sonication, can significantly expand the stacked RGO nanosheets into exfoliated sheets structure [63]. This helps a homogenous and uniform dispersion of RGO nanosheets to bulk PMMA matrix with enhanced electrical conductivity of resulting nanocomposites. The exfoliated nanosheets of RGO are a consequence of large shear force and intense microturbulence/micromixing induced by high intensive ultrasound wave during *in-situ* emulsion polymerization.

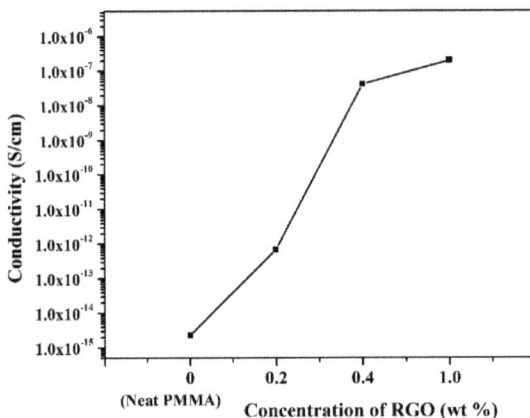

Figure 17: Electrical conductivity of neat PMMA and RGO-based PMMA nanocomposites prepared at different loadings of RGO [61].

Electromagnetic interference shielding (EMI): EMI shielding is used as a protection against harmful electromagnetic radiation emitted through electronic appliances. The electromagnetic radiation and radio waves emitted through nearby devices can interfere with another and reduce its performance. In that case, the polymer nanocomposites with high electrical conducting nanofiller such as RGO is useful in order to prevent these devices. The shielding capacity of electromagnetic radiation can be measured using two different mechanisms, *viz.* reflection and absorption. With these two mechanisms, the overall EMI shielding effectiveness (SE) of nanocomposites can be calculated using the following equations as given below:

$$EMI\ SE = 10\log\left(\frac{P_i}{P_o}\right) \tag{1}$$

Where, P_i and P_o are the incident and transmitted EM power, respectively.

The results of EMI shielding effectiveness obtained by (i) reflection (SE_r), (ii) absorption (SE_a) and (iii) sum of reflection and absorption ($SE_r + SE_a$) i.e. overall shielding effectiveness (SE_o) with neat PMMA and after incorporation of the RGO nanosheets into PMMA matrix are presented in Fig. 18 [61]. The results show that after RGO loading, shielding with absorption is much higher than reflection as compared to the neat PMMA and is estimated highest for 1 wt% RGO. The magnitude of the shielding effectiveness of PMMA-RGO composites with RGO of 1 wt% loading was found to be: $SE_r = 1.44$ dB and $SE_a = 1.83$ dB by absorption which is significantly higher than the neat PMMA and shows no shielding against harmful electromagnetic radiation.

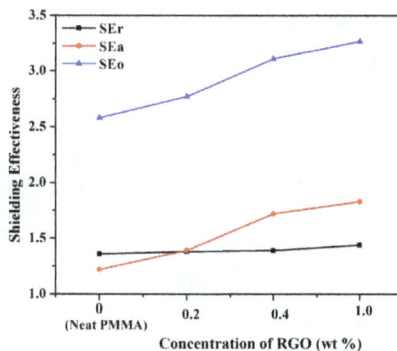

Figure 18: Electromagnetic interference (EMI) shielding properties of RGO-based PMMA nanocomposites prepared at different loadings of RGO [61].

Materials Research Foundations 129 (2022) 233-281   https://doi.org/10.21741/9781644902035-10

Mechanical properties: The mechanical properties of the PMMA/RGO nanocomposites have been evaluated and given in Fig. 19A-D [61]. Fig. 19A depicts the results of tensile strength of pure PMMA and RGO-based PMMA loaded with 0.2, 0.4, and 1.0 wt.% of RGO. The tensile strength was found to be 32.7 MPa for neat PMMA. While incorporation of 0.2 wt% RGO into the matrix of bulk PMMA increased the tensile strength of sample up to 33.6 MPa. For 0.4 wt% RGO loading, the tensile strength shows the highest value of 40.4 MPa. The marked enhancement in tensile strength at low loading (0.4 wt%) of RGO attribute the homogenous dispersion of exfoliated structure of RGO nanosheets into long PMMA chain matrix driven by sonication. Also, RGO with high aspect ratio with strong intermolecular attraction in the form of H-bonding between PMMA chains and RGO nanosheets provide a molecular level distribution of the RGO into PMMA matrix responsible for the increase in tensile strength of resulting nanocomposites. A reduction in tensile strength was observed when RGO loading was 1.0 wt% - which could be due to the agglomerated structure of nanosheets into bulk PMMA chain matrix.

A similar trend in the results of Young's modulus was observed as shown in Fig. 19B. It shows the highest value of 2.6 GPA for RGO loading of 0.4 wt% as compared to 1.13 GPa for neat PMMA. Integration of RGO nanosheets increases the stiffness of the bulk polymer chain that increases Young's modulus. Further increase in nanosheets loading of 1.0 wt.%, a considerable decline in Young's modulus was seen which is due to the agglomeration of RGO nanosheets into the PMMA matrix.

Addition of RGO loading also increases the percentage elongation of nanocomposites as seen from Fig. 19C. The percentage elongation of nanocomposites with 0.2 and 0.4 wt.% RGO loading are 5.94% and 9.3%, respectively, as compared to neat PMMA with only 3.92%. However, a reduction of percentage elongation has been observed with 1.0 wt.% RGO loading. At low loading of RGO, nanosheets are entirely exfoliated and liable to more interfacial interaction with the polymer matrix and increase the nanocomposites properties. However, increasing in nanosheet loading a reduction in percentage elongation can be seen clearly from the results. A similar trend in mechanical properties for PMMA-RGO nanocomposites has been reported by Tripathi et al. [57] and Wang et al. [58].

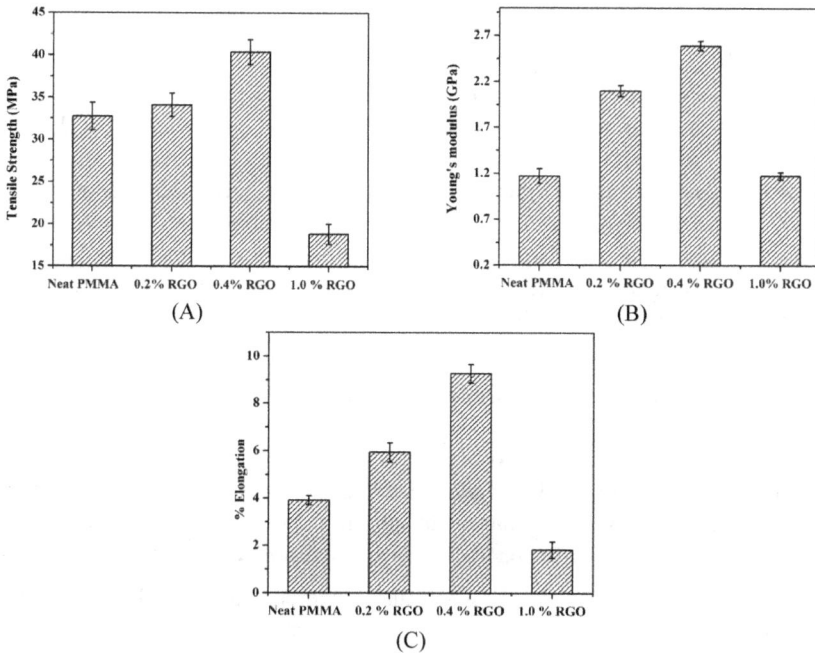

Figure 19: (A) Tensile strength, (B) Young's modulus and (C) percentage elongation results of PMMA and RGO-based PMMA composites prepared at different loadings of RGO: [61].

4.4 PMMA/magnetite nanocomposites

Incorporation of magnetic particles into a polymer matrix composite can open new avenues to make nanocomposites with enhanced electrical and magnetic properties [64–67]. Addition of magnetite nanoparticles (Fe_3O_4) into bulk polymer matrix has attracted considerable interest in designing nanocomposites devices used as a shielding material against harmful EMI radiation generated through electronic gadgets. Among various thermoplastics, PMMA exhibits excellent optical transparency and thereby used as an alternative to glass material in the optical and household application. Other than optical transitivity, additional physical properties of pristine PMMA such as high magnetic saturation, electrical conductivity, high thermal stability and increase in mechanical

Adv. App. of Micro and Nano Clay II – Synthetic Polymer Composites   Materials Research Forum LLC
Materials Research Foundations 129 (2022) 233-281   https://doi.org/10.21741/9781644902035-10

strength can also be augmented via incorporation of magnetite nanofillers and can be widely used in various other applications.

The conventional methods for synthesis of PMMA/magnetite nanocomposites involve the surface modification of magnetic nanofillers for its uniform dispersion and agglomerated free structure into the bulk polymer matrix. The surface modification of magnetite nanoparticles can be achieved either by grafting of long polymeric chains at outer surface of filler particles or treating the surface using silane as a coupling agent [68–69]. In another approach, the use of surfactants can also improve the dispersion behavior and control the agglomeration of nanoparticles in aqueous systems by reducing the inter-particle interaction among the nanomaterials [70].

This section deals about the two-step preparation of PMMA/magnetite nanocomposites through sonochemical *in-situ* emulsion polymerization. First step comprised for the synthesis of magnetite (Fe_3O_4) nanoparticles using co-precipitation technique. In the second step, PMMA/magnetite nanocomposites have been prepared via sonochemical *in-situ* emulsion polymerization with various loading (1, 2 and 5 wt%) of the synthesized magnetite nanoparticles. The change in physical properties of PMMA/magnetite nanocomposites after successful loading of magnetite nanoparticles are discussed below.

Structural morphology: The structural and surface morphology of the PMMA/magnetite nanocomposites with different loading (viz. 1, 2, and 5 wt%) was performed and illustrated in Fig. 20A [71]. The result showed that the magnetite particles are spherical in shape. The bright spots observed in selected area electron diffraction (SAED) pattern (in inset) also confirm its nanocrystalline nature of magnetite nanoparticles [72]. The bright area of the micrograph as shown in Fig. 20B-D [71] confirms the PMMA molecules and the dark spots correspond to the magnetite nanoparticles. With high filler loading of 5 wt%, magnetite nanoparticles show agglomeration and is shown in Fig. 20D.

Magnetic properties: The magnetic properties which are also referred as the hysteresis loop (H–M) curve of magnetite nanoparticles and PMMA/magnetite composite have been studied and measured using vibrational scanning measurement (VSM). The result of H-M curves of magnetite nanoparticles and nanocomposites with PMMA are shown in Fig. 21A-B [71]. Fig. 21A shows the magnetic saturation value (M_s) of magnetite nanoparticles synthesized using sonication is considerably higher (63.37 emu/g) than the magnetic saturation obtained using mechanical agitation (42.16 emu/g). The marked rise of ~ 40% in magnetic saturation value of magnetite nanoparticles synthesized using sonication is attributed to the beneficial facets of the high-intensity ultrasound wave.

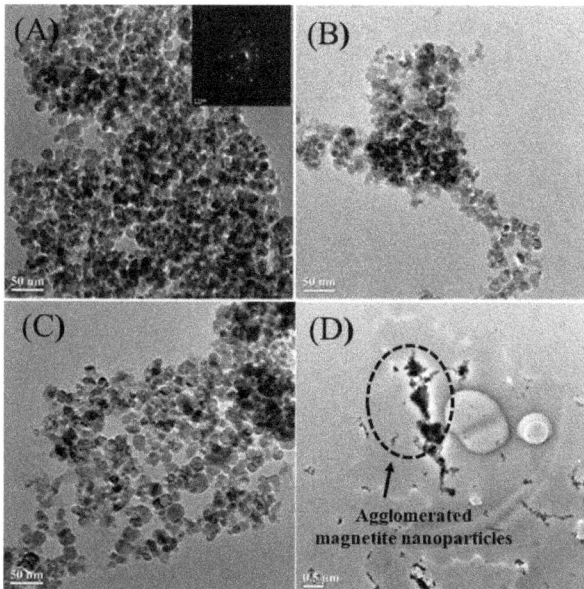

*Figure 20: Results of Transmission Electron Microscopy analysis. (A) Fe₃O₄
nanoparticles with SAED images (inset picture) and PMMA/Fe₃O₄ nanocomposites
loaded with (B) 1.0 wt.%, (C) 2.0 wt.%, and (D) 5.0 wt.% [71].*

Fig. 21B depicts the magnetic saturation value (M_s) of PMMA/magnetite nanocomposites prepared with 1, 2, and 5 wt% magnetite loadings. The M_s value for nanocomposites with 1 wt% magnetite loading is 0.67 emu/g, which is almost 200% times smaller than the M_s value of magnetite synthesized using sonication. The increase in M_s value of the nanocomposites with an increase in nanofiller loading depicts a linear change. For 5 wt% magnetite loading, the M_s value is reported at about 5.12 emu/g. The M_s value of PMMA/magnetite nanocomposites obtained in this study is significantly higher than the M_s value for same magnetite loading as reported previously. For example, Wilson et al. [73] have calculated the highest M_s values of 11.5 emu/g for PMMA/magnetite composites with Fe₃O₄ loading of 10 wt%. The higher M_s value reported by Poddar et al. [71] is attributed to the uniform distribution of the magnetite nanofiller into the PMMA matrix-induced.

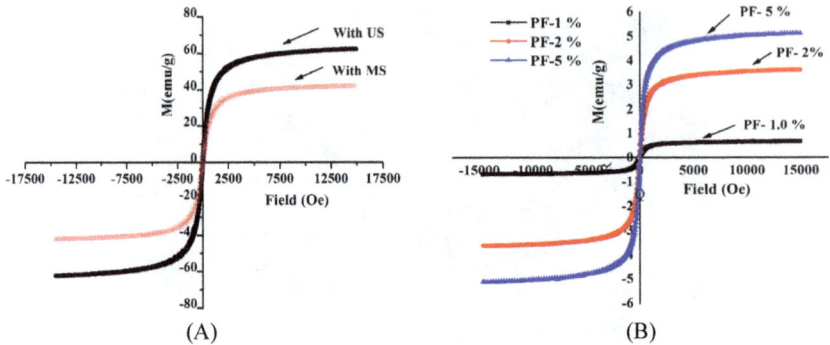

Figure 21: Magnetization curves. (A) Fe₃O₄ nanoparticles synthesized with ultrasound and mechanical stirring, (B) PMMA/Fe₃O₄ nanocomposites synthesized with varying loading of Fe₃O₄. [PF–1%: PMMA/Fe₃O₄ (1.0 wt.%), PF–2%: PMMA/Fe₃O₄ (2.0 wt.%), PF–5%: PMMA/Fe₃O₄ (5.0 wt.%), MS: with mechanical stirring, US: with ultrasonication] [71].

Thermogravimetric study: The TGA results of PMMA/magnetite composites synthesized using high-intensity ultrasound are shown in Fig. 22 [71]. Total weight loss of magnetite particles in the temperature range between 30-600 °C is only 5 wt% which indicates its high thermal resistance against nanoparticles weight loss. A summary of thermal degradation temperature of the TGA curve of neat PMMA and PMMA/magnetite nanocomposites against various weight losses of $T_{5\%}$, $T_{25\%}$, and $T_{50\%}$ are listed in Table 4 [71]. Comparing with neat PMMA, the nanocomposites of PMMA/magnetite for $T_{5\%}$ temperature show a reduction in thermal stability for all loading of nanofillers. This attributes to the loss of residual impurities of $FeCl_3$ remained during nanoparticles synthesis. However, improvement in thermal stability of nanocomposites at $T_{25\%}$ and $T_{50\%}$ were observed. The maximum thermal stability of PMMA/magnetite nanocomposites is reported for 2 wt% loading of magnetite nanoparticles. However, for 5 wt% nanoparticles loading, a significant reduction in thermal stability is reported due to the formation of high agglomerated structures of magnetite nanoparticles into the bulk PMMA matrix.

Variation in thermal stability of nanocomposites corresponding to the rate of maximum degradation temperature or DTG curve is shown in Fig. 23A [71]. Loading of 1 and 2 wt% magnetite nanoparticles increases the inflection point of nanocomposites to 5 and 7 °C as compared to neat PMMA. Increase in inflection points of nanocomposites after nanofiller loading depicts another corroboration of high thermal stability with a uniform dispersion of nanofillers into PMMA matrix. At higher loading of nanofiller of 5 wt%, a reduction in inflection point was observed. The decrease in thermal stability at high nanofiller loading is attributed to the formation of agglomerate, which further weakens the intermolecular attraction between polymer matrix and nanofillers.

Table 4: The TGA analysis results of neat PMMA and PMMA/magnetite composites [71]

Name of Sample	$T_{5\%}$	$T_{25\%}$	$T_{50\%}$	$\Delta T_{25\%}$	$\Delta T_{50\%}$
Neat PMMA	267.4°C	293.2°C	326.0°C	--	--
PMMA + 1 wt.% Fe_3O_4	264.1°C	327.1°C	367.0°C	33.9°C	41.0°C
PMMA + 2 wt.% Fe_3O_4	259.1°C	335.9°C	373.2°C	42.7°C	47.2°C
PMMA + 5 wt.% Fe_3O_4	257.3°C	329.8°C	371.0°C	36.6°C	45.0°C

From Fig. 23B [71] it can be seen that the glass transition temperature (T_g) was increased up to 5.5 and 6.5°C for magnetite loading of 1.0 and 2.0 wt.%, respectively. Successful incorporation of nanofiller loading restricts the segmental motion of polymer matrix via reducing the polymer chain mobility and increase the thermal stability of nanocomposites. However, a further increase in nanofiller loading of 5 wt%, there is a sharp reduction in T_g value of 105 °C. A similar explanation of nanofiller agglomeration with high filler loading is assumed for this as mentioned earlier. The T_g value obtained in these studies is much higher than the T_g value of 103 °C with 22.8 wt% loading of magnetite nanoparticles at explained by Banert et al. [64].

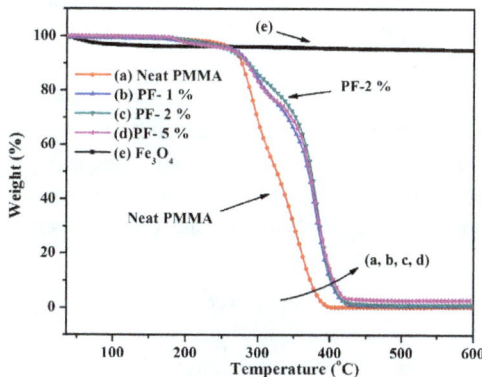

Figure 22: The TGA results of neat PMMA and Fe₃O₄–based PMMA composites synthesized with varying Fe₃O₄ loading [71].

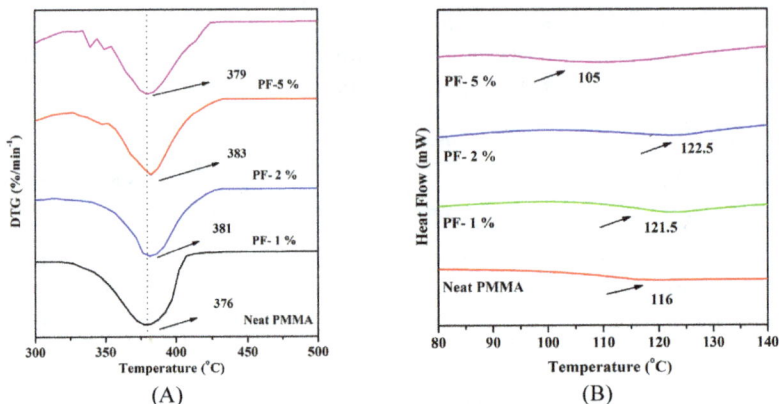

Figure 23: (A) Analysis of differential thermogravimetric (DTG) curve, and (B) glass transition temperature (T$_g$) of neat PMMA and PMMA/magnetite nanocomposites synthesized at different loadings of Fe$_3$O$_4$ nanoparticles [71].

Electrical conductivity: The variation in electrical conductivity of PMMA/magnetite nanocomposites prepared at different loading of magnetite nanoparticles is presented in Fig. 24 [71]. It is observed that there is a substantial enhancement in electrical conductivity of nanocomposites after nanofillers loading. The percolation threshold value, which is an indication of minimum nanofiller loading to change polymer from insulator to electrically conductive, is reported at 1.0 wt.% loading. Above 1.0 wt.% nanofiller loading, the electrical conductivity increases sharply up to a maximum of 2.0×10^{-13} S/cm for 5.0 wt.% loading as compared to the pure PMMA (2.3×10^{-15} S/cm). Increase in electrical conductivity of nanocomposites using magnetite nanofiller is significantly smaller than nanocomposites synthesized using RGO as discussed in the earlier section. The foremost reason for the marginal increase in electrical conductivity is due to spherical morphology of magnetite nanoparticles as compared to nanosheet structures of RGO containing high aspect ratio.

EMI shielding: The electromagnetic interference shielding effectiveness (EMI SE) of PMMA/magnetite nanocomposites was calculated using Eq. 1 and the results are given in Fig. 25 [71]. The results showed that the shielding effectiveness by absorption increases linearly with nanofiller loading. On the other hand, the shielding effectiveness by reflection is not effective and shows a reduction with increase in nanofiller loading. The individual magnitude of shielding effectiveness (SEa, SEr, and SEo) with 5 wt% magnetite nanoparticles loading are SEa = 1.45 dB by absorption, SEr = 1.07 dB by reflection and SE$_o$ =2.52 dB overall shielding.

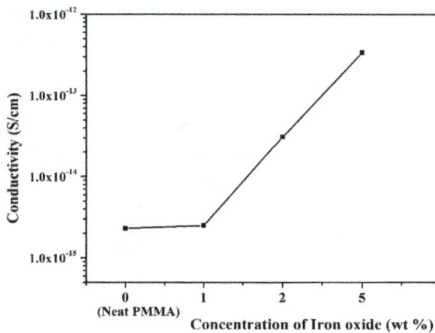

Figure 24: The electrical conductivity of Fe$_3$O$_4$–based PMMA nanocomposites prepared at different loadings of Fe$_3$O$_4$ nanoparticles [71].

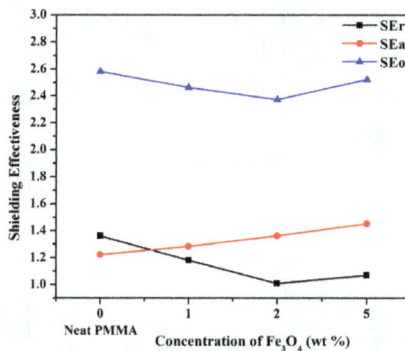

Figure 25: The electromagnetic interference (EMI) shielding properties of Fe$_3$O$_4$–based PMMA composites prepared at different loadings of Fe$_3$O$_4$ nanoparticles [71].

Mechanical properties: Tensile strengths of PMMA and composites of PMMA/magnetite are presented in Fig. 26A [71]. It could be observed that the addition of magnetite nanofiller of 1.0 and 2.0 wt.% increased the PMMA magnetite nanocomposites' tensile strength up to 39.9 MPa and 40.28 MPa, respectively, comparing to the tensile strength of PMMA of

32.74 MPa. The enhancement in the tensile strength is due to the uniform and homogenous distribution of magnetic particles resulted from intense micro-convection generated during sonication. However, nanocomposites at high filler loading of 5.0 wt.% showed a marked reduction in tensile strength. It is quite clear that magnetic nanofillers at higher loading can cause severe agglomeration and consequently, reduction in its tensile strength.

The trend of Young's modulus of PMMA/magnetite nanocomposites shown in Fig. 26B [71] follows a similar trend as tensile strength. A low Young's modulus for PMMA (1.14 GPa) was seen compared to the PMMA/magnetite composites loaded with 1.0 and 2.0 wt% (2.3 and 2.4 GPa, respectively). Further increasing in nanofiller loading, a drastic decline in Young's modulus was reported. The change in Young's modulus parameter with nanofillers loading is explained in the same way as mentioned above for Tensile strength.

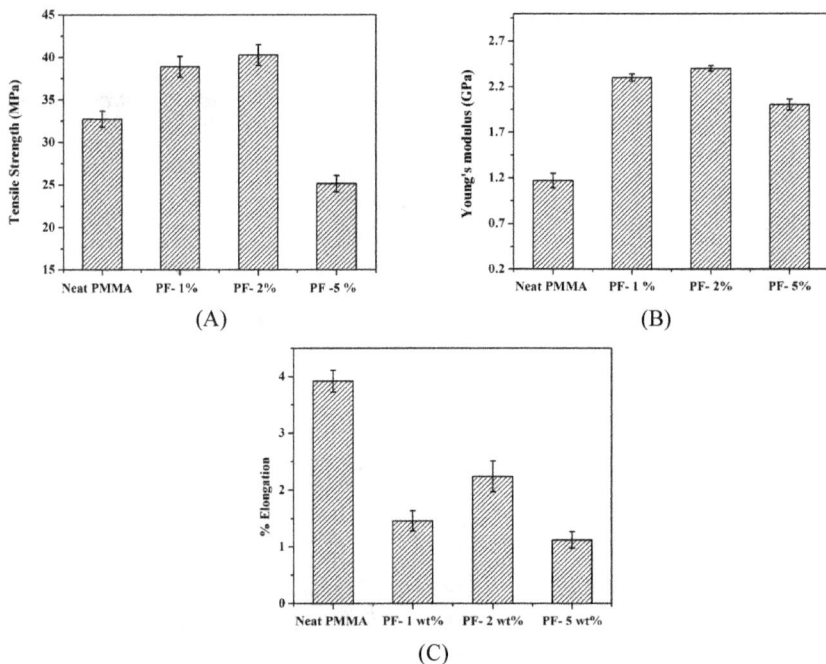

Figure 26: (A) Tensile strength, (B) Young's modulus, (C) Percentage elongation results of PMMA and Fe_3O_4–based PMMA nanocomposites prepared at different loadings of Fe_3O_4 [71].

Adv. App. of Micro and Nano Clay II – Synthetic Polymer Composites   Materials Research Forum LLC
Materials Research Foundations 129 (2022) 233-281   https://doi.org/10.21741/9781644902035-10

The results of the percentage elongation of PMMA/magnetite composites with nanofiller with various loading are shown in Fig. 26C [71]. Addition of 1 wt% magnetite nanofillers caused a significant reduction in percentage elongation value of 1.46% for nanocomposites as compared to high elongation of PMMA of 3.92 wt%. However, a decrease in percentage elongation of up to 2.24% was observed with further increasing in nanofiller loading. The composites with 5.0 wt.% loading showed a minimum percentage elongation of 1.12%. At high filler loading, reduction in percentage elongation essentially confirms the restricted mobility of the long PMMA chains. However, the increase in the elongation percentage for 2 wt% magnetite nanoparticles attributes the improvement in the capability of composite to absorb the deformation energy due to an enhancement in crystallinity of the matrix of polymer with the nanofillers incorporation.

4.5 Cloisite-30B based Poly(MMA-co-BA) nanocomposites

Polymethyl methacrylate (PMMA), a kind of thermoplastic shows excellent optical transparency, high impact strength, and good chemical and environmental stability. However, PMMA shows a major disadvantage of high brittle character that causes its sudden failure under load and limits its application. It has been observed that incorporation of nanofillers (*viz.* clay, RGO, ZnO and magnetite) have increased the PMMA thermal stability, mechanical strength in terms of Young's modulus, electrically conductive and EMI shielding. However, PMMA nanocomposites still show high brittle character with low ductility even after addition of these nanofillers. An alternate way to increase its ductility could be achieved with a blending of another polymer like poly butyl acrylate shows a shallow glass transition temperature (T_g) as compared to PMMA.

An alternative way to increase the ductile behavior of neat PMMA can be achieved through copolymerization of poly butyl acrylate (PBA) polymer into the PMMA matrix. The low T_g of PBA can control the brittle and ductile issues of PMMA after blending into PMMA chains. However, after blending of PBA into PMMA chains, the resulting copolymer poly (MMA-co-BA) further reduces its mechanical and thermal properties owing to PBA's rubbery structure and low T_g. Therefore, the addition of nanofillers into pristine poly(MMA-BA) matrix has become essential in order to increase the ductility of poly(MMA-BA) with enhanced thermal and mechanical properties. Incorporation of Cloisite-30B nanoclay into the copolymer poly(MMA-co-BA) could be the most effective choice to increase the ductility of resulting nanocomposites with enhancing thermal and mechanical properties. The increase in properties like mechanical strength and thermal properties of Cloisite-30B based poly(MMA-co-BA) nanocomposites after addition of Cloisite-30B nanofillers have been discussed below.

Thermogravimetric analysis: The thermal stability of pristine copolymer of poly(MMA–co–BA) and its nanocomposites poly(MMA–co–BA)/Cloisite-30B with different clay amount are presented in Fig. 27 [35]. A summary of the weight loss of nanocomposites due to thermal heating has been summarized in Table 5 [35]. After addition of Cloisite-30B nanoclay into the matrix of poly(MMA–co–BA), thermal degradation temperature for 10% weight loss which corresponding to $T_{10\%}$, increased significantly and was reported to be highest, *i.e.* 30.3°C for 4.0 wt.% clay loading. Similarly, $T_{50\%}$ which corresponds to the thermal degradation temperature of nanocomposites for 50% weight loss showed the maximum enhancement of 15.5° with clay loading of 2.0 wt.%. This significant improvement in the thermal stability of composites was obtained due to the formation of exfoliated nanoclay into the co-polymer matrix. However, for 5 wt% clay loading, reduction in decomposition temperature for $T_{10\%}$ and $T_{50\%}$ is reported. This could be a due to the agglomeration of nano clay as well as phase separation at higher clay loadings.

Table 5: The TGA analysis results of neat poly(MMA-co-BA) and Cloisite-30B clay based poly(MMA-co-BA) composites [35]

Name of Sample	$T_{10\%}$	$T_{50\%}$	$\Delta T_{10\%}$
Neat poly(MMA-co-BA)	289.8°C	357.0°C	--
poly(MMA-co-BA) + 1 wt.% Cloisite 30B	319.2°C	371.8°C	29.4°C
poly(MMA-co-BA) + 2 wt.% Cloisite 30B	319.6°C	372.5°C	28.8°C
poly(MMA-co-BA) +4 wt.% Cloisite 30B	320.1°C	372.7°C	30.3°C
poly(MMA-co-BA) +5 wt.% Cloisite 30B	318.8°C	370.2°C	29.0°C

Nomenclature used in Tables:
$T_{5\%}$ – temperature corresponds to 5% loss
$T_{10\%}$ – temperature corresponds to 10% loss
$T_{50\%}$ – temperature corresponds to 50% loss

DTG curves of Cloisite-30B based poly(MMA–co–BA) nanocomposites are depicted in Fig. 28A [35]. As we can clearly see in the Fig. 28A, with incorporation of clay materials into copolymer matrix, the DTG peak was moved from lower to higher temperature range and confirms the increase in thermal stability of resulting composites. The increase in thermal stability can be confirmed with the increase in the inflection point of composites in comparison to the inflection point of PMMA without Cloisite-30B. The inflection points of poly(MMA–co–BA)/Cloisite-30B composites prepared with 1.0, 2.0, 4.0 and 5.0 wt% clay loadings showed marked rise up to 381, 382, 384 and 374°C, respectively, as compared to the inflection point of PMMA (359°C) without clay. The highest temperature of 25°C was observed for 4.0 wt.% clay loading. With further increase in clay loading (5.0 wt.%), degradation temperature decreased up to 374°C and is due to the agglomerates formation at higher loading.

Another method to evaluate the thermal stability of Cloisite 30B based poly(MMA-co-BA) copolymer composites with varying clay loading was done using DSC analysis and the results are illustrated in Fig. 28B [35]. Pristine copolymer of poly(MMA–co–BA) showed the glass transition temperature (T_g) of 52.1°C. While the addition of clay nanofillers increase the T_g value up to 52.6°, 53.4°, 64.8°C for 1.0, 2.0, and 4.0 wt.% clay loading, respectively. The rise in T_g of the nanocomposites could be due to the additional hindrance incorporated by the clay platelets into the poly(MMA-co-BA) matrix which further slowed down the movement of long polymer chain and increased the T_g. However, the clay amount of 5 wt% loading Tg value decreased up to 53.1°C. The possible consequence of the lowering down of T_g at 5.0 wt.% clay loading could be due to agglomeration of clay into the copolymer matrix.

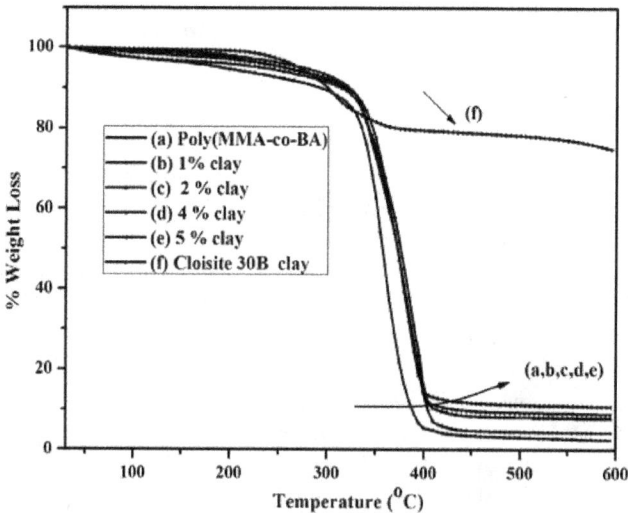

Figure 27: TGA analysis of poly(MMA–co–BA) and Cloisite-30B based poly(MMA–co–BA) composites prepared with different amount of clay [35].

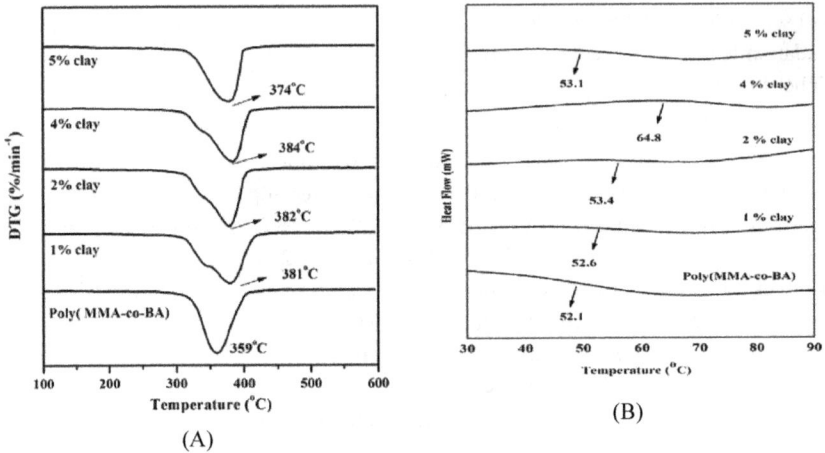

Figure 28: (A) Analysis of differential thermogravimetric (DTG) curve, and (B) glass transition temperature (T$_g$) of poly(MMA-co-BA) and Cloisite-30B based poly(MMA-co-BA) composites [35].

Mechanical Properties: The tensile strength analysis of PMMA, copolymers poly(MMA-co-BA) and Cloisite-30B based poly(MMA-co-BA) composites are given in Fig. 29A [35]. It can be seen from the Fig. 29A that the tensile strength of neat poly(MMA-co-BA) is 16.02 MPa as compared to the high tensile strength of 32.74 MPa for neat PMMA. The low tensile strength of poly(MMA-Co-BA) is obtained due to the blending of PBA polymer which possesses a low glass transition temperature. As discussed earlier, the incorporation of nanofiller in the matrix of polymer restricts the mobility of the polymer chain [74]. Addition of Cloisite-30B nanofillers of 1.0 and 2.0 wt.% increased the tensile strength of poly(MMA-co-BA)/Cloisite-30B nanocomposites up to 18.55 and 20 MPa, while pristine copolymer poly(MMA-co-BA) showed a tensile strength of 16.02 MPa. However, 4 wt% clay loading, a reduction in tensile strength is observed. This could be due to the non-uniform dispersion of clay platelets into the copolymer matrix and caused its agglomeration formation into the bulk copolymer matrix. These agglomerates can form weak spot into the copolymer matrix and reduced the tensile strength of nanocomposites.

The Young's modulus analysis results of copolymer nanocomposites are presented in Fig. 29B [35]. Comparing the Young's modulus of pristine PMMA (1.17 GPa), the pristine co-polymer poly(MMA–co–BA) was found to have 1.02 GPa Young's modulus. The addition of Cloisite-30B nanoclay (1.0 and 2.0 wt.%) into the matrix of co-polymer increase the

Young's modulus up to 1.08 and 1.31 GPa, respectively. However, the Young's modulus was reduced to 1.07 GPa for nanocomposite with clay loading of 4.0 wt.%. Sirapanichart et al. [75] have also reported similar trends for Young's modulus and tensile strength with clay loading for nanocomposites co-polymers of MMA and BA.

The results of percentage elongation of pristine poly(MMA–co–BA) and its Cloisite-30B based nanocomposites are presented in Fig. 29C [35]. Comparing with original PMMA, co-polymer poly(MMA–co–BA) has showed 6 order of magnitude in percentage elongation. Co-polymerization of BA and MMA change the elastic limit of PMMA, polymer which is considered a very hard and brittle material. After nanocomposite formation, the percentage elongation decreased with addition of clay platelets. As reported previously, the interaction between copolymer matrix and nanofiller leads to confinement of dynamics of the polymer chain and it varies proportionately with nanofiller loading which is evidence from the larger dropped in percentage elongation as the clay loading increases.

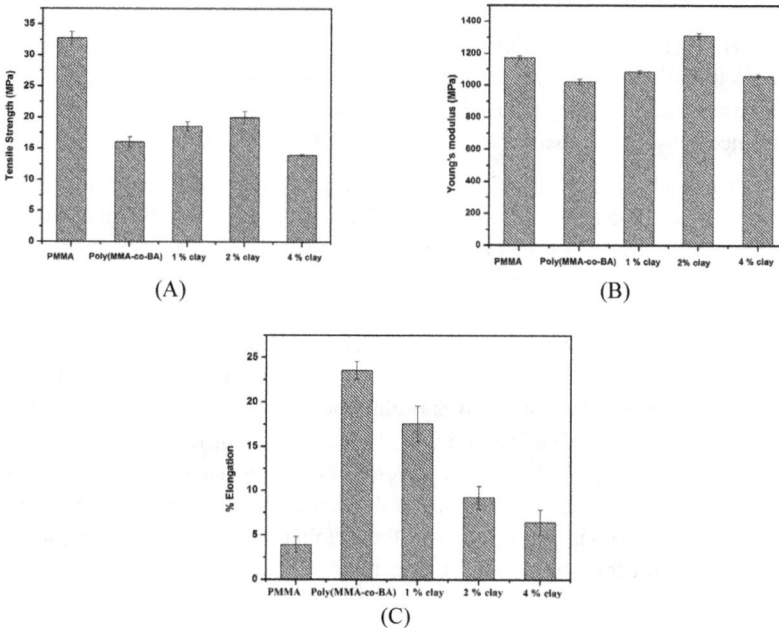

(A)　　　　　　　　　　　　　　　　　(B)

(C)

Figure 29: (A) Tensile strength. (B) Young's modulus. (C) Percentage elongation results of PMMA, poly(MMA-co-BA) and Cloisite-30B based poly(MMA-co-BA) composites. [35].

5.   Overview

The change of properties of nanocomposites with encapsulation of various nanofillers such as Cloisite-30B nanoclay, semiconductor ZnO oxide, reduced graphene oxide (RGO) and magnetite (Fe_3O_4) nanoparticles into polymer matrix at different loadings have been presented. Also, the effect ultrasound wave intensity and its secondary effects of cavitation on the synthesis of polymer nanocomposites have been discussed in detail, including the changes in properties of the polymer nanocomposites. During the synthesis of the nanocomposites, generation of high-intensity ultrasound wave has proven an effective method in uniform and homogenous dispersion of nanofillers into polymer matrix with enhanced physical properties of various PMMA nanocomposites. Addition of Cloisite-30B nanoclay into the PMMA matrix significantly increased the thermal stability and Young's modulus of Cloisite-30B supported PMMA nanocomposites. Also, copolymerization of PBA and PMMA with Cloisite-30B clay loading significantly reduced the brittle characteristics of PMMA with increased in thermal stability of Cloisite-30Bbased poly(MMA-co-BA) nanocomposites. The microturbulence and high energy shock wave produced via ultrasound irradiation enhance the interlayer gallery spacing or d-spacing of clay platelets in the form of the exfoliated structure of resulting nanocomposites. These exfoliations can cause homogenous dispersion of clay layers into the bulk PMMA matrix with enhanced physical properties of Cloisite-30B based PMMA nanocomposites.

On the other hand, ZnO as nanofiller in ZnO-based PMMA nanocomposites increased shielding resistance against UV-radiation in addition to an increase in thermal, mechanical, and electrical properties of PMMA/ZnO nanocomposites. The important aspect of ultrasound assistance technique for the preparation of PMMA/ZnO nanocomposites provides a uniform and homogenous dispersion of nanosize ZnO into the bulk polymer matrix even without surface modification. However, high-intensity ultrasound is also not capable of improving properties due to the agglomeration of nanoparticles at high ZnO loading. Irradiation of ultrasound waves during the synthesis of PMMA/RGO markedly improved the thermal, mechanical and electrical properties of resulting nanocomposites even at low filler loadings ($\leq$ 1 wt%). Addition of iron oxide nanoparticles in the polymer matrix increases its magnetic, thermal, and mechanical properties. These enhanced physical properties primarily attributed to the exfoliation of the graphene sheet in the PMMA matrix with the presence of strong micro-convection induced by high energy ultrasound.

References

[1] Y. Xu, W.J. Brittain, C. Xue, R.K. Eby, Effect of clay type on morphology and thermal stability of PMMA-clay nanocomposites prepared by heterocoagulation

method, Polymer. 45 (2004) 3735-3746.
https://doi.org/10.1016/j.polymer.2004.03.058

[2] T.Y. Tsai, M.J. Lin, Y.C. Chuang, P.C. Chou, Effects of modified Clay on the morphology and thermal stability of PMMA/clay nanocomposites, Mater. Chem. Phys. 138 (2013) 230-237. https://doi.org/10.1016/j.matchemphys.2012.11.051

[3] E. Tang, G.X. Cheng, X. Ma, Preparation of nano-ZnO/PMMA composite particles via grafting of the copolymer onto the surface of zinc oxide nanoparticles, Powder Technol. 161 (2006) 209-214. https://doi.org/10.1016/j.powtec.2005.10.007

[4] M. A. Aldosari, A. A. Othman, E.H. Alsharaeh, Synthesis and characterization of the in situ bulk polymerization of PMMA containing graphene sheets using microwave irradiation, Molecules. 18 (2013) 3152-3167. https://doi.org/10.3390/molecules18033152

[5] R.K. Layek, S. Samanta, D.P. Chatterjee, A.K. Nandi, Physical and mechanical properties of poly(methyl methacrylate)-functionalized graphene/poly(vinylidine fluoride) nanocomposites: Piezoelectric β polymorph formation, Polymer. 51 (2010) 5846-5856. https://doi.org/10.1016/j.polymer.2010.09.067

[6] H. Martinez, L.D. Onofrio, G. Gonzalez, Mossbauer study of a Fe3O4/PMMA nanocomposite synthesized by sonochemistry, Hyperfine Interac. 224 (2013) 99-107. https://doi.org/10.1007/s10751-013-0829-2

[7] V.V. Vodnik, D.K. Bozanic, E. Dzunuzovic, J. Vukovic, J.M. Nedeljkovic, Thermal and optical properties of silver-poly(methylmethacrylate) nanocomposites prepared by in-situ radical polymerization, Eur. Polym. J. 46 (2010) 137-144. https://doi.org/10.1016/j.eurpolymj.2009.10.022

[8] L.Y. Li, S.L. Li, Y. Shao, R. Dou, B. Yin, M. B. Yang, PVDF/PS/HDPE/MWCNTs/Fe3O4 nanocomposites: Effective and lightweight electromagnetic interference shielding material through the synergetic effect of MWCNTs and Fe3O4 nanoparticles, Curr. Appl. Phys. 18 (2018) 388-396. https://doi.org/10.1016/j.cap.2018.01.014

[9] D. Guan, Z. Gao, W. Yang, J. Wang, Y. Yuan, B. Wang, M. Zhang, L. Liu, Hydrothermal synthesis of carbon nanotube/cubic Fe3O4 nanocomposite for enhanced performance supercapacitor electrode material, Mater. Sci. Eng. B: Solid-State Materials for Advanced Technology. 178 (2013) 736-743. https://doi.org/10.1016/j.mseb.2013.03.010

[10] K. Muller, E. Bugnicourt, M. Latorre, M. Jorda, Y.E. Sanz, J.M. Lagaron, O. Miesbauer, A. Bianchin, S. Hankin, U. Bolz, G. Perez, M. Jesdinszki, M. Lindner, Z. Scheuerer, S. Castello, M. Schmid, Review on the processing and properties of polymer nanocomposites and nanocoatings and their applications in the packaging, automotive and solar energy fields, Nanomaterials 7,74 (2017) 1-47. https://doi.org/10.3390/nano7040074

[11] S. Fu, Z. Sun, P. Huang, Y. Li, N. Hu, Some basic aspects of polymer nanocomposites: A critical review, Nano Mater. Sci. 1 (2019) 2-30. https://doi.org/10.1016/j.nanoms.2019.02.006

[12] J. Fawaz, V. Mittal, Synthesis techniques for polymer nanocomposites, first Ed., Wiley-VCH Verlag GmbH & Co. KGaA: 2015.

[13] M. Akashi, H. Ajiro, Template polymerization (molecular templating), Encyclopedia of polymeric nanomaterials, Springer-Verlag Berlin Heidelberg: 2014. https://doi.org/10.1007/978-3-642-36199-9_202-1

[14] Gordon Armstrong, An introduction to polymer nanocomposites, Eur. J. Phys. 36 (2015) 063001 (34pp). https://doi.org/10.1088/0143-0807/36/6/063001

[15] P. Kruus, O. Neill, Polymerization ultrasound and depolymerization by ultrasound, Ultrasonics. 26 (1988) 352-355. https://doi.org/10.1016/0041-624X(88)90035-2

[16] B. A. Bhanvase, D.V. Pinjari, P.R. Gogate, S.H. Sonawane, A. B. Pandit, Synthesis of exfoliated poly(styrene-co-methyl methacrylate)/montmorillonite nanocomposite using ultrasound assisted in situ emulsion copolymerization, Chem. Eng. J. 181-182 (2012) 770-778. https://doi.org/10.1016/j.cej.2011.11.084

[17] E.A.Z. Contreras, C.A.H. Escobar, A.N. Fontes, S.G.F. Gallardo, Synthesis of carbon black/polystyrene conductive nanocomposite. Pickering emulsion effect characterized by TEM, Micron 42 (2011) 263-270. https://doi.org/10.1016/j.micron.2010.10.005

[18] T.J. Mason, J.P. Lorimer, Applied Sonochemistry. The Uses of Power Ultrasound in chemistry and processing, Wiley-VCH Verlag, Weinheim, 2002. https://doi.org/10.1002/352760054X

[19] B.M. Teo, S.W. Prescott, M. Ashokkumar, F. Grieser, Ultrasound initiated miniemulsion polymerization of methacrylate monomers., Ultrason Sonochem. 15 (2008) 89-94. https://doi.org/10.1016/j.ultsonch.2007.01.009

[20] H. Xu, K.S. Suslick, Sonochemical preparation of functionalized graphenes, J. Am. Chem. Soc. 133 (2011) 9148-9151. https://doi.org/10.1021/ja200883z

[21] R. Kuppa, V.S. Moholkar, Physical features of ultrasound-enhanced heterogeneous permanganate oxidation, Ultrason. Sonochem. 17 (2010) 123-131. https://doi.org/10.1016/j.ultsonch.2009.05.011

[22] B.A. Bhanvase, D.V. Pinjari, P.R. Gogate, S.H. Sonawane, A.B. Pandit, Process intensification of encapsulation of functionalized CaCO3 nanoparticles using ultrasound assisted emulsion polymerization, Chem. Eng. Process.: Process. Intensif. 50 (2011) 1160-1168. https://doi.org/10.1016/j.cep.2011.09.002

[23] S. Chakma, V.S. Moholkar, Physical mechanism of sono-Fenton process, AIChE J. 59 (2013) 4303-4313. https://doi.org/10.1002/aic.14150

[24] P. Kruus, D. McDonald, T. J. Patraboy, Polymerization of styrene initiated by ultrasonic cavitation, J. Phys. Chem. 91 (1987) 3041-3047. https://doi.org/10.1021/j100295a080

[25] G. J. Price, D. J. Norris, P. J. West, Polymerization of methyl methacrylate initiated by ultrasound, Macromolecules. 25 (1992) 6447-6454. https://doi.org/10.1021/ma00050a010

[26] B.A. Bhanvase, S.H. Sonawane, New approach for simultaneous enhancement of anticorrosive and mechanical properties of coatings: application of water repellent nano CaCO3-PANI emulsion nanocomposite in alkyd resin, Chem. Eng. J. 156 (2010) 177-183. https://doi.org/10.1016/j.cej.2009.10.013

[27] J.G. Ryu, H. Kim, J.W. Lee, Characteristics of polystyrene/polyethylene/clay nanocomposites prepared by ultrasound-assisted mixing process, Polym. Eng. Sci. 44 (2004) 1198-1204. https://doi.org/10.1002/pen.20114

[28] M. Garcia, G.V. Vliet, M.G.J.T. Cate, F. Chavez, B. Norder, B. Kooi, W.E.V. Zyl, H. Verweij, D.H.A. Blank, Large-scale extrusion processing and characteriza- tion of hybrid nylon-6/SiO2 nanocomposites, Polym. Adv. Technol. 15 (2004) 164-172. https://doi.org/10.1002/pat.458

[29] S.S. Barkade, J.B. Naik, S.H. Sonawane, Ultrasound assisted miniemulsion synthesis of polyaniline/Ag nanocomposite and its application for ethanol vapor sensing, Colloids Surf. A 378 (2011) 94-98. https://doi.org/10.1016/j.colsurfa.2011.02.002

[30] S. V. Krishna, G. Pugazhenthi, Influence of processing conditions on the properties of polystyrene (PS)/organomontmorillonite (OMMT) nanocomposites prepared via solvent blending method, Int. J. Polym. Mater. 60 (2010) 144-162. https://doi.org/10.1080/00914037.2010.504167

[31] G.A. Wang, C.C. Wang, C.Y. Chen, The disorderly exfoliated LDHs/PMMA nanocomposites synthesized by in situ bulk polymerization: The effects of LDH-U on thermal and mechanical properties, Polym. Degrad. Stab. 91 (2006) 2443-2450. https://doi.org/10.1016/j.polymdegradstab.2006.03.008

[32] C. Zeng, L.J. Lee, Poly (methyl methacrylate) and Polystyrene / Clay Nanocomposites Prepared by in-Situ Polymerization, Macromolecules. 34 (2001) 4098-4103. https://doi.org/10.1021/ma010061x

[33] M. Alexandre, P Dubois, Polymer-layered silicate nanocomposites: preparation, properties and uses of a new class of materials. Mater. Sci. Eng., R, 28 (2000) 1-63. https://doi.org/10.1016/S0927-796X(00)00012-7

[34] M.K. Poddar, S. Sharma, V.S. Moholkar, Sonochemical synthesis of PMMA/Cloisite-30B nanocomposites: A mechanistic investigation, Macromolecualr Symposia. 361 (2016) 82-100. https://doi.org/10.1002/masy.201500009

[35] S. Sharma, M. Kumar Poddar, V.S. Moholkar, Enhancement of thermal and mechanical properties of poly(MMA-co-BA)/Cloisite-30B nanocomposites by ultrasound-assisted in-situ emulsion polymerization, Ultrason. Sonochem. 36 (2017) 212-225. https://doi.org/10.1016/j.ultsonch.2016.11.029

[36] M. K. Poddar, K. Vishwakarma ,V. S. Moholkar, Rheological and mechanical properties of PMMA/organoclay nanocomposites prepared via ultrasound-assisted in-situ emulsion polymerization, Korean J. Chem. Eng. 36 (2019), 828-836. https://doi.org/10.1007/s11814-019-0252-8

[37] L.J. Borthakur, D. Das, S.K. Dolui, Development of core-shell nano composite of poly (styrene-co-methyl acrylate) and bentonite clay by ultra sonic assisted mini-emulsion polymerization, Mater. Chem. Phys. 124 (2010) 1182-1187. https://doi.org/10.1016/j.matchemphys.2010.08.055

[38] M.K. Poddar, A Ph.D. thesis on 'Ultrasound assisted synthesis and characterization of polymethyl methacrylate (PMMA) nanocompsoites' (2017), Indian Institute of Technology, Guwahati-Assam, India.

[39] A.B. Morgan, H.D. Harris, Exfoliated polystyrene clay nanocomposites synthesized by solvent blending with sonication, Polymer 45 (2004) 8695-8703. https://doi.org/10.1016/j.polymer.2004.10.067

[40] J. W. Gilman, Flammability and thermal stability studies of polymer-layered silicate (clay) nanocomposites, Appl. Clay. Sci. 15 (1999) 31-49. https://doi.org/10.1016/S0169-1317(99)00019-8

[41] P. Maiti, P.H. Nam, M. Okamoto, N. Hasegawa, A. Usuki, Influence of crystallization on intercalation, morphology, and mechanical properties of polypropylene/clay nanocomposites, Macromolecules. 35 (2002) 2042-2049. https://doi.org/10.1021/ma010852z

[42] M. Kumar, V. Kumar, P. Upadhyaya, G. Pugazhenthi, Fabrication of poly (methyl methacrylate)(PMMA) nanocomposites with modified nanoclay by melt intercalation, Compos. Interface. 21 (2014) 819-832. https://doi.org/10.1080/15685543.2014.961780

[43] H.C. Huang, T.E. Hsieh, Preparation and characterizations of highly transparent UV-curable ZnO-acrylic nanocomposites, Ceramics International. 36 (2010) 1245-1251. https://doi.org/10.1016/j.ceramint.2010.01.010

[44] Y. Zhang, X. Wang, Y. Liu, S. Song, D. Liu, Highly transparent bulk PMMA/ZnO nanocomposites with bright visible luminescence and efficient UV-shielding capability, J. Mater. Chem. 22 (2012) 11971-11977. https://doi.org/10.1039/c2jm30672g

[45] K. Hayashida, Y. Takatani, Poly(methyl methacrylate)-grafted ZnO nanocomposites with variable dielectric constants by UV light irradiation, J. Mater. Chem. C. 4 (2016) 3640-3645. https://doi.org/10.1039/C6TC00882H

[46] B. Kulyk, V. Kapustianyk, O. Krupka, B. Sahraoui, Optical absorption and photoluminescence properties of ZnO/PMMA nanocomposite films, Journal of Physics: Conference Series. 289 (2011) 012003. https://doi.org/10.1088/1742-6596/289/1/012003

[47] R.Y. Hong, J.Z. Qian, J.X. Cao, Synthesis and characterization of PMMA grafted ZnO nanoparticles, Powder Technol. 163 (2006) 160-168. https://doi.org/10.1016/j.powtec.2006.01.015

[48] M.K. Poddar, S. Sharma, V.S. Moholkar, Investigations in two-step ultrasonic synthesis of PMMA/ZnO nanocomposites by in-situ emulsion polymerization, Polymer 99 (2016) 453-469. https://doi.org/10.1016/j.polymer.2016.07.052

[49] A. Das, D.Y. Wang, A. Leuteritz, K. Subramaniam, H.C. Greenwell, U. Wagenknecht, G. Heinrich, Preparation of zinc oxide free, transparent rubber nanocomposites using a layered double hydroxide filler, J. Mater. Chem. 21 (2011) 7194-7200 https://doi.org/10.1039/c0jm03784b

[50] K.F. Lin, H.M. Cheng, H.C. Hsu, L.J. Lin, W.F. Hsieh, Band gap variation of size-controlled ZnO quantum dots synthesized by sole-gel method, Chem. Phys. Lett. 409 (2005) 208-211. https://doi.org/10.1016/j.cplett.2005.05.027

[51] M.A.A. Rahman, S. Mahmud, A.K. Alias, A.F.M. Nor, Effect of nanorod zinc oxide on electrical and optical properties of starchebased polymer nanocomposites, J. Phys. Sci. 24 (2013) 17-28.

[52] J.H. Li, R.Y. Hong, M.Y. Li, H.Z. Li, Y. Zheng, J. Ding, Effects of ZnO nanoparticles on the mechanical and antibacterial properties of polyurethane coatings, Prog. Org. Coat. 64 (2009) 504-509. https://doi.org/10.1016/j.porgcoat.2008.08.013

[53] G. Goncalves, P.A.A.P. Marques, A. Barros-Timmons, I. Bdkin, M.K. Singh, N. Emami, J. Gracio, Graphene oxide modified with PMMA via ATRP as a reinforcement filler, J. Mater. Chem. 20 (2010) 9927-9934. https://doi.org/10.1039/c0jm01674h

[54] K.P. Pramoda, H. Hussain, H.M. Koh, H.R. Tan, C.B. He, Covalent bonded polymer-graphene nanocomposites, J. Polym. Sci. Part A: Polym. Chem. 48 (2010) 4262-4267. https://doi.org/10.1002/pola.24212

[55] K.W. Putz, O.C. Compton, M.J. Palmeri, S.T. Nguyen, L.C. Brinson, High-nanofiller-content graphene oxide-polymer nanocomposites via vacuum-assisted self-assembly, Adv. Funct. Mater. 20 (2010) 3322-3329. https://doi.org/10.1002/adfm.201000723

[56] H. Zhang, Q. Yan, W. Zheng, Z. He, Z. Yu, Tough rraphene polymer microcellular foams for electromagnetic interference shielding, ACS Appl. Mater. Interfaces. 3 (2011) 918-924. https://doi.org/10.1021/am200021v

[57] S.N. Tripathi, P. Saini, D. Gupta, V. Choudhary, Electrical and mechanical properties of PMMA/reduced graphene oxide nanocomposites prepared via in situ polymerization, J. Mater. Sci. 48 (2013) 6223-6232. https://doi.org/10.1007/s10853-013-7420-8

[58] J. Wang, Z. Shi, Y. Ge, Y. Wang, J. Fan, J. Yin, Solvent exfoliated graphene for reinforcement of PMMA composites prepared by in situ polymerization, Mater. Chem.Phys. 136 (2012) 43-50. https://doi.org/10.1016/j.matchemphys.2012.06.017

[59] S.H. Lee, D.R. Dreyer, J. An, A. Velamakanni, R.D. Piner, S. Park, Y. Zhu, S.O. Kim, C.W. Bielawski, R.S. Ruoff, Polymer brushes via controlled surface-initiated atom transfer radical polymerization (ATRP) from graphene oxide. Macromol. Rapid Commun. 31 (2010) 281-288. https://doi.org/10.1002/marc.200900641

[60] M. Fang, K. Wang, H. Lu, Y. Yang, S. Nutt, Covalent polymer functionalization of graphene nanosheets and mechanical properties of composites, J. Mater. Chem. 19 (2009) 7098-7105. https://doi.org/10.1039/b908220d

[61] M.K. Poddar, S. Pradhan, V.S. Moholkar, M. Arjmand, U. Sundararaj, Ultrasound-assisted synthesis and characterization of polymethyl methacrylate/reduced graphene oxide nanocomposites, AIChE J. 64 (2018). https://doi.org/10.1002/aic.15936

[62] L.Z. Guan, L. Zhao, Y.J. Wan, L.C. Tang, Three-dimensional graphene-based polymer nanocomposites: preparation, properties and applications, Nanoscale. 10 (2018) 14788-14811. https://doi.org/10.1039/C8NR03044H

[63] K. Krishnamoorthy, G.S. Kim, S. J. Kim, Graphene nanosheets: Ultrasound assisted synthesis and characterization, Ultrason. Sonochem. 20 (2013) 644-649. https://doi.org/10.1016/j.ultsonch.2012.09.007

[64] T. Banert, U.A. Peuker, Preparation of highly filled super-paramagnetic PMMA-magnetite nano composites using the solution method, J. Mater. Sci. 41 (2006) 3051-3056. https://doi.org/10.1007/s10853-006-6976-y

[65] S. Kirchberg, M. Rudolph, G. Ziegmann, U.A. Peuker, Nanocomposites based on technical polymers and sterically functionalized soft magnetic magnetite nanoparticles: Synthesis, processing, and characterization, J. Nanomater. (2012) 670531. https://doi.org/10.1155/2012/670531

[66] R. Bera, A.K. Das, A. Maitra, S. Paria, S.K. Karan, B.B. Khatua, Salt leached viable porous Fe3O4 decorated polyaniline - SWCNH/PVDF composite spectacles as an admirable electromagnetic shielding efficiency in extended Ku-band region, Composites Part B: Engineering. 129 (2017) 210-220. https://doi.org/10.1016/j.compositesb.2017.07.073

[67] A. Kaushik, R. Khan, P.R. Solanki, P. Pandey, J. Alam, S. Ahmad, B.D. Malhotra, Iron oxide nanoparticles-chitosan composite based glucose biosensor, Biosensors and Bioelectronics. 24 (2008) 676-683. https://doi.org/10.1016/j.bios.2008.06.032

[68] S. Cui, X. Shen, B. Lin, Surface organic modification of Fe3O4 nanoparticles by silane-coupling agents, Rare Metals. 25 (2006) 426-430. https://doi.org/10.1016/S1001-0521(07)60118-1

[69] L.G. Bach, M.D. Islam, J.T. Kim, S. Y. Seo, K.T. Lim, Encapsulation of Fe3O4 magnetic nanoparticles with poly(methyl methacrylate) via surface functionalized thiol-lactam initiated radical polymerization, Appl. Surf. Sci. 258 (2012) 2959-2266. https://doi.org/10.1016/j.apsusc.2011.11.016

[70] F. Nsib, N. Ayed, Y. Chevalier, Dispersion of hematite suspensions with sodium polymethacrylate dispersants in alkaline medium, Colloids Surf. A. 286 (2006) 17-26. https://doi.org/10.1016/j.colsurfa.2006.02.035

[71] M.K. Poddar, M. Arjmand, U. Sundararaj, V.S. Moholkar, Ultrasound-assisted synthesis and characterization of magnetite nanoparticles and poly(methyl methacrylate)/magnetite nanocomposites, Ultrsono. Sonochem. 43 (2018) 38-51. https://doi.org/10.1016/j.ultsonch.2017.12.035

[72] W. Cai, J. Wan, Facile synthesis of superparamagnetic magnetite nanoparticles in liquid polyols, J. Colloid. Interf. Sci. 305 (2007) 366-370 https://doi.org/10.1016/j.jcis.2006.10.023

[73] J.L. Wilson, P. Poddar, N.A. Frey, H. Srikanth, K. Mohomed, J.P. Harmon, S. Kotha, J. Wachsmuth, Synthesis and magnetic properties of polymer nanocomposites with embedded iron nanoparticles, J. Appl. Phys. 95 (2004) 1439-1443. https://doi.org/10.1063/1.1637705

[74] M. Tanniru, Q. Yuan, R.D.K. Misra, On significant retention of impact strength in clay reinforced high density polyethylene (HDPE) nanocomposites, Polymer 47 (2006) 2133-2146. https://doi.org/10.1016/j.polymer.2006.01.063

[75] S. Sirapanichart, P. Monvisade, P. Siriphannon, J. Nukeaw, Poly(methyl methacrylate-co-butyl acrylate)/organophosphate-modified montmorillonite composites, Iran. Polym. J. 20 (2011) 803-811.

Keyword Index

About the Editors

Dr. Amir Al-Ahmed is working as a Research Scientist-II (Associate Professor) in the Interdisciplinary Research Center for Renewable Energy and Power Systems (IRC-REPS), at King Fahd University of Petroleum & Minerals (KFUPM), Saudi Arabia. He graduated in chemistry from the Department of Chemistry, Aligarh Muslim University (AMU), India. Then completed his M.Phil. (2001) and Ph.D. (2004) in Applied Chemistry from the Department of Applied Chemistry, AMU, India, followed by three consecutive postdoctoral fellowships in South Africa and Saudi Arabia. During this period, he worked on various multidisciplinary projects, in particular, conducting composites, electrochemical sensors, nano-materials, polymeric membranes, electro-catalysis and solar cells. At present, his research activity is fundamentally focused on the 3rd generation solar cell devices, such as, low band gap semiconductors, quantum dots, perovskites, and tandem cells. At the same time, he is also working on energy storage technologies, such as, heat storage, evaluation of electricity storage devices and dust repellent coating for PVs. He has worked on different NSTIP, KACST and Saudi Aramco funded projects in the capacity of a principle and co-investigator. Dr. Amir has eight US patents, over 60 journal articles, invited book chapters and conferences publications. He has edited ten books with Trans Tech Publication, Springer, Elsevier, Materials Research Forum LLC, and several other books are in progress. He is also the Editor-in-Chief of an international journal "Nano Hybrids and Composites" along with Professor Y. H. Kim.

Dr. Inamuddin is working as Assistant Professor at the Department of Applied Chemistry, Aligarh Muslim University, Aligarh, India. He obtained Master of Science degree in Organic Chemistry from Chaudhary Charan Singh (CCS) University, Meerut, India, in 2002. He received his Master of Philosophy and Doctor of Philosophy degrees in Applied Chemistry from Aligarh Muslim University (AMU), India, in 2004 and 2007, respectively. He has extensive research experience in multidisciplinary fields of Analytical Chemistry, Materials Chemistry, and Electrochemistry and, more specifically, Renewable Energy and Environment. He has worked on different research projects as project fellow and senior research fellow funded by University Grants Commission (UGC), Government of India, and Council of Scientific and Industrial Research (CSIR), Government of India. He has received Fast Track Young Scientist Award from the Department of Science and Technology, India, to work in the area of bending actuators and artificial muscles. He has also received the Sir Syed Young Researcher of the Year Award 2020 from Aligarh Muslim University. He has completed four major research projects sanctioned by University Grant Commission, Department of Science and Technology, Council of Scientific and Industrial Research, and Council of Science and Technology, India. He has published 200 research articles in international journals of repute and nineteen book chapters in knowledge-based book editions published by renowned international publishers. He has published 150 edited books with Springer (U.K.), Elsevier, Nova Science Publishers, Inc. (U.S.A.), CRC Press Taylor & Francis Asia Pacific, Trans Tech Publications Ltd. (Switzerland), IntechOpen Limited (U.K.), Wiley-Scrivener, (U.S.A.) and Materials Research Forum LLC (U.S.A). He is a member of various journals' editorial boards. He is also serving as Associate Editor for journals (Environmental Chemistry Letter, Applied Water Science and Euro-Mediterranean Journal for Environmental Integration, Springer-Nature), Frontiers Section Editor (Current Analytical Chemistry, Bentham Science Publishers), Editorial Board Member (Scientific Reports-Nature), Editor (Eurasian Journal of Analytical Chemistry), and Review Editor (Frontiers in Chemistry, Frontiers, U.K.). He is also guest-editing various special thematic special issues to the journals of Elsevier, Bentham Science Publishers, and John Wiley & Sons, Inc. He has attended as well as chaired sessions in various international and national conferences. He has worked as a Postdoctoral Fellow, leading a research team at the Creative Research Initiative Center for Bio-Artificial Muscle, Hanyang University, South Korea, in the field of renewable energy, especially biofuel cells. He has also worked as a Postdoctoral Fellow at the Center of Research Excellence in Renewable Energy, King Fahd University of Petroleum and Minerals, Saudi Arabia, in the field of polymer electrolyte membrane fuel cells and computational fluid dynamics of polymer electrolyte membrane fuel cells. He is a life member of the Journal of the Indian Chemical Society. His research interest includes ion exchange materials, a sensor for heavy metal ions, biofuel cells, supercapacitors and bending actuators.

www.ingramcontent.com/pod-product-compliance
Lightning Source LLC
Chambersburg PA
CBHW071333210326
41597CB00015B/1433